Hazardous Materials Warning Labels

DOMESTIC LABELING

EXPLOSIVE A — EXPLOSIVE B — EXPLOSIVE C — BLASTING AGENT — POISON GAS — FLAMMABLE GAS — NON-FLAMMABLE GAS — CHLORINE

OXYGEN — FLAMMABLE LIQUID — FLAMMABLE SOLID — DANGEROUS WHEN WET — OXIDIZER — ORGANIC PEROXIDE — POISON — IRRITANT

BIOMEDICAL MATERIAL — RADIOACTIVE I — RADIOACTIVE II — RADIOACTIVE III — CORROSIVE — DANGER Cargo Aircraft only — MAGNETIZED Magnetized Material — Package Orientation Markings

Handling Labels

General Guidelines on Use of Labels

- Labels illustrated above are normally for *domestic shipments*. However, some air carriers *may* require the use of International Civil Aviation Organization (ICAO) labels.

- Domestic Warning Labels *may* display UN Class Number, Division Number (and Compatibility Group for Explosives only.) Sec. 172.407(g).

- Any person who offers a hazardous material for transportation MUST label the package, if required. [Sec. 172.400(a)].

- Label(s), when required, must be printed on or affixed to the surface of the package near the proper shipping name. [Sec. 172.406(a)].

- When two or more different labels are required, display them next to each other. [Sec. 172.406(c)].

- Labels may be affixed to packages (even when not required by regulations) provided each label represents a hazard of the material in the package. [Sec. 172-401].

- The Hazardous Materials Tables, Sec. 172.101 and 172.102, identify the proper label(s) for the hazardous materials listed.

UN Class Numbers

Hazardous materials class numbers associated with the hazard classes.

Class 1—Explosives

Class 2—Gases (Compressed, Liquefied or dissolved under pressure)

Class 3—Flammable liquids

Class 4—Flammable solids or Substances

Class 5—Oxidizing Substances

Class 6—Poisonous and infectious Substances

Class 7—Radioactive Substances

Class 8—Corrosives

Class 9—Miscellaneous dangerous Substances

INTERNATIONAL LABELING

Substance liable to Spontaneous Combustion — Poisonous Substance — IRRITANT — Poisonous Substance — HARMFUL STOW AWAY FROM FOODSTUFFS * — INFECTIOUS SUBSTANCE — Infectious Substance

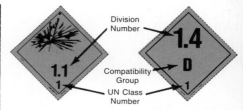

Division Number — Compatibility Group — UN Class Number — 1.1 — 1.4 D

EXAMPLES OF INTERNATIONAL LABELS

- These are examples of International Labels not presently used for domestic shipments.

- Most of the domestic labels (illustrated above) *may* be used Internationally.

- Text, when used Internationally *may* be in the language of the country of origin.

- Text is *mandatory* on Radioactive Material, St. Andrews Cross,* and Infectious Substance labels.

EXAMPLES OF EXPLOSIVE LABELS

- The NUMERICAL DESIGNATION represents the CLASS or DIVISION.

- ALPHABETICAL DESIGNATION represents the COMPATIBILITY GROUP (for Explosives Only)

- DIVISION NUMBERS and COMPATIBILITY GROUP combinations can result in over 30 different "Explosives" labels (see IMDG Code/ICAO).

For complete details, refer to one or more of the following:

- Code of Federal Regulations, Title 49, Transportation. Parts 100-199. [All Modes]

- International Civil Aviation Organization (ICAO) Technical Instructions for the Safe Transport of Dangerous Goods by air. [Air]

- International Maritime Organization (IMO) Dangerous Goods Code. [Water]

- Canadian Transport Commission (CTC) Regulations. [Rail]

U.S. Department of Transportation

Research and Special Programs Administration

Materials Transportation Bureau
Washington, D.C. 20590

CHART 8
JANUARY 1985

THE COMMON SENSE
APPROACH TO
HAZARDOUS
MATERIALS

**Other Related Fire Service Books
Available from FIRE ENGINEERING**

About the Author

Frank L. Fire is Director of Marketing at Americhem, Inc. in Cuyahoga Falls, Ohio. He is also a recognized expert in hazardous-materials chemistry and incident analysis. Mr. Fire does extensive teaching and advising on these subjects at the National Fire Academy, the Akron Fire Department, and at the University of Akron where he was previously Coordinator of the Fire Science Program.

He has had numerous articles published in *Fire Engineering* on hazardous materials and firefighting, and has written the section on plastics in *The Chemistry of Hazardous Materials* as taught by the National Fire Academy. Mr. Fire is a member of the National Fire Protection Association, the Society of Plastics Engineers, and the Society of the Plastics Industry, where he is on the Coordinating Committee on Fire Safety. He has a B.S. in Chemistry and M.B.A. from the University of Akron.

THE COMMON SENSE APPROACH TO

HAZARDOUS MATERIALS

Frank L. Fire

FIRE "Leading the fire service since 1877"
ENGINEERING
a publication of Technical Publishing,
a company of the Dun & Bradstreet Corporation New York

Printed in the United States of America

Library of Congress Cataloging-in-Publication Data

Fire, Frank L., 1937-
    The common sense approach to hazardous materials

    Includes index.
    1. Hazardous substances. I. Title.
T55.3.H3F57 1985   604.7   85-81203
ISBN 0-912212-11-X

To my wife, Marlene, whose support and editing got me through.

To Joe Lentini, who started it all.

And to my fourth son Christopher, who would never forgive me if I didn't mention him.

Contents

List of Illustrations

List of Tables

Preface

This book is written for the first responder to a hazardous-materials incident. If that first responder is a firefighter, then a lot of reactions will have to be changed, because you do not go storming into a hazardous-materials incident as you do at a structure fire. As a chief officer, you do not order "Attack, Attack, Attack!" as you ordinarily do at a structure fire. As an incident commander, you do not immediately throw water at the problem, as you do at a structure fire. Your previous training in attacking fires can kill you at a hazardous-materials incident. Instead of reacting as you have in the past, you are going to have to come to an immediate halt, determine that you do in fact have a hazardous-materials incident, identify what that material is, and then use common sense in handling the incident. This book does not teach incident command. It is merely a primer on hazardous materials for the first responder.

If the first responder is not a firefighter, but a peace officer or a property protection specialist, the response in the hazardous-materials incident will be different from that in the other types of incidents emergency personnel have been trained to handle. Emergency personnel are expected to use common sense in handling hazardous materials. This book does not teach property protection nor how to enforce the peace. This book provides an introduction to the hazardous-materials problem by presenting the foundation needed to go further in the study of hazardous materials, such as "hands-on" courses and incident-command courses. The hazardous-materials problem is not a simple one, and one course will not produce an expert in the field.

The "Common Sense Approach to Hazardous Materials" is the approach everyone should take when it is necessary to approach hazardous materials. Actually, the common sense approach is the approach everyone should use in undertaking any endeavor. It is easily defined and easily understood by everyone. Quite simply, the common sense approach to anything is nothing more than breaking down a seemingly impossible task into very small pieces, each of which is easily handled. As each piece is handled and assimilated, it is put together with the piece handled just before this one, and so on. Before you know what is happening, you have mastered whatever it was that seemed so impossible.

Each person reading this *preface* has accomplished some seemingly impossible task in the past (at least, it seemed impossible to *someone*). Think about it. Remember how impossible it was to learn to ride a bike? To read? To drive a car? To get through training? Stop a minute and analyze all those things you have accomplished in your lifetime up to this point. All of them were done in the same way—the common sense way. You accomplished one small task at a time, continuing until the entire job was done. Think about those attempts at something that failed. You probably tried to take on too much at once. You did not break it down into easily handled parts.

You remember the old sayings, don't you? "It's hard by the yard, and a cinch by the inch." "You can't eat an elephant in a single bite." "A journey of a thousand miles begins with a single step." There are a lot more, cornier than these. Corny, *but true*. Vince Lombardi took rawboned rookies and taught them a complex system of playing football—and winning. Legend has it that he started each year's first training session by holding a football aloft and proclaiming, "Gentlemen, this is a football." Legend also has it that a lineman once interrupted him by protesting, "Slow down, Coach!"

You, too, may feel that the material in this book moves too fast; if you do, go back and read it again. This book is written with Lombardi's philosophy of "Back to Basics" in mind. It begins with a chapter on the chemistry of hazardous materials; this information is *really* basic, and it is presented in a thoroughly basic manner. It is not intended to give anyone a start toward a degree in chemistry, nor is it intended to "snow" anyone. It is presented so that you will understand the basic language and the basic principles of how and why hazardous materials behave as they do. Without this basic beginning, you will always *react* to hazardous materials, rather than *act*. Things will happen to you, rather than you having control. This rather meager background in chemistry will prepare you for the next few chapters, which explain the makeup of the chemicals that surround us. This background will give you the "edge" that you need in controlling a hazardous-materials incident; it is also the foundation for the information to come.

Chapters Two, Three, Four, and Five take you through the types of compounds that you will be facing, how they are held together (which, of course, will tell you how they come apart), and how they differ from one another. These chapters focus on what these chemicals consist of, and knowing the types of compounds that make up classes of hazardous materials will give you tips on how to handle them.

Chapter Six looks into the various theories of fire and dissects them, because knowing just how fires burn will give you the clues needed to extinguish them. This chapter also covers pyrolysis and bond energies.

These first six chapters are the fundamentals. It is common sense to attack a huge study such as hazardous materials by taking very small beginning steps; these first six steps cover all the background and teach you all the language you will need to understand the rest of the text. This foundation will lead you, one step at a time, through each class of hazardous materials.

Students of hazardous materials complain that they learn placarding and labeling and even identification of classes of hazardous materials, but they do not know the names of these materials they will be facing. The common sense approach provides the names of the most common hazardous materials in each class, so that you will be able to place names with hazards. It is not possible for any book to name *all* the chemicals that pose a hazard to people and/or the environment (there are literally tens of thousands of them), but the chemicals named in this book represent the materials involved in 99 percent of the incidents to which you may ever respond. It is not difficult to become acquainted with the 75 or more that represent 90 percent of the problems you will face; you can use reference books to handle the rest.

With Chapter Seven, we begin the study of individual classes of hazardous materials, and this is carried through to Chapter Eighteen, the last in the book. These last twelve chapters may be used as the entire text for a thirteen-week quarter, if the instructor feels there is not enough time to present all the material. It is recommended, however, that the credit hours awarded for this course be of a sufficient number to allow the entire book to be covered. In a semester system, there should be no time problem in covering the entire text.

Do not allow yourself to be "psyched out" by the enormity of the problem presented by hazardous materials. Take it one step at a time, and you will be amazed at the relative ease with which you absorb the material. You will still need refresher courses, plus courses that use "hands-on" techniques in handling hazardous-materials incidents. Take every course available to you in this field. This study is necessary because there is no solution that is "right" in every situation.

Introduction

The subject of hazardous materials is *not* complicated. What *is* complex is the myriad of chemical compounds and other substances that make up the *total* list of hazardous materials; what is downright stupefying are the tongue-twisting, mind-boggling names of some of these materials. Do not be intimidated by the topic. Do not think you have to memorize lists of thousands of chemicals, and try to pronounce each name correctly, because you don't.

The goal of this book is *recognition*. It is written so that when you do encounter a hazardous material, you will *recognize* it as a hazardous material, and you will also recognize the hazard class to which it belongs. Therefore, if you concentrate on the hazards of the top one hundred or so chemicals, you will be able to recognize, probably, 90 percent of the hazardous materials that will be involved in an incident to which you must respond. Knowing the hazard class to which *these* belong will prepare you to handle all members of that class safely.

The list of the most likely chemical substances to be involved in an incident is not hard to produce, once you know what chemicals are manufactured in the largest quantity in the United States, and which are most often shipped. With the exception of radioactive materials, these substances will be the ones encountered most often. They have such terrifying names as oxygen, nitrogen, ammonia, and so on. Of course, sulfuric and nitric acids are on the list, but these are not difficult to memorize. There are *some* substances with difficult names, but those can be handled.

The term firefighter is used extensively in this book, but it has several meanings. It can mean a fireman or a female firefighter, and the word "he" will be used throughout the text, applying to both male and female members of the fire services. The person can be paid, volunteer, or a hybrid of the two; he may be a public or private firefighter or a member of an industrial brigade. The term should apply to *anyone* who is a first responder to a hazardous-materials incident, whether that person is a policeman, deputy sheriff, state highway patrolman, disaster worker, or a lab technician who has volunteered to help with a spill in the laboratory. That person may be an insurance underwriter, the manager of a chemical plant, or a dispatcher or worker in a truck terminal. The key definition includes anyone who has any responsibility to respond to, and take part in, the handling of a hazardous-materials incident.

The hazardous-materials problem is not a new one. It has been with us since the advent of the flammable liquids that we use to fuel our automobiles and has been escalating at an alarming rate ever since. With the standard of living increasing in this country, and everyone wanting to increase his or her living standards to keep up with the neighbors, the demand for new products is increasing rapidly.

Inherent in this increased demand for new products are the increases in demand for the raw materials that are used to make these products. More often than not, some of the raw materials used along the line are hazardous in one or more ways. In addition to these raw materials, the manufacturing processes themselves may be hazardous. The firefighter of today must be aware of these new hazards and new hazardous materials and must be prepared to learn how to handle them safely.

The firefighter must learn the characteristics of these new hazardous materials and how they can harm him, his colleagues, the taxpayers who pay his salary, any and all innocent bystanders, exposed property, and the environment. Little will be said in this book about the exposed property and the environment, but this fact does not indicate that the protection of these entities is not important. Protection of the first responder is the goal of this book; protection of property and the environment are other topics for other books.

The emphasis in this book will be on the number-one exposure in any and all fire situations and hazardous-materials incidents—the firefighter. The firefighter is always the number-one exposure in any situation to which he is expected to respond. The problem is that the firefighter himself (or herself) does not believe it, and if he or she does, he or she does not act that way.

The firefighter *must* accept the position of number-one exposure, or he will not be effective in doing his job. If he does not properly protect himself, he *will* be injured or killed and will not be able to effect life rescue. Even if no life is threatened, the injured or dead firefighter becomes a liability in the incident, rather than an asset. Attention that should be paid to gaining control of the incident is focused instead on the rescue of a firefighter who did not protect himself properly.

With ten to twenty thousand new chemicals developed each year and one to two thousand of these entering the stream of commerce, the firefighter may object to the statement that he should learn the hazards and characteristics of each chemical—and well he should object. No one, chemists included, can master the hazards and characteristics of every chemical used today, along with the thousands of new ones added annually. But it *is* possible to learn the *general* hazards and characteristics of eight or nine classes of hazardous materials; as long as you can place a particular chemical within a particular hazard class, you will gain a head start on handling the incident safely.

This book is designed to teach the hazards of each class of hazardous materials. It consists of eighteen chapters, and the suggestions for its use include different instructions for different academic session lengths. Minimum class time should be four hours per week. For a fifteen-week semester, the first ten chapters should be covered on a one-chapter-per-week basis, with the last eight chapters covered at a rate of two per week. The second option is to cover the first chapter the first week, cover the next four outside the classroom as reference reading, and cover the remaining chapters on a one-a-week basis.

On a thirteen-week (quarter-system) basis, the first six chapters may be covered at a rate of one a week, then two a week for the next six weeks. A second option is to cover the first chapter the first week, read the next four chapters outside class, cover one chapter per week for the next nine weeks, and two chapters each for the last two weeks.

On either academic system, the speed with which the material is covered may be adjusted to each class, determined by the amount of material any particular group can assimilate. The most important material covering the hazard classes is in the last thirteen chapters, but the first five are an important foundation upon which to build. The first chapter is, by far, the most important, as it provides the language the student will need to understand as the material on hazard classes is being presented. The next four chapters are purely expansion on Chapter One, chemistry, and mastery of these chapters will make the rest of the book easy.

A word about the chemistry in this book—it is not difficult, nor is it presented in the manner that it would be in a college chemistry course. In other words, a teacher of chemistry, even of high school chemistry, might scoff at the manner in which the material is presented and might even have a valid argument that some of the statements are not "one hundred percent" scientifically correct. The presentation of the chemistry in this book is intentionally different, however, changed ever so slightly to make its understanding easier, in a manner that non-chemistry students will understand, with very little work. The instructor should insist on that work in this area, for it will make the rest of the book relatively simple.

A very simple Periodic Table, similar to the one used in this book, should be kept on display at all times throughout the course. It should include no more information than the table presented in this book. There is no need for a Periodic Table complicated with unnecessary information.

The chemistry presented in this book is designed to be the absolute minimum that a firefighter will need to understand what each hazard class is and what he can expect in general from each member of that class, in both a fire and a non-fire situation. Much of the material must be memorized, and, once it is, simple deductions can be made to explain chemical actions and reactions.

No attempt is made to list *every* hazardous chemical in existence, since any attempt to do so would be immediately outdated. Nor is any attempt made to list all the physical and chemical properties of the hazardous materials included. What is attempted is the mention in the text of every commercially valuable chemical that could be classed as a hazardous material, so that the student may become familiar with those chemicals that he has the highest probability of encountering at an incident. Listed at the end of each chapter on hazard classes are the names of additional substances that possess the same hazard as that covered in the chapter. These tabulations too are not designed to be all-inclusive but can be used as reference guides to check the inclusion of a substance in a particular hazard class. The criteria used for inclusion in the lists include regulations for transportation. Since most hazardous-materials incidents involve transportation, a transportation accident will most probably provide the contact with such materials. No attempt has been made to include in the lists chemicals that are intermediates in chemical processes in manufacturing plants.

Many hazardous materials possess more than one hazard, but they are described in the chapter that best represents the *principal* hazard for that particular chemical. They are also mentioned in the other chapters that describe other hazards that they possess and also appear in the list at the end of those chapters. What is important to a first responder is that he be aware that a substance may have multiple hazards. It is also important to know that every member of that hazard class may not have *all* the possible hazards listed for that class, but that each member of the class *might* have more than one hazard.

This book is intended to provide the informational background needed to handle a hazardous-materials incident. What the firefighter must do to supplement this background is to take part in hands-on training exercises that will give him experience in handling those incidents. The firefighter has trained long and hard to handle structural fires, and he has succeeded in learning that subject well. He has the opportunity to supplement that training by responding to structural fires in his response district. The Fire Services must learn to accept the fact that the training received and reinforced in fighting structural fires is, in most cases, *totally inadequate* to handle hazardous-materials incidents safely. In many tragic instances, it has been dead wrong.

It is the hope of the author that this book will contribute toward the proper education and training of the firefighter in the safe handling of hazardous-materials incidents. The firefighter is still the number-one exposure and must be protected at all costs.

THE COMMON SENSE APPROACH TO HAZARDOUS MATERIALS

1

The Chemistry of Hazardous Materials

Introduction

The first thing the student of fire science must do is to get over the feeling of sheer terror when he sees the word "chemistry." In studying the chemistry of hazardous materials, the goal is not to make you a chemist nor an expert on hazardous materials. Instead, the goal is simply one of recognition. You will learn just enough chemistry to achieve this goal, and no more.

You will not be taught as much chemistry as a junior in high school, let alone a freshman in college. As a matter of fact, we take a few liberties in this book and make statements that might not be absolutely correct scientifically; we are safe in making those statements because, again, we are not interested in producing chemists, but firefighters and other first responders who will have *enough* education in chemistry to *recognize* a hazardous material and then know where to go and what to do to get the proper information to handle the situation. This chapter, like the others that will follow, will introduce words that may appear strange to you. The words will be defined as we move through each subject, however, and you will find them gathered in a glossary at end of each chapter.

One of the major complaints of firefighters and other first responders is that while many of them understand what oxidizing agents, corrosives, flammable solids, and other classes of hazardous materials are, they are never told what actual materials make up each class. This complaint is handled here by specifically listing all the materials in each hazard class at the end of the section dealing with that class. Remember that there are literally tens of thousands of chemicals in use; the lists presented in each chapter will contain all, or very nearly all, of the hazardous materials with which you may come into contact. It will be your responsibility to find the specific characteristics of any particular substance by checking several of the reference books that should be available at the scene.

The most common of the materials in each class, and therefore the most likely to be shipped through or used in your jurisdiction, will be discussed briefly in the section covering that specific hazard, and the rest will simply be listed by chemical name. Do not be intimidated by some of the long, complicated chemical names. You will quickly notice that you are already familiar with the names of the most common materials in each hazard class, and that the strange-sounding, unfamiliar materials are not commonly encountered.

Remember, the goal you should be working toward is recognition; that is, recognition of the fact that you are dealing with a hazardous material. You cannot solve the problem of what to do with the hazardous material if you do not realize that you are dealing with one. If you are not aware of the problem, you will in all probability quickly become part of it, rather than part of the solution.

The chemistry portion of this book exists only to teach you the language of hazardous materials, which quite naturally must include more than just the words "oxidizer," or "corrosive," or "poison," or whatever the name of the hazard class. If you grasp the basics, and that is all that will be presented, you will have a head start on the hazardous-materials problem simply because you understand the actions and reactions in which the materials will be involved. If you know what sequence of events the hazardous materials will follow in any given incident, you will then be able to decide at what stage of the incident you will intervene. In other words, by knowing what these chemicals will do under certain circumstances, you will increase your chances of successfully interrupting the incident and bringing it to a safe conclusion.

The effort you make to learn the names and the ways of naming certain classes of chemicals will enable you to associate that name with a particular hazard, which should then lead you to a logical method of interrupting the incident. It should also tell you when the best attack is *no* attack; that is, to evacuate and cordon off the area and allow the incident to run its course, with no attempt on your part to intervene in and interrupt the incident. This response will take great discipline on your part, simply because your training has probably conditioned you to rush in and try to return the environment to normal as soon as possible. Handling a hazardous-materials incident is different enough from a structure fire to warrant stopping far enough *away* from the site to try to identify just what hazardous material is involved in the incident and then after, *and only after*, consultation with the proper references and resources, to make a decision as to whether or not to continue your approach, and whether or not to mount some sort of attack. In other words, the old joke that the two most valuable pieces of equipment in a hazardous-materials incident are a pair of running shoes and a pair of powerful binoculars is really not a joke at all.

Your decision to become involved in the incident should be made on the basis of whether or not you can actually get the situation *safely* under control, or if you will be needlessly endangering the lives of your men. The only way you can ever make this decision *intelligently* is to educate yourself in the area which is probably your weakest, and that is undoubtedly the *chemistry* of hazardous materials. This book should help you correct that weakness.

Now take a deep breath, relax, and read on with an open mind. You will be surprised how easy it is.

Chemistry

Chemistry is the science of matter, energy, and reactions. You are already familiar with all three of these things, as all three are required to exist in a fire. This is demonstrated in figure 1, which shows the chemical reaction involving the burning of methane. Methane, oxygen, carbon dioxide, and water all have mass, and energy is released from the reaction as heat and light. We repeat that the only chemistry to be presented here and in the following pages is the absolute minumum needed to understand the properties of the substances that will be discussed.

$$CH_4 + 2O_2 \rightarrow CO_2 + 2H_2O$$

Figure 1

Methane + Oxygen   Yields   Carbon Dioxide + Water.

Matter is anything that has mass and occupies space. The matter we are concerned with is material that can burn, support combustion, or stop combustion. It may be toxic (poisonous), corrosive, water- or air-reactive, or radioactive. Its natural state may be solid, liquid, or gaseous. It may be inert or highly reactive. It may be extremely hot or incredibly cold. It may be metallic or non-metallic. But whatever it is, it is matter.

For some reason, probably because it is invisible, many people do not consider air as matter. Air is not *always* invisible, as many people who live in highly polluted areas will attest. Air *does* have mass, as proved by the fact that it weighs 14.7 pounds for every square inch of area that it covers, measured at sea level. All gases have mass, even though some of them rise when released in air, and actually lift a balloon or some other container they may be enclosed in at the time. Therefore, the first thing that you may have to look at differently is the fact that gases are matter, just like liquids and solids.

Matter comes in two forms, pure substances and mixtures of pure substances. Pure matter includes elements and compounds separately, while mixtures are physical combinations of different elements and compounds, or of both. Examples of pure substances that are elements are gold, silver, sodium, iron, carbon, helium, and neon, plus 96 more. Pure substances that are compounds include sodium chloride (common table salt), sucrose (table sugar), pure water, carbon monoxide, and carbon dioxide.

Mixtures are substances such as gasoline and kerosene (mixtures of compounds), air (mixture of compounds and elements), brass or bronze (mixtures of elements), and wood (mixtures of compounds). The difference between a mixture and a pure substance is important and will become more apparent later. In addition, the difference between pure substances and mixtures of substances will be defined, based on the smallest particles of each substance that can still be identified as that substance. The information will make a little more sense to you then.

Basically, then, all matter is made up of elements, compounds, mixtures of elements, mixtures of compounds, and mixtures of elements *and* compounds. Since all compounds are made up of elements bound together chemically, all matter is really made up of elements and the different ways they combine. The way elements combine to form compounds, compounds react with elements and other compounds, and energy is absorbed or released in such reactions—all of these make up the science of chemistry. We will pursue the question of exactly what elements and compounds are a little later in this chapter and in Chapter Two.

X   A reaction is something that happens to the substance; it may be chemical or physical. Both types of reactions are important, but the more important is the chemical reaction. Whenever anything burns, a chemical reaction is occurring. The chemical combination of anything with oxygen is called *oxidation*, and if heat and light are generated by this chemical combination, the reaction is called *combustion*. The rusting of iron is an example of *oxidation*, but the reaction is so slow that no light is evolved, and no apparent heat is given off; it really is, but it is so slow it is almost immeasurable. Nevertheless, since the old material has reacted with something, and changed its *chemical* make-up, then a chemical reaction took place. When wood or paper (or indeed, any class A or class B material) burns, the chemical reaction is exactly the same as the rusting reaction, only very much faster. In this case, the oxidation manifests itself as fire. This reaction, fire, is several orders of magnitude faster than the rusting reaction, and if a reaction is several orders of magnitude faster than a fire, the resulting oxidation reaction is called an *explosion*.

If all that changes is the *physical* make-up of the material (that is, it evaporates, or freezes, or condenses, or melts, or is cut in half), it is a physical reaction, not a chemical reaction.

(CAN CHANGE IT BACK)
(CHEMICAL YOU CANNOT CHANGE IT BACK)

Therefore, in the common, everyday, run-of-the-mill fire (if there were such an event) all the things that make up chemistry occur. *Matter* (the fuel and oxygen) is *reacting* together, with the liberation of *energy* (heat and/or light). That statement is all there is to chemistry.

The Elements

The forerunners of chemists were known as alchemists. They were not interested in matter, energy, and reactions but were interested in, among other things, producing gold, a fairly rare metal upon which man had placed some value, from some other metal (lead, for example), which was much more common and upon which man placed no great value. The constant failure of these men led others to wonder why they were failing, and these curious men began to look at matter, what caused it to change form, and how the energy that evolved could be used. These men were the first chemists, and they realized that the elements of the alchemists (earth, air, and fire, among others) were not really basic materials, but were made up of other materials, or, in the case of fire, the manifestation of chemical change of matter from one substance to another.

As these first chemists began investigating matter, their objective was to break matter down into its simplest parts, so they could better understand it. When they were able to find a material that could not be further broken down into simpler matter by chemical means, they called it an *element*. As more and more elements were discovered, a Russian chemist decided to try to list all that were known in some order; he discovered that as he arranged the elements by their increasing *atomic weights*, there were some rather remarkable repetitions of chemical properties among the elements. For instance, he discovered that lithium, sodium, and potassium were remarkably similar chemically, as were beryllium, calcium, and barium, and fluorine, chlorine, bromine, and iodine. Why they are similar in their chemical properties will be covered later. In any event, what the Russian chemist Mendeleev had constructed was a crude form of the Periodic Table of the Elements, so named for the *periodicity*, or regular repetition of chemical properties in several elements. His only mistake was that he arranged the elements in order of their increasing atomic weights, rather than in their increasing *atomic numbers*. To define what these terms are, we are going to take a Disney-like trip into the *atom*.

Atomic Structure

As we mentioned earlier, elements are substances that cannot be broken down into simpler substances by chemical means. They are the building blocks of the universe. Everything in creation is made up of elements, and differences among the different things that exist are due to the *different* way various elements will react with, and combine with, other elements. You may select a substance to try to identify its chemical composition, which means you are interested in what elements have combined together, and *how* they have combined together, to make up whatever substance you are trying to analyze. The major facts to remember about elements are that they are basic substances and are *not* made up of other substances, and that when they react *chemically* with other elements or compounds, they no longer remain elements but become a part of some new substance. The key here is that it must be a *chemical* reaction, not a physical reaction, for the element to lose its identity and become an unrecognizable, integral part of another substance.

The smallest particle of an element is known as an atom, the smallest part of an element that can still be identified as that element. It cannot be broken down into anything smaller and still be recognized as an atom of a particular element. The only way an atom can be destroyed is by physical means (that is, nuclear physics). We are not about to touch on that subject until, of course, we reach the topic of radioactivity in Chapter Sixteen. The implication here is that when an atom of an element reacts with another atom of another element or a with the molecule of a compound, the atom is not destroyed; its chemical reaction has caused it to change its appearance as it becomes part of another compound.

The atom itself is made up of two particles (subatomic particles) called a *nucleus* and an *electron* (or electrons). The nucleus is the center of each atom and contains all the positive charge of the atom and essentially all of its weight. The electron is a particle that orbits the nucleus like a planet around the sun, has a negative electrical charge of -1, and is almost weightless. (Although the electron does weigh something, it is so light compared to other particles that we refer to it as essentially weightless.) Whatever the positive charge of the nucleus of any given atom of any given element is, it will be balanced by the appropriate number of electrons. This fact is what leads to the natural law that states all atoms are electrically neutral. Let us now probe even deeper, into the nucleus, to see where the atom gets its weight and positive charge.

The nucleus is also made up of two particles (nuclear particles); they are called the *proton* and the *neutron*. The proton is a nuclear particle with a positive electrical charge of 1 and an atomic weight of 1. The neutron has no electrical charge (it is electrically neutral) and has an atomic weight of 1. Since the electron has essentially no weight, and both the proton and neutron each weigh 1 atomic mass unit (a.m.u.), you can see that essentially all the weight of the atom will be in the nucleus. Since all atoms are electrically neutral, and each electron has a negative electrical charge which is equal to, but opposite in charge to, that of a proton, there must be the same number of electrons in orbit around the nucleus as there are protons in the nucleus. How the electrons arrange themselves in orbit around the nucleus determines the chemistry of each element. Before we examine the chemical behavior of these elements because of this electronic configuration, let us look at those elements that will be important to this book.

Elements Important to the Course

There are 103 known elements (or 102, or 104, or 106), depending on which chemistry book you read. The first 92 exist in nature, and the rest are man-made. We are not interested in all 103 elements in studying hazardous materials, although they all certainly are important. We will look at 48 of them, with about half of those being the constituents of most hazardous materials.

It is at this point we will become involved with elemental *symbols* and chemical *formulas*, which are nothing more than a method of shorthand that saves a considerable amount of time and space. It is not difficult to learn, but the only way you can do so is to memorize it. The 48 elements in table 1 are the ones important to this course, and their *symbols* are listed next to them.

You will notice that the symbols are of two kinds: one-letter and two-letter symbols. All the one-letter symbols are capitalized and, in all of the two-letter symbols, the first letter is *always* capitalized, and the second is *always* lower case. You will also notice that the symbols are not always identical to the first or first and second letters of the element's name. This is because the symbol may be derived from either the English or the Latin name (and, in one case, the Greek). The only way to master the list of 48 symbols is to memorize them.

Table 1 is the list of 48 elements and their symbols.

Let us look at a few of these elements and describe them briefly. (See figure 2.) You must realize that in most cases, entire *books* have been written about many of these *individual* elements, so in no way can an element be properly described in a few lines. We do this as a method of introduction to elements in general, however, and many of these elements will turn up later, as hazardous materials themselves, or as integral parts of them.

Table 1 / 48 Elements and Their Symbols

Hydrogen	H	Helium	He	Rubidium	Rb	Strontium	Sr
Lithium	Li	Beryllium	Be	Iodine	I	Xenon	Xe
Boron	B	Carbon	C	Cesium	Cs	Barium	Ba
Nitrogen	N	Oxygen	O	Radon	Rn	Chromium	Cr
Fluorine	F	Sodium	Na	Manganese	Mn	Iron	Fe
Magnesium	Mg	Neon	Ne	Cobalt	Co	Nickel	Ni
Aluminum	Al	Silicon	Si	Copper	Cu	Zinc	Zn
Phosphorus	P	Sulfur	S	Gold	Au	Silver	Ag
Chlorine	Cl	Argon	Ar	Platinum	Pt	Tin	Sn
Potassium	K	Calcium	Ca	Antimony	Sb	Mercury	Hg
Arsenic	As	Selenium	Se	Lead	Pb	Radium	Ra
Bromine	Br	Krypton	Kr	Uranium	U	Plutonium	Pu

Figure 2
Atomic Model of
Lithium with Key.

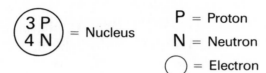

P = Proton
N = Neutron
◯ = Electron

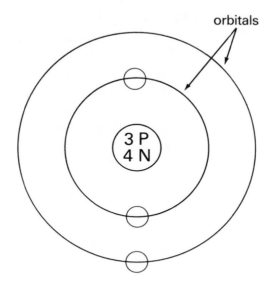

orbitals

Lithium
Li
Atomic Number 3
Atomic Weight 6.941

Figure 3
Atomic Model of
Hydrogen.

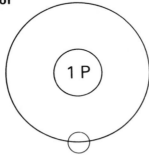

Hydrogen
H
Atomic Number 1
Atomic Weight 1.008

Figure 4
Atomic Model of
Helium.

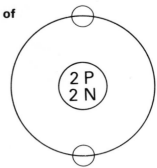

Helium
He
Atomic Number 2
Atomic Weight 4.003

Hydrogen (figure 3) is the smallest of all the elements; that is, its atom is the smallest of all the atoms. It is a colorless, odorless, tasteless gas, and it is the lightest of all the gases. It burns with a very hot, almost invisible flame (you will learn later that hydrogen, like oxygen, nitrogen, fluorine, and chlorine, actually does not exist as a gas made up of atoms, but rather a gas made up of molecules).

Helium (figure 4) is an inert gas, colorless, odorless, and tasteless, and it does absolutely nothing chemically. It is very much lighter than air (although not as light as hydrogen) and does not burn.

Sodium is a very soft, light metal, silvery in color when freshly cut, but usually it is a dull gray. It will float on water and will react violently with it. It is usually stored under kerosene (with which it does not react) or in a vacuum-packed can.

Potassium looks and acts chemically just like sodium, and even a chemist would have a difficult time telling them apart. It is stored in the same way as sodium.

Lead is a very heavy, gray, soft metal, though not as soft as sodium or potassium. Lead is toxic.

Oxygen is a colorless, odorless, tasteless gas that is essential to life. *Oxygen will not burn*, but it *will* support combustion. It makes up approximately 21 percent of our atmosphere.

Nitrogen is a colorless, odorless, tasteless gas that is inert at normal temperatures. The gas makes up at least 78 percent of the atmosphere.

Chlorine is a yellowish-greenish gas with a sharp, pungent, choking odor and an acrid taste. It is toxic, corrosive, and will support combustion.

Fluorine is a yellowish gas with exactly the same chemical properties as chlorine.

Gold is a heavy, yellow metal that is extremely malleable (can be hammered into very thin sheets) and is considered very valuable.

Iodine is a grayish non-metal with a purplish sheen to it. Iodine gives off reddish vapors that are toxic, corrosive, and will support combustion.

Mercury is a heavy, silvery, shiny, *liquid* metal.

Bromine is a corrosive, toxic, brownish liquid that gives off reddish-brown fumes, which are toxic, corrosive, and will support combustion.

Sulfur is a yellow non-metallic solid that burns, giving off choking, toxic fumes.

Magnesium is a silver-white metal that burns with a blinding white light.

The Periodic Table

The *Periodic Table of the Elements* (table 2) is the instrument by which all the known elements are arranged in order of their increasing *atomic number*. We will be interested only in the long vertical *groups* that begin with hydrogen, beryllium, boron, carbon, nitrogen, oxygen, fluorine, and helium, but we will not be interested in *all* the elements in those groups. The elements in the groups beginning with those listed above are known as the *representative*, or *main group* elements, while all the others are known as the *transitional* elements. You should recognize, however, that table 2 is divided in another way. We have used a heavy line to separate the *two* types of elements. All the elements to the left and below the line are *metals*, while those elements to the right and above the line are *non-metals*. It will be extremely important for you to know these differences. It will also be very easy for you to memorize the few non-metals, which, of course, means that all the remaining elements are metals.

Remember, the Periodic Table of the Elements is now organized correctly by order of increasing atomic number, rather than by increasing atomic weight. In doing so, the periodicity of chemical properties appears very strongly in certain parts of the Table. For example, the group IA elements headed by lithium (figure 5) are known as a "family" or group of elements as the *alkali metals*. Although hydrogen is shown at the top of group IA, since it is a non-metal, it is not considered a part of the "family" of elements below it. Just to the right of the alkali metals in group IIA are a group or family of elements known as the *alkaline earth* metals.

Over to the right, in the next-to-the-last column marked group VIIA, and headed by the element fluorine, is the group or family of elements known as the *halogens*. And finally, in the last column on the right is group VIIIA, headed by helium. These elements are known as the *inert*, or *noble*, gases. These four groups all contain elements whose chemical properties are exceedingly similar to other elements in the same group or family. Why this is so will be covered later. At present, it is sufficient if you recognize that the chemistries of sodium and potassium are very similar to each other, and that both are similar to lithium, cesium, rubidium, and francium. Beryllium (figure 6), barium, calcium, magnesium, strontium, and radium, are all very similar to each other in their chemical properties, just as fluorine, chlorine, bromine, iodine, and astatine are all similar to each other. In group VIIIA, the chemistry exhibited by the inert gases helium, neon, argon, krypton, xenon, and radon is very similar in that they exhibit *no*

Table 2 / Periodic Table of the Elements

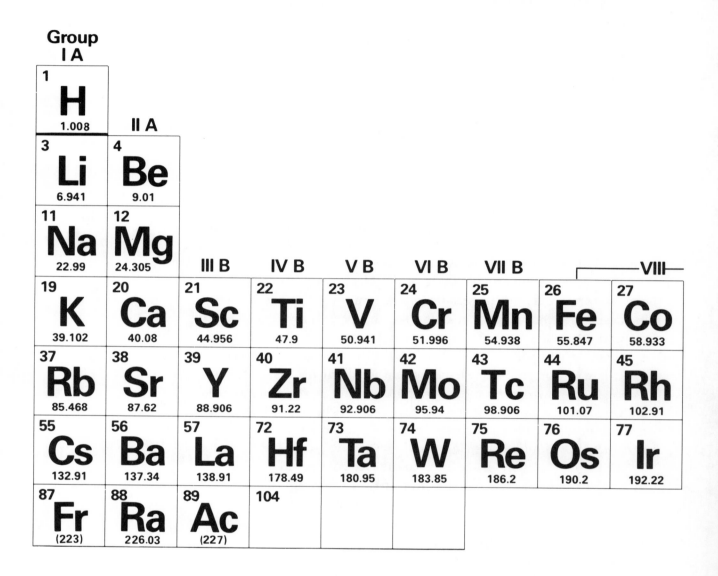

Group
I A

1								
H 1.008								

II A

3 **Li** 6.941	4 **Be** 9.01

11 **Na** 22.99	12 **Mg** 24.305

III B **IV B** **V B** **VI B** **VII B** —VIII—

19 **K** 39.102	20 **Ca** 40.08	21 **Sc** 44.956	22 **Ti** 47.9	23 **V** 50.941	24 **Cr** 51.996	25 **Mn** 54.938	26 **Fe** 55.847	27 **Co** 58.933
37 **Rb** 85.468	38 **Sr** 87.62	39 **Y** 88.906	40 **Zr** 91.22	41 **Nb** 92.906	42 **Mo** 95.94	43 **Tc** 98.906	44 **Ru** 101.07	45 **Rh** 102.91
55 **Cs** 132.91	56 **Ba** 137.34	57 **La** 138.91	72 **Hf** 178.49	73 **Ta** 180.95	74 **W** 183.85	75 **Re** 186.2	76 **Os** 190.2	77 **Ir** 192.22
87 **Fr** (223)	88 **Ra** 226.03	89 **Ac** (227)	104					

Lanthanide series

58 **Ce** 140.12	59 **Pr** 140.91	60 **Nd** 144.24	61 **Pm** (147)	62 **Sm** 150.4	63 **Eu** 151.96	64 **Gd** 157.25

Actinide series

90 **Th** 232.04	91 **Pa** 231.04	92 **U** 238.03	93 **Np** 237.05	94 **Pu** 239.05	95 **Am** (243)	96 **Cm** (247)

					VIII A
					2 **He** 4.0026

III A	IV A	V A	VI A	VII A	
5 **B** 10.81	6 **C** 12.01	7 **N** 14.007	8 **O** 15.999	9 **F** 18.998	10 **Ne** 20.179
13 **Al** 26.98	14 **Si** 28.086	15 **P** 30.974	16 **S** 32.06	17 **Cl** 35.453	18 **Ar** 39.95

	I B	II B						
28 **Ni** 58.71	29 **Cu** 63.546	30 **Zn** 65.37	31 **Ga** 69.72	32 **Ge** 72.59	33 **As** 74.92	34 **Se** 78.96	35 **Br** 79.904	36 **Kr** 83.80
46 **Pd** 106.4	47 **Ag** 107.87	48 **Cd** 112.40	49 **In** 114.82	50 **Sn** 118.69	51 **Sb** 121.75	52 **Te** 127.6	53 **I** 126.90	54 **Xe** 131.3
78 **Pt** 195.09	79 **Au** 196.97	80 **Hg** 200.59	81 **Tl** 204.37	82 **Pb** 207.2	83 **Bi** 208.98	84 **Po** (210)	85 **At** (210)	86 **Rn** (222)

65 **Tb** 158.93	66 **Dy** 162.50	67 **Ho** 164.93	68 **Er** 167.26	69 **Tm** 168.93	70 **Yb** 173.04	71 **Lu** 174.97
97 **Bk** (245)	98 **Cf** (248)	99 **Es** (254)	100 **Fm** (253)	101 **Md** (256)	102 **No** (254)	103 **Lw** (257)

Figure 5
Atomic Model of
Lithium.

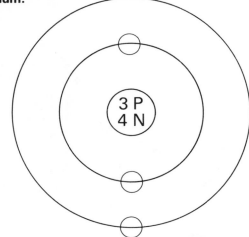

Lithium
Li
Atomic Number 3
Atomic Weight 6.941

Figure 6
Atomic Model of
Beryllium.

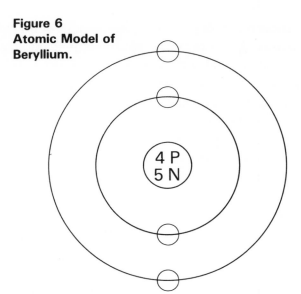

Beryllium
Be
Atomic Number 4
Atomic Weight 9.012

chemical properties at all; that is, they are all gases that do not react with anything, chemically. The other three groups are highly active chemically, while the rest of the elements in the Periodic Table are just "so-so" in their chemical activity. Just now, though, the important thing for you to recognize is that *representative* elements in the same column have chemical properties similar to each other.

We have mentioned several times that the elements on the Periodic Table are listed in order of their increasing atomic numbers. It is now time to define just what atomic numbers and atomic weights are. As we mentioned before and as shown in the drawings of the several elements (beginning with figure 2 on through figure 15), atoms are made up of a nucleus and one or more electrons in orbit around the nucleus (subatomic particles). The nucleus itself contains protons and neutrons (nuclear particles). Remember that both protons and neutrons have mass equal to one atomic mass unit each, and that the orbiting electrons have essentially *no* mass; the atomic weight of the element, therefore, must be the same as the total number of protons and neutrons in the nucleus. The atomic number, on the other hand, is simply the number of protons in the nucleus.

The number of protons in the nucleus of *every* atom of a particular element is *always* the same. You cannot change the number of protons in the nucleus and still have the atom remain that of the element you began with. For example, carbon is atomic number 6. Look at the Periodic Table and you will see a large number 6 either to the left and below this element's symbol, or, in the case of some other Periodic Tables, it will appear above the symbol. In any case, on the same Table, it will always appear in the same position in the box containing the symbol for each element, and it will always be a whole number. The "6" in carbon's case means that there are six protons in the nucleus. If you were somehow to add another proton to the nucleus, the atom would no longer be an atom of carbon but rather an atom of nitrogen; look at nitrogen in the Table and note the number "7" in its box. There is no way, chemically *or* physically, that you can change the number of protons in an atom and have it remain the same element. Carbon *always* has six protons in *each* and *every* atom of carbon on Earth (and everywhere else)!

On the other hand, there are some atoms of carbon that have more than six *neutrons* in the nucleus. This does not make the atom some element other than carbon; it just makes the atom different in its atomic weight. As a matter of fact, the type of carbon known as carbon-14 has eight neutrons in its nucleus, and it is radioactive. Most of the elements have atoms with differing numbers of neutrons in the nucleus, but, just because they do, they are not *chemically* any different from any other atom of the same element. Remember, for any element, the number of protons in the nucleus is *always* the same, but there may be some atoms of an element that have different numbers of neutrons in the nucleus. The name given to an atom of this type is an *isotope*. Not all isotopes are radioactive.

The most stable form of the element is that listed in the Periodic Table; however, you will notice that the atomic weight is not a whole number, but a decimal. This is really a *weighted* average of the atomic weights of all the atoms of a particular element. You can tell how many neutrons there are in the nucleus of the most stable form of the element by looking at the atomic weight, subtracting *from* it the atomic number, and rounding to the nearest whole number. (Refer to table 3.) For example, carbon's atomic weight is listed at 12.01 (again, you will see very slight differences on different Tables). Since the atomic number is 6, then 12.01 minus 6 equals 6.01. Rounding to the nearest whole number gives the answer as six neutrons in the nucleus of the most stable form of carbon, which might then be called carbon-12 (we will not attach numbers to the elements, *unless* we mean the

**Figure 7
Atomic Model of
Boron.**

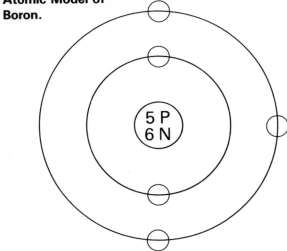

**Boron
B
Atomic Number 5
Atomic Weight 10.81**

**Figure 8
Atomic Model of
Carbon.**

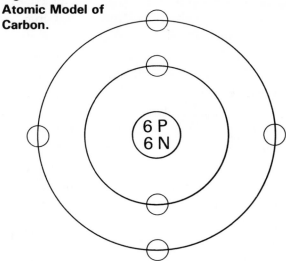

**Carbon
C
Atomic Number 6
Atomic Weight 12.011**

**Figure 9
Atomic Model of
Nitrogen.**

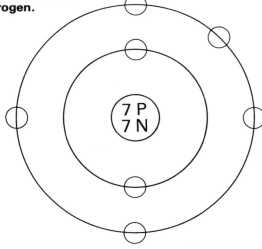

**Nitrogen
N
Atomic Number 7
Atomic Weight 14.007**

**Figure 10
Atomic Model of
Oxygen.**

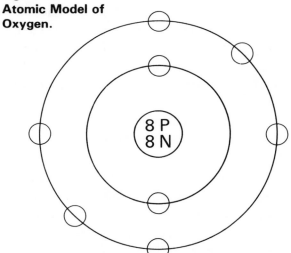

**Oxygen
O
Atomic Number 8
Atomic Weight 15.999**

**Figure 11
Atomic Model of
Fluorine.**

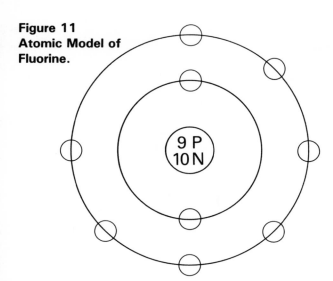

**Fluorine
F
Atomic Number 9
Atomic Weight 18.998**

**Figure 12
Atomic Model of
Neon.**

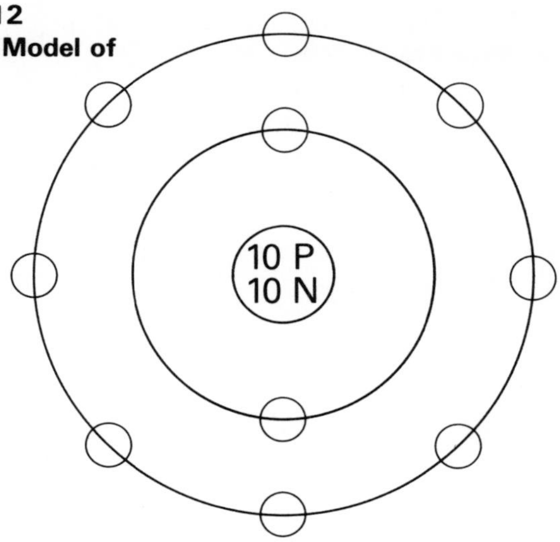

**Neon
Ne
Atomic Number 10
Atomic Weight 20.179**

**Figure 13
Atomic Model of
Sodium.**

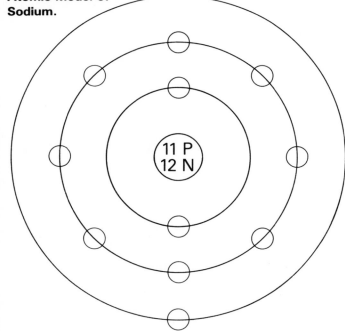

**Sodium
Na
Atomic Number 11
Atomic Weight 22.99**

**Figure 14
Atomic Model of
Chlorine.**

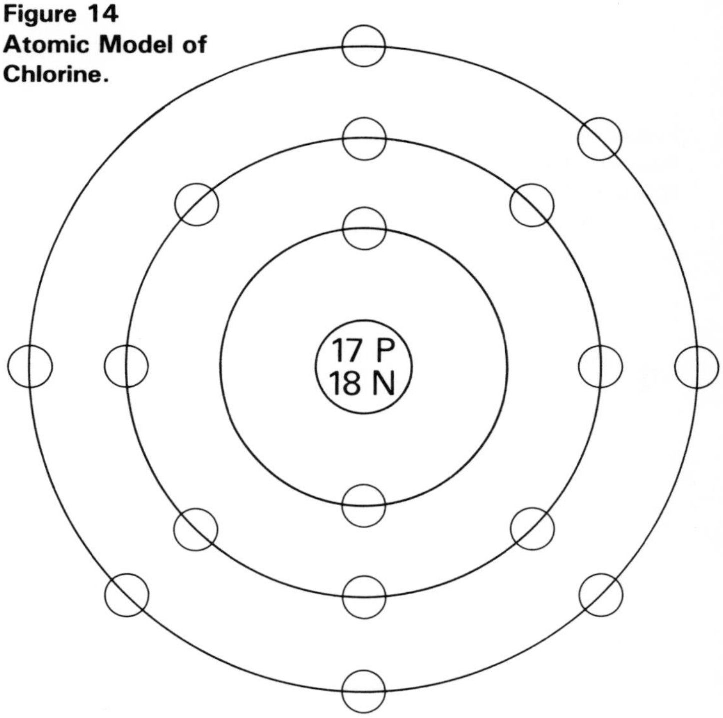

**Chlorine
Cl
Atomic Number 17
Atomic Weight 35.453**

**Figure 15
Atomic Model of
Argon.**

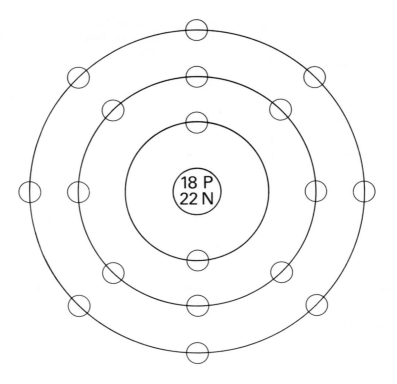

**Argon
Ar
Atomic Number 18
Atomic Weight 39.948**

isotopes). For sodium, there are 12 neutrons in the nucleus. Do you see why? For calcium, there are 20 neutrons. Right? And for chlorine, there are 18 neutrons. Therefore, by looking at the Periodic Table, you can tell (if you have memorized the chemical symbols) the atomic number of any element, its atomic weight, and how many protons and how many neutrons it has in its nucleus. You can also describe all of the chemistry of an element; that is, how it will combine chemically with other elements. How you do that is the topic of Chapter Two.

The difference between metals and non-metals is more than appearance, although most metals are easily recognized. Metals (we are discussing elements, not *alloys*, which are physical mixtures of the elemental metals, created by melting them and then stirring them together) are usually crystalline in nature and have a "metallic" luster, are good conductors of heat and electricity, may be hammered into sheets or drawn into wire, and, more technically, are materials whose oxides form hydroxides (bases) with water. They are all solid, with one exception, mercury, which is a liquid.

Non-metals, on the other hand, are the elements that are not metals. They are usually amorphous (non-crystalline) in nature, are usually poor conductors of heat and electricity, and cannot be drawn into wire or hammered into sheets. Chemically, their oxides form acids with water, rather than bases as the metallic oxides do. They are all solids or gases, with one exception, bromine, which is a liquid.

Electronic Configuration

The physics of an element is contained in its nucleus, but the chemistry of an element is contained in its electrons, or, more correctly, how those electrons align themselves when they orbit the nucleus.

Since all atoms are electrically neutral, protons have an electrical charge of $+1$, and electrons have an electrical charge of -1, each atom must have as many positive charges as it does negative charges. Therefore, it is easy to determine how many electrons an atom has in orbit around the nucleus by looking at its atomic number. If carbon has six protons in the nucleus, it must have six electrons in orbit. The same principle is true for every element. Sodium has 11 electrons in orbit, potassium has 19, and krypton has 36. Remember, you can tell how many electrons there are because the atom requires the same number of electrons as protons to become electrically neutral. The atomic number is defined as the number of protons in the nucleus, not how many electrons there are in orbit.

We have said quite a few times that the chemistry of an element is determined by the electronic configuration of the atoms, and that is true. It is the configuration of the electrons in the *last* orbit (the electrons in the outer ring) that is the true determinant. Look at the drawings you have of different elements. Each has a nucleus, but in some there are differing numbers of "rings," or orbitals, around the nucleus. Hydrogen and helium have only one ring, while lithium through neon each have two rings, and sodium through argon have three rings. This is due to a physical law that the electrons obey as they occupy each ring. The rules are:

1. There may be one or two electrons in the first ring, but no more than two.
2. The second ring may have one through eight electrons, but no more than eight.

Table 3 / Atomic Numbers and Atomic Weights of the Elements

Name	Symbol	Atomic Number	Atomic Weight	Name	Symbol	Atomic Number	Atomic Weight
Actinum	Ac	89	(227)	Erbium	Er	68	167.26
Aluminum	Al	13	26.98	Europium	Eu	63	151.96
Americium	Am	95	(243)	Fermium	Fm	100	(253)
Antimony	Sb	51	121.75	Fluorine	F	9	18.998
Argon	Ar	18	39.95	Francium	Fr	87	(223)
Arsenic	As	33	74.92	Gadolinium	Gd	64	157.25
Astatine	At	85	(210)	Gallium	Ga	31	69.72
Barium	Ba	56	137.34	Germanium	Ge	32	72.59
Berkelium	Bk	97	(245)	Gold	Au	79	196.97
Beryllium	Be	4	9.01	Hafnium	Hf	72	178.49
Bismuth	Bi	83	208.98	Helium	He	2	4.0026
Boron	B	5	10.81	Holmium	Ho	67	164.93
Bromine	Br	35	79.904	Hydrogen	H	1	1.0080
Cadmium	Cd	48	112.40	Indium	In	49	114.82
Calcium	Ca	20	40.08	Iodine	I	53	126.90
Californium	Cf	98	(248)	Iridium	Ir	77	192.22
Carbon	C	6	12.01	Iron	Fe	26	55.847
Cerium	Ce	58	140.12	Krypton	Kr	36	83.80
Cesium	Cs	55	132.91	Lanthanum	La	57	138.91
Chlorine	Cl	17	35.453	Lawrencium	Lr	103	(257)
Chromium	Cr	24	51.996	Lead	Pb	82	207.2
Cobalt	Co	27	58.933	Lithium	Li	3	6.941
Copper	Cu	29	63.546	Lutetium	Lu	71	174.97
Curium	Cm	96	(247)	Magnesium	Mg	12	24.305
Dysprosium	Dy	66	162.50	Manganese	Mn	25	54.938
Einsteinium	Es	99	(254)	Mendelevium	Md	101	(256)

3. The third ring may have one through eight electrons, but no more than eight (except where the transition elements begin).

4. Each succeeding ring will finish the period (row) with eight electrons in the last ring.

Looking at the drawings of the elements again, notice the similarity that occurs in the electronic configuration of the first two elements in the alkali metals (group IA). Although sodium and lithium have different numbers of electrons in orbit, the similarity is that both have one electron in the outer ring. If you were to continue drawing the electronic configuration of all the elements, when you came to potassium and drew in its 19 electrons, if you followed the rules of

Name	Symbol	Atomic Number	Atomic Weight	Name	Symbol	Atomic Number	Atomic Weight
Mercury	Hg	80	200.59	Samarium	Sm	62	150.4
Molybdenum	Mo	42	95.94	Scandium	Sc	21	44.956
Neodymium	Nd	60	144.24	Selenium	Se	34	78.96
Neon	Ne	10	20.179	Silicon	Si	14	28.086
Neptunium	Np	93	237.05	Silver	Ag	47	107.87
Nickel	Ni	28	58.71	Sodium	Na	11	22.99
Niobium	Nb	41	92.906	Strontium	Sr	38	87.62
Nitrogen	N	7	14.007	Sulfur	S	16	32.06
Nobelium	No	102	(254)	Tantalum	Ta	73	180.95
Osmium	Os	76	190.2	Technetium	Tc	43	98.906
Oxygen	O	8	15.999	Tellurium	Te	52	127.6
Palladium	Pd	46	106.4	Terbium	Tb	65	158.93
Phosphorus	P	15	30.974	Thallium	Tl	81	204.37
Platinum	Pt	78	195.09	Thorium	Th	90	232.04
Plutonium	Pu	94	239.05	Thulium	Tm	69	168.93
Polonium	Po	84	(210)	Tin	Sn	50	118.69
Potassium	K	19	39.102	Titanium	Ti	22	47.9
Praseodymium	Pr	59	140.91	Tungsten	W	74	183.85
Promethium	Pm	61	(147)	Uranium	U	92	238.03
Protactinium	Pa	91	231.04	Vanadium	V	23	50.941
Radium	Ra	88	226.03	Xenon	Xe	54	131.3
Radon	Rn	86	(222)	Ytterbium	Yb	70	173.04
Rhenium	Re	75	186.2	Yttrium	Y	39	88.906
Rhodium	Rh	45	102.91	Zinc	Zn	30	65.37
Rubidium	Rb	37	85.468	Zirconium	Zr	40	91.22
Ruthenium	Ru	44	101.07				

filling orbitals, you would place two electrons in the first ring, eight electrons in the second ring, eight electrons in the third ring, and finally one electron in the outer ring. This similarity (that is, having the same number of electrons in the outer ring as other elements in the same group) is responsible for the so-called "family effect" that causes each element in a group to have chemical properties similar to other elements in that group. The family effect is most prevalent in groups IA, IIA, VIIA, and VIIIA. It is very weak in other groups, but it is still there.

Looking at the drawings of the electronic configuration for fluorine (figure 11) and chlorine (figure 14) will show that they also share a similarity, that of having the same number of electrons in the outer ring, seven, and therefore possessing similar chemical properties. Probably the most interesting elements are the members of group VIIIA, the noble gases. Again, they all have the same number of electrons in the outer ring and therefore have similar chemical properties, but, in this case, similar chemical properties means they do not do anything (hence, the reference to nobility). This inactivity is because these elements and their electronic configuration satisfy the octet rule.

The octet rule states that all atoms must strive to reach an ultimate state of stability; that is, they *must* attempt to reach a state where the atom is totally satisfied to be totally stable and need not enter into any chemical reactions to become stable. Nature says that all atoms must have eight electrons in the outer ring to reach this state of stability. Once this state occurs, the atom will not react chemically with anything else. You can see that the elements of group VIIIA have reached this state, and that fact explains why they will not react chemically with anything. Helium, which sits atop this group, looks like an exception to the octet rule, and it is. Since it (and hydrogen) follow rule number 1 of filling orbitals with electrons, and helium has only two electrons while hydrogen has one, they follow a special case of the octet rule, called the duet rule. The duet rule states that an element reaches its ultimate state of stability when it has two electrons in the outer ring. Therefore, helium truly does belong in group VIIIA.

It is most important for you to remember the family effect, the octet rule, and the duet rule, because these phenomena explain why every chemical reaction that can or will occur, will happen. All the chemistry you will ever need is tied up in these very simple principles.

The Kinetic Molecular Theory

The kinetic molecular theory is a fancy name for a rather simple premise. The word *kinetic* means moving, *molecular* has to do with molecules, and a *theory* is a supposition set forth by someone who has made certain observations (scientific, it is hoped) about some phenomena, and, on the basis of this supposition, is able to make certain predictions and explanations. A theory will exist until someone (or some event) disproves it. Very simply stated, the kinetic molecular theory states that *all* molecules are in constant motion, as long as the temperature of the matter under observation is at a temperature above absolute zero (− 459.67°F., or − 273.16° C.). At absolute zero, according to the theory, all molecular motion (and hence all life) ceases.

The theory says that, at any given temperature, all molecules of a particular type will be vibrating (moving) at a certain rate, and as the temperature of the matter involved is increased (that is, as more energy is applied to and absorbed by the matter), the speed of the vibration or motion of the individual molecules will increase. Conversely, as the temperature of the matter involved drops (that is, as energy is given up by the matter), the motion or vibration of the molecules will decrease. Using a familiar substance, water, the following scenario should demonstrate the operation of the theory in a simple manner.

At room temperature, water is in its natural state of matter, liquid, and its molecules are moving at a constant rate (in actual fact, they are sliding over each other in the characteristic manner of molecules in the liquid state). As energy in the form of heat is applied to the water, the individual molecules of water begin to move faster and faster, and the rate of the molecules that are escaping through the surface of the liquid (evaporation) speeds up. This fact explains why warm or hot water evaporates faster than cold water. As the water absorbs more and more energy (that is, the water gets warmer and warmer), the molecules move faster and faster, and, of course, the evaporation rate increases as the more energetic water molecules escape into the atmosphere; these escaped molecules are known as water vapor. Finally, at the boiling point of water, the molecules are moving at the fastest rate possible (at atmospheric pressure). At this point, the molecules possess enough energy so that tremendous numbers of molecules are escaping, and, if more and more energy is applied, all the molecules will undergo the phase change from liquid to vapor; at the boiling point of water, this water vapor is known as steam. Let us assume that this is what has happened, so we now have a room full of hot vapor, with the individual water molecules moving as fast as they can.

If we reverse the heating process (that is, if we withdraw energy, or heat, from the molecules), the molecules begin to slow down, the intermolecular forces that attract molecules to each other begin to take over again, and the vapor undergoes a phase change back to a liquid (it is said to condense), with the molecules sliding over each other as they did before. If we continue to withdraw energy, the motion of the molecules will continue to slow down, and much of the molecules' freedom of movement is threatened by the previously mentioned intermolecular forces. As more and more energy is removed, the liquid water cools down to its freezing point; *at* the freezing point, the molecules have slowed down so much that they become locked to each other, undergoing a change to the third state of matter, solid, better known (in the case of water) as ice. Even in the solid state, the molecules still possess enough energy to have motion. In this case, the molecules are not sliding over each other as they were in the liquid state or in random, free motion, as in the case of the vapor state but instead are vibrating to some degree. To overcome the forces holding the molecules in the solid state, all that must be done is to apply heat energy. As energy is applied to the ice, the molecules begin to vibrate faster and faster, until they possess enough energy to overcome the forces holding them in the locked position and begin to slide over each other as liquid molecules (that is, the ice melts).

Whether we use water as the example, or start with a piece of wood (made up mostly of cellulose, a long-chain molecule in a constant state of vibration, which, when enough heat energy is applied, begins to vibrate so rapidly that molecular fragments break off as gases), or a bar of metal, the principle is the same: the molecules move faster as heat energy is absorbed, and slower as the energy is withdrawn. The kinetic molecular theory explains many physical reactions that occur in nature and explains every situation that leads to a fire! These will be called to your attention as we progress through the next several chapters.

Glossary

Absolute Zero: The temperature at which all molecular motion ceases: −459.67°F. or −273.16°C.

Alkali Metals: The elements of group IA.

Alkaline Earth Metals: The elements of group IIA.

Alloy: A physical mixture of two or more metals.

A.M.U.: Atomic mass unit: the unit of weight used to state atomic or molecular weights of atoms or molecules. The proton and the neutron each weigh 1 a.m.u.

Atom: The smallest particle of an element that can still be identified as the element.

Atomic Number: The number of protons in the nucleus.

Atomic Weight: The total number of protons and neutrons in the nucleus. A fractional atomic weight is due to the averaging of the weights of all the isotopes of the element.

Chemical Reaction: A chemical change which occurs when two or more substances are brought together, and energy is either absorbed or liberated; the types of chemical reactions are oxidation, reduction, ionization, combustion, polymerization, hydrolysis, and condensation.

Chemistry: The science of matter, energy, and reactions.

Combustion: The rapid chemical combination of a substance with oxygen, usually accompanied by the liberation of heat and light.

Compound: A chemical combination of two or more elements, either the same elements or different ones, that is electrically neutral.

Duet Rule: A special case of the octet rule; the rule which states that to reach stability, an atom (or ion) must have two electrons in its only orbital.

Electron: A subatomic particle that has essentially no weight and has an electrical charge of -1.

Element: A pure substance that cannot be broken down into simpler substances by chemical means.

Energy: The capacity for doing work.

Family Effect: The fact that each element in the "family" or group has very similar chemical properties, because of the number of electrons in the outer ring.

Halogens: The elements of group VIIA: fluorine, chlorine, bromine, iodine, and astatine.

Inert Gases: The elements of group VIIIA: helium, neon, argon, krypton, xenon, and radon; also known as the noble gases.

Isotope: A form of the same element having identical chemical properties but a different number of neutrons in the nuclei of its atoms.

Kinetic Molecular Theory: A theory which states that all molecules are in constant motion at any temperature above absolute zero. Furthermore, as energy is absorbed by the molecules, their speed of motion or vibration will increase, and as energy is withdrawn, molecular motion will decrease.

Main Group: Those elements of the groups with an A after the Roman numeral: the vertical columns headed by hydrogen, beryllium, aluminum, carbon, nitrogen, oxygen, fluorine, and helium; also known as the representative elements.

Matter: Anything that has mass and occupies space.

Mixture: A substance made up of two or more compounds, physically mixed together; a mixture may also contain elements and compounds mixed together.

Molecular Formula: The use of chemical symbols to show what atoms and how many of them are present in a compound.

Molecule: The smallest particle of a compound that can still be identified as the compound; two or more atoms bound together chemically by covalent bonds and electrically neutral.

Neutron: A nuclear particle that has an atomic weight of one and is electrically neutral.

Noble Gases: The elements of group VIII: helium, neon, argon, krypton, xenon, and radon; also known as the inert gases.

Nucleus: A subatomic particle that contains all the positive charge and essentially all the weight of the atom.

Octet Rule: The rule which states that to reach stability, each atom must have eight electrons in its outermost ring; the fulfillment of the octet rule is what causes ionization.

Orbital: The "ring" around the nucleus that contains the orbiting electrons.

Oxidation: The chemical combination of a substance with oxygen.

Periodic Table: A systematic arrangement of all the known elements by their atomic numbers, which demonstrates the periodicity, or regular repeating, of chemical properties of the elements.

Proton: A nuclear particle with an atomic weight of 1 and an electrical charge of $+1$.

Rings: The paths around the nucleus in which the orbiting electrons travel.

Symbol: The chemical shorthand used instead of the full name of an element.

Valence Electrons: The electrons in the atom's outermost ring; the configuration of these electrons determines the chemistry of each element.

2

Compounds

Introduction

So far, all of our discussions have concerned elements, the basic building blocks of the universe. We have also mentioned how Nature hates to allow *anything* to be in an unstable condition. As a matter of fact, Nature just *will not* allow anything to be unstable, thereby forcing whatever is unstable to find a way to become stable. Now, if you will recall a previous section where it was stated that all atoms *must* strive to reach a state of stability, and that state was eight electrons in the outer ring (with the sole exception of a few elements that strive to have two electrons in the outer ring), how is it that oxygen, which has six electrons in the outer ring, can exist in this seemingly unstable state?

The answer is, of course, that it does not. Oxygen cannot and does not exist in the elemental state; that is, oxygen does not exist in nature as an element. It cannot, because oxygen is too active chemically to exist by itself; furthermore, there is a law that states all atoms must have eight electrons in the outer ring (the octet rule), and oxygen has six. This fact, that oxygen has six electrons in its outer ring, is *what makes oxygen active chemically*. It is this lack of stability or rather this striving *for* stability that produces chemical activity in any chemical species, whether it is an element, a compound, or a molecular fragment. Therefore, oxygen does not exist in our atmosphere as O, but rather as O_2. The reason it does is simple. In striving to reach its state of ultimate stability, oxygen will do *whatever* it must in order to get eight electrons in the outer ring. As a result, if any oxygen does exist ever so briefly as O, it immediately reacts with another O to form O_2. Picture the oxygen structure in figure 10; it clearly shows six electrons in the outer ring. Now, bring this atom together with another atom of oxygen (again, with six electrons in the outer ring), and you have the structure shown in figure 16. Look carefully, and you will see how the two atoms of oxygen have combined, and how they *both* now believe they each have eight electrons in the outer ring! They have both reached stability, and atmospheric oxygen does indeed exist as O_2.

This is a highly important concept to grasp, but you already had some forewarning of it. Most, if not all of you, from time to time have

**Figure 16
Molecular Model of
Oxygen.**

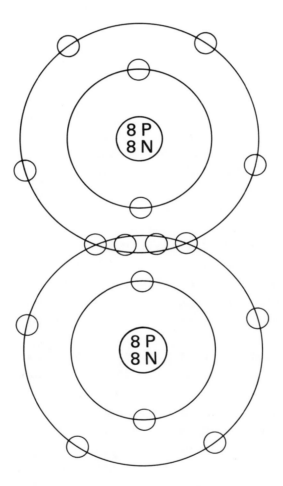

**Oxygen
O_2
Atomic Weight 31.998**

referred to oxygen as O_2; you may have known nothing about the octet rule or any other fundamental rule of chemistry, but you did know that O_2 meant the oxygen in the air we breathe. The fact that O_2 is a *compound* and not the element oxygen should not bother you. What may be confusing is that only some of the gases exist as molecules (the unit particles of a compound). Remember, only certain gases behave in this way, and these are mentioned in the next paragraph. The gases that exist as monatomic (one atom in the molecule) rather than diatomic (two atoms in the molecule) are the noble, or inert, gases of group VIIIA. It is not necessary for these gases, helium, neon, argon, krypton, xenon, and radon, to combine with another of their own atoms to form a molecule to reach stability because they are already stable; helium has two electrons in its only ring and the others have eight in the last ring. As a matter of fact, they are so stable that they will not react with *anything*! This is why they are called inert.

The octet rule is the driving force behind the formation of molecules. If an atom does not have eight electrons in the outer ring (or two, in certain cases), it *must* react with something to reach stability. Some atoms do this by forming molecules, which are the smallest parts of a compound that can still be identified as that compound; some atoms do it by forming ions, which will be explained later in this chapter. When a molecule is formed, the atom must react with another atom in order to form it. This process is not necessary in the case of the inert gases, since they have already reached stability and should be considered *both* as atoms and as monatomic molecules. Therefore, if by some chance there existed an atom of oxygen, it would be very unstable because of the six electrons in its outer ring, it would *have* to react with something, and it does. If there is another atom of oxygen nearby, it forms the compound oxygen, whose molecular formula is O_2. The important concept for you to understand is that the striving for stability, the striving of an element to achieve eight electrons in the outer ring (or two, in those few special cases) is what causes a chemical reaction to take place. If all the atoms of all the elements had full outer rings, there would be no possible chemical reactions and hence no life.

Hydrogen, an element with only one electron in its outer (and only) ring, will behave in the same way. When it combines with another atom of hydrogen, each atom thinks it has two electrons in the outer ring, and therefore by the duet rule both are satisfied. Hydrogen does not exist as H, but rather as H_2. It does not exist as H because of the lone electron in its outer ring, whose presence drives it to seek out another atom with which it can combine to provide the additional electron it needs for stability. (See figure 17, for examples.) The same is

Figure 17
Molecular Model of
Hydrogen Fluoride.

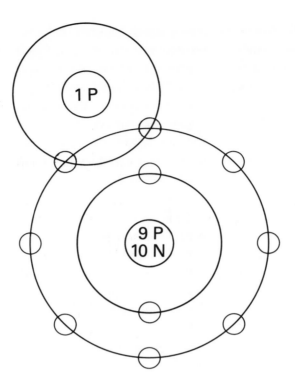

Hydrogen Fluoride
HF
Molecular Weight 20.006

true for all the elemental gases except the noble gases. Nitrogen exists as N_2, fluorine exists as F_2, and chlorine exists as Cl_2. This is so because nitrogen has five electrons in its outer ring, and fluorine and chlorine each have seven. These atoms will combine or react with atoms of other elements to enable each of them to achieve eight electrons in their outer rings.

Now consider the possibility of the combination of hydrogen and oxygen. Oxygen needs to add two electrons to its six to reach stability, while hydrogen needs to add one to satisfy the duet rule. It is easy to see how hydrogen will attach itself to oxygen in order to make it seem as if hydrogen has two electrons in its outer ring, and it is therefore satisfied. That leaves oxygen with only seven electrons in the outer ring, however, and it still needs one more. Therefore, *another* hydrogen atom will attach itself to the oxygen, and while the hydrogen atom is satisfied with what it thinks are *its* two electrons, oxygen is satisfied because it now believes it has eight electrons in its outer ring. And presto, H_2O, or water. (See figure 18 for examples of other compounds.)

This kind of chemical attachment has a special name, *covalent bonding*. We will have much more to say about it in Chapter Three later, but first we want to look at the other common form of chemical bonding, ionic bonding. We will then return to covalent bonding.

At this point, a distinction must be drawn between the kind of instability an element shows when its atoms do not satisfy the octet or duet rules (eight or two electrons in the outermost ring), and the instability of a *compound*. We have already mentioned that each atom of each element will strive to have eight electrons in its outermost ring (or two, in a few instances), so that it can reach the state of stability represented by the inert gases in group VIIIA, which are totally satisfied as they are and will enter no chemical reactions. This condition is one form of stability and is used only when discussing electronic configurations, and why elements will react *chemically*. The other form of stability relates to compounds, and their inclination to *break down* into their components, usually under the slightest provocation, usually with the evolution of a hazardous substance (a toxic gas, a fuel, or an oxidizer), and usually at the worst possible time.

An example would be hydrogen peroxide, H_2O_2, which will decompose, liberating oxygen. The decomposition can be slow or violent, depending on the concentration of the hydrogen peroxide and the nature of the reason for its breakdown (heat, shock, or contamination). Another, more familiar example of the instability of a compound is contaminated nitroglycerine, which will decompose explosively when heated or shocked, even if very slightly.

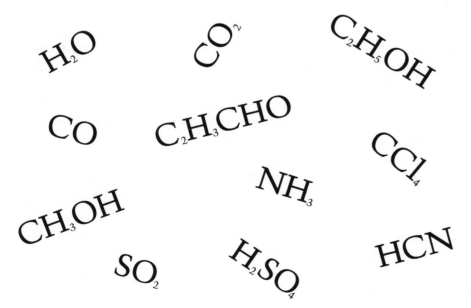

**Figure 18
Compounds.**

On the other hand, examples of stable compounds include water, carbon dioxide, and silicon dioxide, SiO_2, which is sand. These compounds are so stable that they do not enter into chemical reactions under normal conditions and, in fact, are so stable that they are all used as fire-extinguishing agents. This does *not* mean, however, that they *cannot* be broken down. We will see (in later chapters, when we discuss combustible metals), that there are occasions where these materials *will* break down, but such extraordinary circumstances do not mean that the materials are usually unstable.

From this point on, then, unless reference is made directly to an element, or to an atom of that element, stability and instability will refer to the *chemical* stability of a compound or an ion. Later, in Chapter Four, we will introduce the radical, or functional group, which may also be unstable. Stability means that the compound will resist entering into a chemical reaction of one sort or another; instability means that the compound (and, in some cases, the mixture) will undergo some chemical change very easily, usually when you do not want it to. Not *all* compounds that are unstable are hazardous, but, since the subject we are studying is hazardous materials, you may assume that any reference to an unstable substance or any mention of a material as unstable means that something bad might happen to the material, the surroundings, and *you*, at any time.

Ionic Bonding

When elements that are metals react chemically with elements that are non-metals, the resulting chemical compound is called a *salt*, and the type of bonding that occurs is called *ionic*. It is important to remember that ionic bonding takes place *only* between a metal and a non-metal. This is another good reason for you to be able to look at the Periodic Table and tell the difference between metals and non-metals. (Eventually, you will have the differences memorized.) Ionic bonding is quite different from covalent bonding in that while it becomes obvious to the observer that there is a *sharing* of electrons in covalent bonding, what happens in ionic bonding is different. Just now, though, you are expected to recognize that when a compound contains a metal and a non-metal, the compound is ionic.

When an atom of a metallic element comes near an atom of a non-metallic element, an amazing thing happens: one or more electrons actually leave the orbit (the outer ring) of the metal atom and jump into the orbit (again, the outer ring) of the non-metal atom. The number of electrons that leave the metal atom are equal to the number of electrons in its outer ring (the same number, incidentally, as the group it belongs to); that is, all the alkali metals lose one electron, all the alkaline earth metals lose two electrons, and aluminum, the only metal in group IIIA that is important to this course, loses three electrons. The number of electrons that jump into the outer ring of the non-metallic element depends on just how many electrons are needed to fill up, or total, eight electrons in the outer ring. Thus, the non-metals of Group VA will accept three electrons, the non-metals of Group VIA will accept two electrons, and all the halogens (group VIIA) will accept one electron. By accepting these electrons, the non-metals now have eight electrons in the outer ring, while the metals also have eight electrons in the outer ring (or two, to satisfy the duet rule). In any case, both metal and non-metal are now satisfied that they are in a state of stability.

It will soon become obvious to you that if an atom of aluminum, the metal in group III, loses its three electrons to a non-metal, and the atom of the non-metal that has been brought near can accept only *one* electron, that it will be necessary for three of the atoms of the non-metal to participate in the reaction. If you can grasp this situation, you will have no problems understanding the formation of ions, or the writing of the chemical formulas for the ionic compounds.

What is important for you to know is that once an electron or electrons are lost or gained by an atom, *it is no longer an atom*. Since all atoms are electrically neutral, and these particles that were once atoms now have more or fewer electrons than protons, it must be obvious that the particles that were once atoms now carry an electrical charge; therefore they can no longer be atoms. They are now *ions*. An ion is a charged particle formed by an atom gaining or losing electrons. You will see later in this chapter that an ion may also be formed from a group of atoms, bound together chemically, which have collectively gained or lost electrons.

These charged particles called ions have electrical charges equal to the number of electrons gained or lost. For example, when the alkali metals lose their one electron in the outer ring, they become ions that carry an electrical charge of $+1$; the charge is positive because the electron has a charge of -1 and the proton is $+1$; since one electron is now gone, there is one more proton in the nucleus than there are electrons in orbit. When the alkaline earth metals lose two electrons, they become ions with a charge of $+2$; when aluminum loses its three electrons in the outer ring, it becomes an ion with a $+3$ charge. On the other hand, when the halogens gain one electron, they become ions with a − 1 charge. Do you see why? The same reasoning will show that the non-metals of group VIA will produce ions with a − 2 charge, while − 3 charged ions will be produced when the non-metals of group V gain their three electrons. These negatively charged ions are called *anions*, while the positively charged ions are called *cations*.

Look at an example and see what happens. When sodium, an alkali metal, is placed in a container that has an atmosphere of 100 percent chlorine, and some energy, *always* necessary to begin a chemical reaction, is introduced, a violent chemical reaction takes place. If we started with 22.99 grams of sodium and 35.453 grams of chlorine, we would now have 58.443 grams of sodium chloride, common table salt. In the process of the chemical reaction, one electron from each of an incredible number of sodium atoms leaped from its outer ring to the outer ring of an equal number of chlorine atoms, creating a compound, written as NaCl. The exact amounts of the *reactants*, sodium and chlorine, were selected because these weights expressed in grams correspond to their atomic weights, and if one atom of sodium reacts with an atom of chlorine, two ions are attracted to each other to form the compound, sodium chloride. In other words, if the ratios of reactants are kept equal to their atomic weights, there will be no reactant left over. If you used equal *weights* of the reactants, there would be an excess of one left over after the reaction.

To prove that the compound is ionic (that is, made up of ions rather than atoms), you can run a simple experiment. Place electrodes leading from the positive and negative ends of a dry-cell battery in pure (distilled) water, and, using the proper meter, measure the flow of electricity through the water. The reading will be zero! Now add a few grams of common table salt; instantly, as the salt dissolves, a current will begin to flow. This is due to the presence of positively and negatively charged particles in the water, with electrons flowing from one electrode, along the countless cations to anions to the other electrode. The definition of electricity is the flowing of electrons. If you try the same experiment with sucrose (common table sugar), you will get a reading of zero even after dissolving the sugar in the water, because sucrose is covalently bonded. Sodium chloride is ionically bonded, and when it is dissolved in water, the cations and anions separate, thus becoming able to carry an electrical charge. While the material is in solid form, the ions are held very tightly, but they could still conduct electricity if they were melted, rather than if they were dissolved. This is because the electrical charges on the ions are opposite, and opposite charges attract. So the force that holds a solid ionic compound together is electrostatic in nature and opposite in charge. This force is extremely strong, as evidenced by the fact that sodium chloride and other ionically bonded compounds are exceedingly stable (not able to be broken down into their components very easily). When they are dissolved in water, however, this electrostatic force is overcome, and the ions actually do separate. As the water is evaporated, however, the electrostatic attraction becomes stronger and the ions come closer and closer to each other; as the last molecules of water are evaporated, the solids re-form and once again are held together very strongly.

The Naming of Ionic Compounds

When there are only two elements in the compound, even though there may have been more than one atom of each kind in the beginning of the reaction, the compound is known as a *binary* compound. When one element is a metal and the other element is a non-metal, the compound is ionic. The naming of ionic compounds is quite simple. The positively charged (metal) ion is always named first, and its name is the same as the name of the element or atom; for example, when the element is sodium, the resulting ion is the sodium ion; if it is calcium, it is the calcium ion. The negative (non-metal) ion is always named second, and the name of the element is changed to end in *-ide*; for example, oxygen becomes the oxide ion, chlorine becomes the chloride ion, and sulfur becomes the sulfide ion. Therefore you have compounds with names like sodium fluoride, potassium chloride, lithium bromide, aluminum oxide, calcium sulfide, barium nitride, and so on. If you consider eleven different metals (lithium, sodium, potassium, rubidium, cesium, beryllium, magnesium, calcium, strontium, barium, and aluminum) and ten different non-metals (hydrogen, carbon, nitrogen, phosphorus, oxygen, sulfur, fluorine, chlorine, bromine, and iodine), you should now be able to name one hundred and ten different binary ionic compounds. You can do this by starting with lithium and put it together with the anion (the negative ion formed by the non-metal), remembering to change the last syllable of the non-metal so that it ends in -ide. For example, lithium hydride, lithium carbide, lithium nitride, lithium phosphide, lithium oxide, lithium sulfide, lithium fluoride, lithium chloride, lithium bromide, and lithium iodide. Next, do this with sodium, then potassium, and so on. With ten ionic compounds named for each of eleven metals, you will reach a total of one hundred and ten. It is as simple as that. When we get to complex ions, there will be more compounds to name.

When an atom loses or gains an electron to become an ion, the process is called *ionization*. When this change occurs, the chemical symbol for the atom becomes the chemical formula for the ion. That is, Na for the sodium atom becomes Na^{+1} for the sodium ion, Ca for the calcium atom becomes Ca^{+2} for the calcium ion, and Al for aluminum becomes Al^{+3} for the aluminum ion. On the other hand, Cl for the chlorine atom becomes Cl^{-1} for the chloride ion, O becomes O^{-2} for the oxide ion, and N becomes N^{-3} for the nitride ion. The $+1$ charge is the same for all the alkali metals, and the $+2$ charge is the same for all the alkaline earth metals. The -3 charge is the same for any non-metal ion in group V, the -2 charge is the same for the ions formed by the non-metals of group VI, and the -1 charge is the same for all the halogens. Since the inert gases are already at stability, they will not gain or lose electrons to form ions. It is important for you to know the chemical formula for the ion (which includes the electrical charge by size and sign), so you can write the chemical formula for the ionic compound. The chemical formula *never* shows the charges. Since all compounds are electrically neutral, however, it is important to know the formula of the ion so you will have the proper number of cations and anions in the formula to balance it electrically.

Chemical Formulas for Ionic Compounds

Like atoms, all chemical compounds are electrically neutral. If ions are charged particles (which they are), and if they are electrostatically attracted to one another to form ionic compounds (which they are), and if all chemical compounds are electrically neutral as stated above, then there *must* be a balancing of cations and anions to leave an electrical charge of 0, and there is. That is, in every ionic compound, there will be an equal number of positive and negative charges. If any alkali metal combines with any halogen, it will *always* be at a one-to-one ratio, so there will be one cation and one anion, as in sodium chloride, NaCl. Similarly, if an alkaline earth metal combines with a non-metal from group VIA, or if aluminum combines with a non-metal from group VA, it will always be on a one-to-one basis. Obviously it takes one ion with a $+1$ charge to balance one ion with a -1 charge, as well as an ion with a $+2$ charge to balance an anion with a -2 charge, and so on. It is just slightly trickier when the ions that combine carry unequal charges. For example, if barium combined with the chloride ion, it would take two chloride ions to balance one barium ion. If sodium combined with the oxide ion, it would take two sodium ions to balance one oxide ion. And if aluminum combined with the oxide ion, it would take two aluminum ions to balance the charge of *three* oxide ions. Therefore, the chemical formulas would be $BaCl_2$, Na_2O, and $AlCl_3$. Can you see why formulas like LiI, BaS, and AlN are correct? Again, notice that the electrical charges of the ions are *not* shown, even though they *are* present. They are *understood* to be there, but, since the compounds are electrically neutral (that is, the *total* positive charges of the cations balance out the *total* negative charges of the anions), there is no reason to show them.

Knowing that all chemical compounds are electrically neutral, and that metals combine with non-metals to form salts, which are ionic, and that the positive charges and the negative charges *must* balance each other, you should now be able to write out correctly the chemical formulas for the one hundred and ten binary ionic compounds you learned to name in the preceding section.

Notation

Please remember that whenever we have discussed compounds, ions, and elements, we have used the term "chemical formula" for the shorthand used to represent compounds and ions, and "chemical symbol" for the shorthand used to represent elements. You must learn to recognize the difference between O and O_2. O is the symbol for the element oxygen while O_2 is the chemical formula for the compound oxygen, which is what actually exists in our atmosphere. Similarly H is the chemical symbol for the element hydrogen, while H_2 is the chemical formula for the compound hydrogen.

Formulas are also used to designate ions. Na is the symbol for the element sodium, while Na^{+1} is the formula for the sodium ion. Al is the symbol for the element aluminum, while Al^{+3} is the formula for the aluminum ion, as Cl is the symbol for the element chlorine, while Cl^{-1} is the formula for the chloride ion. It is important to remember that symbols represent only the element or an atom of the element, while the formula represents an ion of a compound that may have been formed. The Periodic Table contains only symbols, one in each box. The symbol stands for *both* the element and for one atom of the element.

Complex Ions

So far, all the ionic compounds we have discussed have been binary; that is, made up of two elements, even if there were more than one atom of each element involved in the ionization. It is easy to see that LiCl is a binary compound, but it takes a little more work to understand that Al_2O_3 and K_2O are binary. Even though Al_2O_3 has five ions and K_2O has three, each is made up of the atoms of only two elements, which makes the compound binary.

If oxidizers, poisons, and explosives were not hazardous materials, we would be tempted to skip the topic of complex ions, but such is not the case. We stated earlier that ions were charged particles, the result of an atom losing or gaining an electron or electrons. In the case of a complex ion, atoms of more than one element are bound together chemically, and this group of atoms has collectively lost or gained one or more electrons, making this a special type of ion. There are complex ions that contain only one element but have more than one atom of that element gaining electrons; the most common example is the peroxide ion, which is made up of two atoms of oxygen bound together chemically; these two atoms of oxygen have *collectively* gained two electrons, thus giving the ion an electrical charge of -2.

An example of a complex ion is the nitrate ion, whose formula is NO_3^{-1}. Note that this is not a compound, since all compounds are electrically neutral, and this collection of atoms has a charge of -1. The chemical bonding of the nitrogen and oxygen is a type that is far beyond the scope of this course, so we will not discuss it here. Just accept the fact that three atoms of oxygen and one atom of nitrogen are bound together chemically, and this collection of four atoms has collectively gained one electron from somewhere. This means that the electronic configuration of three oxygens and one nitrogen was almost satisfied that it was stable, except that it lacked one electron; when it accepts this electron, it accepts the -1 charge that accompanies it. Thus the nitrate ion NO_3^{-1} will act as a unit in all chemical reactions in which it participates, except for the reactions that will decompose the ion. When it is near a potassium ion, K^{+1}, the electrostatic attraction of opposite charges will attract the two ions together to form the ionic compound potassium nitrate, KNO_3. In this case, the nitrate ion is known as an *oxyion*, because it is an ion that contains oxygen, acts just an any other anion (non-metallic ion) would, and reacts with a metal. The result is the same as in the case of the binary ionic compounds where a metal and a non-metal combine; a salt—in this case, an oxysalt—is formed. You can plainly see from the formula,

KNO$_3$, that this compound contains three elements and is therefore not a binary compound. An important thing to remember is that the anion will act like *any* anion, so that the complex ion may be thought of as a non-metal ion, even though it clearly is not.

KNO$_3$ is not an isolated case of a complex ion combining with a metal ion to form a salt, because there are many types of oxyions. The type of ion that NO$_3$$^{-1}$ is, the complex anion known as an oxyanion, is extremely important to the study of hazardous materials since it is an *oxidizer*, in whatever combination into which it enters. In other words, the nitrate ion can combine with any metal ion to form an *oxysalt*. The alkali metals (potassium in the above example) form LiNO$_3$ and NaNO$_3$, among others. The alkaline earth metals (whose ions have a +2 charge) form Be(NO$_3$)$_2$, Ca(NO$_3$)$_2$, and Ba(NO$_3$)$_2$, and so on. Aluminum forms Al(NO$_3$)$_3$. Using the eleven metal ions, including the seven above, you can name and write the formulas for eleven metal nitrates.

Another oxyion is the *nitrite* ion, NO$_2$$^{-1}$. Using the eleven metal ions important to this course, you should be able to form, name, and write the correct chemical formulas for eleven metal nitrites.

The nitrate and nitrite ions are related in that they both contain nitrogen and oxygen (albeit in different ratios), they are both oxidizers, and both anions have an electrical charge of −1. There are several other oxyions that act as oxidizers; the largest group of these are the chlorine-containing oxyions. They are the *perchlorate* ion, ClO$_4$$^{-1}$, the *chlorate* ion, ClO$_3$$^{-1}$, the *chlorite* ion, ClO$_2$$^{-1}$, and the *hypochlorite* ion, ClO^{-1}. They *all* have a −1 charge on the anion, and they all combine with metal ions in the same way all other ions do, because of electrostatic attraction. Typical compounds include potassium perchlorate, KClO$_4$, calcium chlorate, Ca(ClO$_3$)$_2$, aluminum chlorite, Al(ClO$_2$)$_3$, and sodium hypochlorite, NaClO. Using the eleven metal ions and the four chlorine-containing oxyions, you should be able correctly to name and write the formulas for forty-four oxysalts.

The chlorine- and nitrogen-containing oxyions are the most common complex anions that are oxidizers. There are a few others that present the same problem. They include the permangamate ion, MnO$_4$$^{-1}$, the chromate ion, CrO$_4$$^{-2}$, and the peroxide ion, O$_2$$^{-2}$. Please notice the difference between the peroxide ion, O$_2$$^{-2}$, and the oxide ion, O^{-2}. The peroxide ion contains two oxygen atoms that have collectively gained two electrons bound together, while the oxide ion started as just one oxygen atom that gained two electrons. This makes the peroxide ion much more hazardous.

The peroxide ion is definitely a complex ion, but it is not a true oxyion, since there is no other element represented aside from oxygen. It is usually mentioned at the same time that oxyions are discussed, however, because of its oxidizing power. All the peroxides are hazardous because of this power. The peroxide ion is not particularly stable; it will break down with little provocation to release oxygen to the atmosphere or to a fuel, whether one already burning, or one just waiting for an ignition source.

Any chemical whose first name is a metal, *any* metal, and whose last name is nitrate, nitrite, perchlorate, chlorate, chlorite, hypochlorite, permanganate, chromate, or peroxide is an oxidizing agent and will contribute oxygen to a fire or potential fire that will increase its severity. As a matter of fact, the knowledge that *any* compound whose chemical name ends in -ate or -ite contains oxygen within the compound is a valuable tool. This fact does *not* mean that any compound containing oxygen within its chemical makeup is an oxidizer. We know that water and carbon dioxide have a lot of oxygen in the molecule and neither of them are oxidizers. Even oxyions like the sulphate ion, SO_4^{-2}, the sulphite ion, SO_3^{-2}, the phosphate ion, PO_4^{-3}, the phosphite ion, PO_3^{-3}, the bicarbonate ion, HCO_3^{-1}, and the carbonate ion, CO_3^{-2}, which all have a lot of oxygen in the ion, are not oxidizers under normal conditions. Your job is to commit to memory those complex ions that *are* oxidizers and to be constantly on the alert for them.

There are a couple of other anions which are not oxyions that are hazardous; they are the cyanide ion, CN^{-1}, and the hydroxide ion, OH^{-1}. The cyanide ion you will recognize as deadly poisonous, but you might not be familiar with the hydroxide ion. We will cover this complex ion later when we discuss corrosives in Chapter Thirteen.

So far, all the complex ions we have covered have been anions; that is, ions with a negative charge. These complex anions behave exactly like non-metal ions in their reactions with metals; that is, they all combine with metals ionically to form salts, with the exception of the hydroxide ion. There are several more complex anions, but they are relatively rare. We will mention them only if we meet them as hazardous materials. There is, however, one important complex *cation* (positively charged ion), the ammonium ion, NH_4^{+1}. This cation acts exactly like any metal ion in that it will combine ionically with anions in the same ratio as the alkali metals do, because it carries the same +1 electrical charge as the group IA metal ions. Therefore, ionic compounds can be formed, called ammonium chloride, NH_4Cl, ammonium nitrate, NH_4NO_3, ammonium sulphate, $(NH_4)_2SO_4$, and so on. The ammonium ion is very important in commerce and is present in many extremely hazardous materials.

If you remember that all chemical compounds are electrically neutral, it must by now be painfully obvious that for you to be able to put these 17 anions together with the 12 (11 metals plus ammonium) cations to be able correctly to name and write the formulas for the resulting 204 ionic compounds, you will not only have to memorize the formulas for the individual ions, but also the *valence*, or electrical charge of each. It is easy to determine the valence for the simple ions; all you do is to glance at the Periodic Table, and the number of valence electrons (those electrons in the outer ring) will indicate the combining power (or valence) of each ion. The number of electrons in the outer ring for the alkali metals of group IA is equal to the valence, or electrical charge on the ion. For all these metals the valence is $+1$. For the alkaline earth metals, the valence is $+2$. For aluminum, the valence is $+3$. Carbon seldom enters into ionic reactions, but its valence can be $+4$ or -4. For the non-metals of group VA, there are five valence electrons; the atom must gain three to reach stability, and doing so will give the ion a -3 charge, so its valence is -3. The non-metals of group VIA need two electrons to reach stability, so their valences are -2. The valence of the halogens is -1. Ionic formulas are simply the putting together of cations and anions in the proper ratio so that their opposite charges balance each other, and the resulting compounds are electrically neutral.

This process may seem complicated, but it really is not. What you must remember is that *all* compounds are electrically neutral; if the compound for which you are attempting to write a formula is ionic, then the positive charges and the negative charges *must* be equal to each other (to balance each other), so that the compound will have no electrical charge; that is, the charge on the compound will be zero. Therefore, once you have memorized the electrical charges on all the ions, it should not be difficult to put cations and anions together so the resulting electrical charge is zero! Just remember that you cannot have half-ions; for example, trying to balance the aluminum ion, Al^{+3}, with one and one-half sulfate ions, SO_4^{-2}; what is needed are *three* sulfate ions to balance *two* aluminum ions, to give the formula $Al_2(SO_4)_3$, for aluminum sulfate. For calcium phosphate three calcium ions, Ca^{+2}, are needed to balance two phosphate ions, PO_4^{-3}, in order to end with the correct formula, $Ca_3(PO_4)_2$.

In summing up ionic compounds, you have been presented with 11 metallic ions and one complex cation for a total of 12 cations, and 12 simple anions plus 17 complex anions for a total of 29 anions. This means that with a little work and a lot of memorization you will be equipped to name correctly and to write the formulas for 348 ionic compounds! Remember that the ionic charges—the electrical charges on *both* the cation and the anion—must balance each other so that the total charge on the compound is *zero* (electrically neutral).

Of course, there is no need for you to do this on demand, but you had better be able to recognize which compounds are hazardous and which are not. Be advised that, as a rule, ionic compounds are solids, and most are soluble in water. Nearly all are non-flammable, so, with a few notable exceptions, you need not worry about their burning. Some ionic compounds are corrosive, however; some *will* burn; some will explode; some are oxidizing agents; some are air-reactive; some are water-reactive; and some are deadly poisons. The study of ionic compounds may seem like a nuisance to you, but it cannot compare with the pain involved if you ignore their hazards!

One last thing about ions—they do not exist free in nature. For every positively charged ion in a compound, there *must* be a negatively charged ion next to it, in the proper ratio so that the compound is electrically neutral. The only time that ions separate from each other is in water, but then they are all still there, very close to each other. If you could actually count the number of ions present (an incredibly large, mind-boggling number; if you are interested, pick up any chemistry textbook, and look up Avogadro's number in the index), you would find the same ratio of cations to anions, so that the solution is electrically neutral. When the process of ionization takes place, cations *and* anions are formed, always in ratios that balance each other. You can not make positively charged ions without creating negatively charged ions; the electrons thrown off by the metals have to go *somewhere*. Mother Nature will see that everything remains in balance.

Glossary

Anion: A negatively charged ion.

Binary Compound: A compound that contains the atoms or ions of two elements; it may contain more than one atom or ion of an element, but only two elements may be represented in the compound.

Cation: A positively charged ion.

Chemical Reaction: A chemical change which occurs when two or more substances are brought together, and energy is either absorbed or liberated; the types of chemical reactions are oxidation, reduction, ionization, combustion, polymerization, hydrolysis, and condensation.

Complex Ion: Two or more atoms, bound together chemically, that have collectively gained or lost one or more electrons and are now electrically charged.

Compound: A chemical combination of two or more elements, either the same elements or different ones, that is electrically neutral.

Covalent Bond: The sharing of two electrons between the atoms of two non-metallic elements.

Formula: The shorthand used to signify what elements or ions are present in a compound, and how many of each; although the compound may be ionic, the charges on the ions are never shown in a formula.

Ion: An atom, or group of atoms, bound together chemically, that have gained or lost one or more electrons, and are electrically charged according to how many electrons were gained or lost.

Ionic Bond: The electrostatic attraction of oppositely charged ions to each other.

Ionization: The process by which an atom or group of atoms bound together chemically gain or lose one or more electrons and become an ion.

Oxyanion: A complex anion containing oxygen and at least one other element.

Oxysalt: An ionic compound containing a metal and an oxyanion.

Reactants: The chemical compounds that are brought together in a chemical reaction.

Salt: An ionic compound formed by a metal and a non-metal.

Symbol: The chemical shorthand used instead of the full name of an element.

Valence Electrons: The electrons in the atom's outer ring; the configuration of these electrons determines the chemistry of each element.

3

Covalent Bonding

Introduction

Back in chapter one, we discussed how compounds could exist in nature if they had to follow the octet rule. Most of the representative elements, those elements in the long columns or groups of the Periodic Table are too *active* to exist in nature as pure elements. Their electronic configuration is such that they have fewer than eight electrons in their outer or valence shell or ring. What is meant by an active element is that the drive to satisfy the octet rule by having a "full" outer ring is so strong that the element cannot just "lie" there or float about in the atmosphere without attempting to satisfy that drive and comply with the octet rule. What the element does, of course, is to react chemically with something else, to give up or gain the electrons required to fulfill the obligation of having the "right" number of electrons in the last ring.

There are many elements that are not very active (of course, the inert, or noble, gases of group VIIIA are totally inactive and form no chemical compounds at all), and the so-called "transition" elements, those in the short groups in the middle of the Periodic Table, are, as a category, examples of elements, all metals, that are not nearly as active chemically as the representative elements. Gold is a good example of a fairly inactive element; the proof is that gold can be found (or at least *used* to be found) in a free and elemental state as a "vein" inside rock, or as nuggets, lying on the ground or in river beds. This is not to say that the other transition elements are found free in nature. Gold is simply an extreme case. The other metals are found as their "ores," usually oxides and sulfides, easily separated from the other ion and thereby produced as the pure element, as in iron production, from iron ore, which is iron oxide. Iron, if left to lie in the open as the pure metal, will very slowly convert back to iron oxide by slowly reacting with oxygen in the air, the process we call rusting. Iron, therefore, *can* exist in nature for some period of time, before fulfilling its need to satisfy the octet rule. Many of the other transition elements have chemical activities somewhere between iron and gold.

On the other hand, sodium is a perfect example of a metallic element that is "too active" to exist in nature as the pure element. With one electron in its outer ring, it has an overpowering tendency to get rid of that electron, and thereby "uncover" the next ring (that now becomes the outer ring), which has eight electrons in it, and which would satisfy the octet rule. Usually, metals that need to react to reach stability do so with oxygen to form the corresponding oxides, which are the usual composition of ores mined to produce a pure metal. Sodium is one of the most abundant metals in the earth's crust; chemically, you can substitute lithium and potassium anywhere that sodium is mentioned. This is also true of cesium and rubidium, although they are relatively rare and are almost never found. While there may be some sodium oxide in the earth, sodium is *so* active that it reacts with almost anything available that will accept its lone electron. That is why there are so many sodium compounds found in nature, particularly in the ocean, since almost all sodium compounds are soluble in water. Chlorine, another abundant and active element, goes through the same reactions as sodium, except that chlorine needs to gain one electron to satisfy the octet rule, so it is a perfect match for sodium; this explains why sodium chloride is so common in the ground and in the ocean.

You will recall that there are two common ways the octet rule is satisfied: ionic bonding and covalent bonding. Remember, you can substitute lithium, potassium, rubidium, and cesium wherever sodium is mentioned, and fluorine, bromine, and iodine wherever chlorine is mentioned, simply because they belong to the same "family" of elements as sodium and chlorine; therefore their chemical reactions are the same. The previous description of how some elements solve the problem of satisfying the rule is a description of *ionic bonding*. Before we leap into covalent bonding, which, by the way, is the way *most* compounds are bonded, make sure you remember the important rules that control how elements react chemically.

1. All atoms are electrically neutral; that is, each atom has as many electrons in orbit around the nucleus as there are protons in the nucleus. Since electrons have an electrical charge of -1, and protons have an electrical charge of $+1$, these particles balance each other out and produce a net electrical charge of zero.

2. The group number indicates how many electrons are in the outermost ring, called the valence ring. These electrons are called valence electrons.

3. Because of nature's laws, all atoms strive to be in their most stable condition, which is eight electrons in the outer ring (the octet rule), or, in a few instances, two electrons in the outer ring (the duet rule, a special case of the octet rule).

4. The most stable condition is represented by the electronic configuration of the noble gases (the elements of group VIIIA).

5. Each atom *will do* whatever is necessary to reach its most stable condition. It will do this in one of three ways:

   a. It will give up an electron or electrons.
   b. It will accept an electron or electrons.
   c. It will share an electron or electrons with another atom or atoms.

6. To be able to tell what an element will do, you must be familiar with its electronic configuration, especially the valence electrons, because in the representative elements the chemistry they will go through is represented by those electrons in the outer ring. Knowing which elements are metals and which are non-metals tells you if they form ionic or covalent compounds. By looking at a molecular formula you can determine the type of bonding. If a metal has combined with a non-metal, it is an ionic compound. If the compound is a combination of non-metals, it is covalently bonded.

7. All electrons have a tendency to "pair up." It is not necessary for you to understand fully why the pairing up occurs, but it will be important to know what the configurations of those valence electrons are. Remember that the number of valence electrons is equal to the group number in the Periodic Table.

In Chapter Two we discussed ionic bonding, which is the gaining or losing of electrons by the atoms to form electrically charged particles called ions; a compound is formed by the electrostatic attraction of oppositely charged ions. But, just before we did that, we showed how atmospheric oxygen exists, along with hydrogen, nitrogen, chlorine, and fluorine. Refer to the molecular model of oxygen (figure 16) in Chapter One. We also discussed water, and in figure 17 you have a drawing of the hydrogen fluoride molecule. These compounds all represent covalent bonding, and that is the next topic.

The seven important rules that control how elements react chemically are valid for covalently bonded compounds as well as for ionic compounds. The atoms of the non-metallic elements that enter into covalent reactions are the same as those which enter into ionic reactions, but the number of elements that form covalently bonded compounds is considerably fewer. You will remember that ionic compounds are formed *only* when *metals* combine chemically with non-metals. Covalent compounds, on the other hand, are formed only when *non-metals* react with non-metals. The non-metals that are important to this study are hydrogen, carbon, oxygen, sulfur, phosphorus, nitrogen, fluorine, chlorine, bromine, iodine, helium, neon, argon, krypton, and xenon; since the last five elements are noble gases and do not form *any* type of compounds, we are really only concerned with ten elements. It is not asking much of yourself to memorize the position of ten elements on the Periodic Table; once you have done that, you will be able to write the molecular formulas of an infinite number of covalently bonded compounds. Furthermore, since covalently bonded compounds make up the vast majority of chemicals that exist in the world today, you can plainly see that covalent compounds are going to represent the vast majority of hazardous materials with which you will come in contact. It is in this area, then, that we will spend a considerable amount of time.

The Covalent Bond

The covalent bond is defined as the sharing of a pair of electrons between two atoms. Covalent bonding is the sharing of one or more pairs of electrons; that is, the covalent bond itself (what is actually holding the atoms together) is a pair of electrons shared by the two atoms so that each atom thinks both electrons in the pair belong to *it*. Covalent bonding may consist of the sharing of one, two, or three pairs of electrons; in the case where there are two or three pairs of electrons being shared, each atom again believes that *all* the electrons being shared are in *its* orbit.

Before we begin the process of describing how you can tell the manner in which these bonds are formed, please remember that we are dealing *only* with non-metals and only with those that most often form covalent bonds. There are only ten, and some are more frequently involved than others. For example, by far the most common elements involved in covalent bonding are carbon, hydrogen, and oxygen. Chlorine is the next most common, followed by nitrogen, fluorine, bromine, iodine, sulfur, and phosphorus, in that order. Probably 90 percent of the hazardous materials consist of carbon, hydrogen, and oxygen alone, and the other 10 percent include the other seven elements, plus carbon, hydrogen, and oxygen. Every so often, another non-metal will appear in our studies, as will some exceptions to what appear to be firm rules, but the vast majority of our work will be centered around compounds made up solely of carbon, hydrogen, and oxygen.

Remember, too, that the ionic bond and the covalent bond are about as different as they can be. In the case of the ionic bond, there is an actual transfer of an electron from the last orbit of the metal to the last orbit of the non-metal. You end with two different and distinct particles, the positive cation and the negative anion, and the force that holds them together is the electrostatic attraction of those opposite electrical charges. This attraction causes the ions to "stack up" on each other; for this reason the ionic compounds are solids. The electrical charge holding the ions together is negated by melting the solid or by dissolving it in water, but once the compound cools or some remains after evaporation, it re-solidifies because of the electrostatic attraction, and the solid is back to its original form. Regardless of the fact that an ionic compound contains electrically charged particles, however, the compound, like *all* compounds, is electrically neutral. Nature will not allow *anything* to exist in an electrically charged condition for very long.

On the other hand, the covalent bond does not create two particles, but one. The force holding the atoms together is not electrical in nature, since all atoms are electrically neutral. The force that *is* the covalent bond is the actual linking of orbitals or rings so that the electrons from both atoms are free to move in the last ring of either atom. Putting it another way, the atoms that will react to form a covalent bond actually intertwine their outer orbitals with each other, so that the valence electrons from each atom are free to move in the valence orbital of either atom, thus resulting in a "full" outer ring for both atoms. Thus two atoms of oxygen may come together to form atmospheric oxygen, the six valence electrons from each oxygen atom are free to travel in the valence orbital of each other atom, and since the rule is that there cannot be more than eight electrons in the outer ring, you will never find 9, 10, 11, or 12 (the total valence electrons of both atoms) in the last ring. There are actually eight electrons in the last ring of both atoms, so that both atoms are satisfied that they have reached stability, and therefore act as if they have.

A word is appropriate here on the "coming together" of two atoms of oxygen to form atmospheric oxygen, or two or more atoms of any non-metals to form a covalent compound. Chemical reactions do not "just happen" when two elements or two compounds or an element or a compound are brought into intimate contact with each other. Energy is always required to start a chemical reaction, and it must come from somewhere. Even hypergolic chemicals and water- and air-reactive chemicals that seem to react spontaneously will draw the energy needed to begin the reaction from the surrounding environment. Any new compound formed from the reaction will have some of that energy now stored in the bonds created when the compound was created. In other words, it took energy of some kind to force the atoms of elements together into a new compound, and some of that energy is now stored in those covalent bonds, ready to be released in a new reaction. We will talk about "bond energies" in a later section of this chapter.

How Non-Metals Combine

Remember that non-metals are those elements to the right and above the line on the Periodic Table (table 2). These are the only elements capable of forming covalent bonds, and they do so in an easily predictable manner. You must be familiar with their positions in the Periodic Table, and therefore know the numbers of the groups in which they reside. This, of course, will tell you the number of electrons in the outer ring, and combining this information with the octet rule will tell you how many electrons in the outer ring they will need to reach stability; since they will not accept or give up electrons (non-metals will *only* accept electrons and will do so *only* in reactions with metals), they *must* reach stability by *sharing* electrons. How many they will share is determined by how many they need to fill the valence ring.

The best way to illustrate this is to use what is called the "dot" method of showing electronic configuration. In this method, each dot stands for one electron; the only electrons that are shown as dots are the valence electrons (those electrons in the outer ring), and the chemical symbol stands for the nucleus *and all the other electrons* of that atom. The dot method is displayed for the most common elements as follows:

$$H\cdot \quad \cdot\overset{\displaystyle\cdot}{\underset{\displaystyle\cdot}{C}}\cdot \quad \cdot\overset{\displaystyle\cdot\cdot}{\underset{\displaystyle\cdot\cdot}{O}}\cdot \quad \cdot\overset{\displaystyle\cdot\cdot}{\underset{\displaystyle\cdot\cdot}{S}}\cdot \quad :\overset{\displaystyle\cdot\cdot}{\underset{\displaystyle\cdot}{N}}\cdot \quad :\overset{\displaystyle\cdot\cdot}{P}\cdot \quad \cdot\overset{\displaystyle\cdot\cdot}{\underset{\displaystyle\cdot\cdot}{F}}: \quad \cdot\overset{\displaystyle\cdot\cdot}{\underset{\displaystyle\cdot\cdot}{Cl}}: \quad \cdot\overset{\displaystyle\cdot\cdot}{\underset{\displaystyle\cdot\cdot}{Br}}: \quad \cdot\overset{\displaystyle\cdot\cdot}{\underset{\displaystyle\cdot\cdot}{I}}:$$

The only way for a covalent bond to be formed would be if electrons had a tendency to form pairs, and in fact they do. (*Why* they do is far beyond the scope of this book.) Note that some electrons are pictured as " . " while others are shown as " : " or " . . ". Where two electrons are shown close together on the same side of the symbol, they stand for electrons that are already paired. You can see therefore that the tendency for electrons to pair is so great that they will do so within their own electronic configuration, if possible. The electrons that have already "paired up" are not available for covalent bonding, which, of course, is the *sharing* of a pair of electrons. The only electrons available for covalent bonding are those electrons shown as single dots. The number of electrons available for covalent bonding is equal to the number of electrons needed to reach stability. Therefore, from the above illustration, hydrogen, fluorine, chlorine, bromine, and iodine all need one electron to complete the outer ring, so they can form only one covalent bond. Oxygen and sulfur have two unpaired electrons (they need two electrons to reach stability), so they can, and will, form two covalent bonds. Nitrogen and phosphorus both have three unpaired electrons (and need another three to reach the total of eight), so they will form three covalent bonds. Carbon, because of its unique structure of four electrons in the outer ring (which is quite symmetrical), has four unpaired electrons and therefore, will create four covalent bonds; there is only one exception to this—carbon monoxide—and it is deadly. You can see that using dots to represent the valence electrons allows you to visualize how electrons form pairs, both with electrons from other atoms and with their own electrons. (You can also use x's and o's to show which atom contributes which electron to which pair.)

Once you understand where the electrons are and how they pair up, it is much easier to switch to the "dash" method of depicting covalent bonds (see figure 19). In this manner, you ignore all the paired electrons in the atom and show a "–" wherever there is an unpaired electron. Whenever an unpaired electron exists, the bond (or more correctly, half a bond) is "dangling" from the atom. Once it pairs with another unpaired electron from another atom, the dash then appears between the two atoms, and it is *then* that the dash truly stands for a covalent bond. To repeat, the dash represents an unpaired electron when shown with *one* atom and represents a covalent bond when it is shown between two atoms. Thus, H· becomes H–, ·Ö· becomes –O–, ·Ç· becomes —C—, ·N̈: becomes –N, and :C̈l· becomes Cl–. The symbol for the element now represents the nucleus of the atom *and all the electrons in orbit,* except for the unpaired electrons. If you want to depict any of the common non-metals that participate in covalent bonding by the dash method (and we will be doing so to a great extent), all you have to do is look at the position on the Periodic Table of the element in question to determine the number of dashes you will have to use to represent unpaired electrons—which, of course, will tell you how many covalent bonds will be formed. When you locate the group

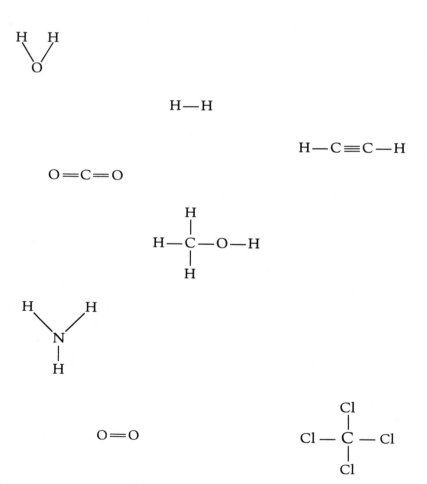

Figure 19
Covalent Bonds.

in which the element resides, just subtract the group number from eight; the answer will be the number of unpaired electrons each atom of the element has, which is also the number of dashes to draw and the number of covalent bonds the atom will form. This "rule of thumb" will enable you to understand the way hydrocarbon compounds and their derivatives are constructed, and these compounds will probably represent 95 percent of all the hazardous materials with which you should be familiar. The other covalently bonded compounds that we will discuss will follow the same rules, but the bonding will appear to be exceptions. These can be discussed separately; they include the explanations for carbon monoxide, sulfur dioxide, sulfur trioxide, and several other oxygen-containing gases. Just accept the fact that the chemical formulas, CO, SO_2, and SO_3 are correct for the compounds just named. The chemistry required to explain how these compounds are bonded can take up entire chapters and in some cases entire books. Needless to say, we do not have to be concerned with it at this time. If you *do* want to pursue the subject, look up "coordinate covalent" bonding in any college chemistry book.

Hydrocarbons

The element carbon, as mentioned before, has a unique electronic configuration in that its four valence electrons form a symmetrical design. Symmetry is not seen in the electronic configuration of any elements other than the noble gases, where the outer ring is completely full. The symmetry, in this case, is really a half-full outer ring, but with the electrons positioned where they are, carbon may, under certain circumstances, act as if it has a *full* outer ring. This symmetry of electrons in the outer ring gives carbon some unique properties. Since an element wants to gain or lose electrons in the number necessary to fulfill the octet rule, all metals find it easier to give up one, two, or three electrons (depending on whether they have one, two, or three valence electrons) than to accept five, six, or seven electrons. Conversely, non-metals find it easier to accept one, two, or three electrons (depending on whether they have five, six, or seven valence electrons) than to lose five, six, or seven electrons. Right in the middle of the metals and non-metals, however, are the elements of group IVA, which find it just as easy to give up four electrons as to accept four electrons—and carbon sits atop group IVA. Carbon has four electrons in the outer ring, and all of them are unpaired. This means carbon can form four covalent bonds, and it does so in *all* hydrocarbon compounds. The element carbon is the basis of organic chemistry, and that makes carbon the basis of all life.

Organic chemistry is the chemistry of compounds that are part of living things, or things that were once alive. Since man has learned to synthesize, or make them, organic chemistry also includes all the man-made chemicals that have identical chemistry in nature. Organic chemistry is based on hydrocarbon chemistry; a hydrocarbon is a compound made up of just two elements, hydrogen and carbon. Carbon is the central element in all hydrocarbon compounds and all hydrocarbon derivatives. A hydrocarbon derivative is a hydrocarbon that has had a functional group substituted for one or more of the hydrogen atoms. We will discuss substitution, derivatives, and functional groups in later chapters. Now we will concentrate on carbon and its place in covalent bonding of hydrocarbons.

Carbon, as stated earlier, has four valence electrons; these four electrons are arranged symmetrically in the outer ring. Since all four of these electrons are unpaired, they are all available for covalent bonding. Let us examine how this occurs.

The carbon atom can be drawn like this, using the dash method of depicting the covalent bond: $-\overset{\displaystyle |}{\underset{\displaystyle |}{C}}-$. If we bring in an atom of hydrogen, which looks like $-H$, we will have a combination of carbon and hydrogen that looks like $-\overset{\displaystyle |}{\underset{\displaystyle |}{C}}-H$. This drawing shows hydrogen as satisfied that it has paired its lone unpaired electron, but it leaves carbon with three unpaired electrons still seeking to pair up. If we bring another hydrogen atom in, we now have $H-\overset{\displaystyle |}{\underset{\displaystyle |}{C}}-H$, but this still leaves two unpaired electrons. Another hydrogen atom makes the diagram $H-\overset{\displaystyle H}{\underset{\displaystyle |}{\overset{\displaystyle |}{C}}}-H$, which still leaves one unpaired electron. Bringing one more hydrogen atom in produces $H-\overset{\displaystyle H}{\underset{\displaystyle H}{\overset{\displaystyle |}{\underset{\displaystyle |}{C}}}}-H$, which finally satisfies all four hydrogen atoms and also the carbon atom. The resulting *hydrocarbon* compound has a molecular formula of CH_4, and is known as methane, the principal ingredient of natural gas, and the simplest of all the hydrocarbons. Whenever you see a formula that shows the covalent bonds as dashes, it is known as a *structural formula*. Whenever you see a formula written as CH_4, or some similar formula showing symbols and subscripts, it is known as the *molecular formula*. The molecular formula shows the elements in the compound and how many of each there are, while the structural formula shows how they are arranged in the molecule and the covalent bonds between the atoms. Stated yet another way, the structural formula is a configuration of all the elements in a compound, showing the bond locations as dashes, and the molecular formula is the writing of a formula showing the type and number of atoms present in the molecule.

We have shown how a simple hydrocarbon is structured by showing the structural formula of methane, CH_4. In this example (figure 20), carbon with four unpaired electrons combines chemically with hydrogen that has one unpaired electron by using one of its unpaired electrons to pair up with the lone (and therefore unpaired) electron of a hydrogen atom. Let us move on to show how carbon combines with oxygen to form carbon dioxide. You must realize that

Figure 20
Structural Diagram of Methane.

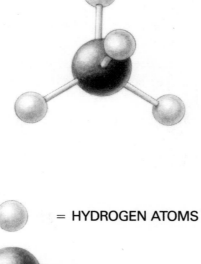

○ = HYDROGEN ATOMS

● = CARBON ATOM

▬ = COVALENT BOND

we are not showing a chemical reaction but instead are depicting a stepwise formation of the compound. The actual reaction occurs with amazing speed.

Carbon, you will recall, is pictured structurally as $-\overset{|}{\underset{|}{C}}-$, while oxygen is $-O-$. When an atom of oxygen is brought to the carbon, the result is $-\overset{|}{\underset{|}{C}}-O$. This shows carbon with three unpaired electrons, and oxygen with one. If another oxygen atom is brought in, the result could be $O-\overset{|}{\underset{|}{C}}-O$. Since we know the correct formula is CO_2, the structure could not possibly be $-O-\overset{|}{\underset{|}{C}}-O-$, because this would mean the *compound* would have two unpaired electrons, which is impossible. Again, a law of nature rules that all electrons in a covalent *compound* will be paired. This means that *all* the valence electrons of all the atoms that are reacting to form the compound will be "used up" by pairing, and that there will be no unpaired electrons left over; it means the structure must be something else, and it is. The correct structural formula for CO_2 is $O=C=O$. We now have the four covalent bonds that we know carbon will form, two between the carbon and each of the oxygens. All the oxygen's electrons are now paired; each atom of oxygen thinks it has eight electrons in its outer ring, as does the carbon, so the octet rule is satisfied for all the elements in the compound, making it the correct structural formula. These two covalent bonds shown between each carbon and oxygen are known as *double* bonds. Double bonds are quite common in covalent chemistry, and when they appear between two *carbon* atoms, it makes the compound more reactive, particularly at the point in the structure where the double bond occurs. The formation of double bonds is an example of the lengths to which the atoms of an element will go to satisfy the laws of nature, but they do not stop there.

There are also *triple* bonds, which would be the sharing of three pairs of electrons. In some fairly rare instances, atoms of elements will react together chemically in such a way that there would be a tremendous deficiency in the number of electrons in the outer ring of both atoms. The formation of the triple bond is the solution to this problem. With six electrons now being shared between them, there will be enough to satisfy whichever atom has the electron deficiency (even both). If both atoms are carbon, a hazard arises. If the carbon-to-carbon double bond is reactive (and it is), the carbon-to-carbon triple bond is even more reactive. As a matter of fact, it is *so* reactive that it is extremely unstable; instability in this case means that it will decompose rather easily, with the release of a great amount of energy. We will explore this reaction further when we consider acetylene, the most common of the chemical compounds that contains a carbon-to-carbon triple bond. Double and triple bonds are known collectively as *multiple* bonds.

With the information just presented to you, you should be able now to draw the structural formulas for water, H_2O, ammonia, NH_3, for fluorine, F_2, and for chlorine, Cl_2.

Do not concern yourselves with the apparent exceptions to the rules. They are *not* exceptions but rather a slightly different form of covalent bonding called *coordinate covalent bonding*. In coordinate covalent bonding, one element's atoms do not always donate as many electrons to the covalent bonding process as the other element's atoms do. Again, do not concern yourself with this form of covalent bonding; it is in no way as important to the study of hazardous materials as is the covalent bonding of elements in organic chemistry.

Glossary

Activity: A reference to chemical activity, the speed with which an element will seek out ways to satisfy the octet rule.

Covalent Bond: The sharing of two electrons between the atoms of two non-metallic elements.

Covalent Compound: A compound containing atoms bonded together by covalent bonds.

Dash Method: A method of showing the configuration of an atom's unpaired electrons by drawing a dash for every unpaired electron; the dash, when used around the symbol of one element, is *not* a covalent bond.

Dot Method: A method of showing the configuration of an atom's valence electrons.

Double Bond: A state in which two atoms are sharing two pairs of electrons between them.

Hydrocarbon: A covalent compound containing *only* hydrogen and carbon.

Hypergolic: The property of reacting with another substance immediately upon coming in contact with that substance; both materials must be hypergolic. An example is materials used in a solid-fuel rocket engine, which can be counted on to react and burn, producing thrust, without depending upon an ignition device that might fail.

Ionic Compound: A chemical combination of oppositely charged ions, held together by the electrostatic attraction of opposite charges.

Metals: The elements to the left of and below the line on the Periodic Table (table 2) in this textbook.

Multiple Bond: A double or triple covalent bond.

Non-Metals: The elements to the right of and above the line on the Periodic Table (table 2) in this textbook.

Orbital: The "ring" around the nucleus that contains the orbiting electrons.

Paired Electrons: Electrons that are near to each other will tend to be attracted to each other in pairs; paired electrons form a covalent bond only when they are shared by two atoms.

Representative: The elements of the long groups in the Periodic Table.

Transition: The elements of the short groups in the Periodic Table.

Triple Bond: A state in which two atoms are sharing three pairs of electrons between them.

Unpaired Electrons: The electrons that are ready to react or pair up with electrons of another atom to form a covalent bond.

Valence Electrons: The electrons in the atom's outermost ring; the configuration of these electrons determines the chemistry of each element.

Valence Ring: The outermost ring of an atom.

4

Hydrocarbons

Introduction

Hydrocarbons, you will recall, are defined as compounds containing only hydrogen and carbon. Since a hydrocarbon is a chemical combination of hydrogen and carbon, both of which are non-metals, hydrocarbons *must* be convalently bonded, and they are. Hydrogen has only one electron in the outer ring and therefore will form only one covalent bond, by donating that one electron to the bond. Carbon, on the other hand, occupies a unique position in the Periodic Table, being halfway to stability with its four electrons in the outer ring. None of these electrons are paired, so carbon uses all of them to form covalent bonds. Carbon's unique structure makes it the basis of organic chemistry, which is the study of compounds that were once a part of living things. The way carbon combines with hydrogen and other non-metals in convalent compounds makes it one of the most important elements to life, and, unfortunately, to the study of hazardous materials.

Carbon not only combines covalently with other non-metals, but also with itself. This fact may not seem remarkable to you, considering that oxygen also reacts with itself to form O_2, hydrogen reacts with itself to form H_2, nitrogen reacts with itself to form N_2, fluorine reacts with itself to form F_2, and chlorine reacts with itself to form Cl_2. Forming a *diatomic* molecule, however, is the extent of the self-reaction of the elemental gases, while carbon has the ability to combine with itself almost indefinitely! Although the elemental gases form molecules when they combine with themselves, the carbon-to-carbon combination must include another element or elements, generally hydrogen. This combination of carbon with itself (plus hydrogen) forms a larger molecule with every carbon atom that is added to the *chain*. When the chain is strictly carbon-to-carbon with no *branching*, the resulting hydrocarbon is called a *straight-chain* hydrocarbon. Where there are carbon atoms joined to carbon atoms to form side branches off the straight chain, the resulting compound is known as a *branched* hydrocarbon, or an *isomer*. These compounds will be defined later in the chapter, as we get to them.

The carbon-to-hydrogen bond is always a single bond, since no matter how badly carbon wants to form four covalent bonds, hydrogen can form *only* one. While the resulting bond between carbon and hydrogen is always a single bond, carbon *does* have the capability and *will* form double *and* triple bonds between itself and other carbon atoms, and/or any other atom that has the ability to form more than one bond. When a hydrocarbon contains only single bonds between carbon atoms, it is known as a *saturated* hydrocarbon; when there is *at least* one double or triple bond between two carbon atoms anywhere in the molecule, is is an *unsaturated* hydrocarbon. When determining the saturation or unsaturation of a hydrocarbon, only the carbon-to-carbon bonds are considered, since the carbon-to-hydrogen bond is *always* single.

Let us look at the phenomena of longer and longer *chains* of carbon atoms by looking at *analogous* series of hydrocarbons. An analogous series is defined as an orderly grouping of compounds with each succeeding or preceding compound being some "unit" different from the compound under consideration, with the "unit" being constant throughout the series. This description sounds complicated, so we will consider the first analogous series of hydrocarbons and see what the "unit" or difference is between each of the compounds in the series.

Straight-Chain Hydrocarbons: The Alkanes

The first analogous series of hydrocarbons we will study is the series of compounds known as the *alkanes*. In this series, the names of all the compounds end in *-ane*. Many of them will sound familiar to you because you have encountered them before in your work. The first compound in this series is one previously discussed, methane. Methane's molecular formula is CH_4 and its structural formula is

$$H-\overset{\overset{\displaystyle H}{|}}{\underset{\underset{\displaystyle H}{|}}{C}}-H.$$ Methane is a gas and is the principal ingredient in the mixture of gases known as natural gas. The next compound is this series is **ethane**, whose molecular formula is C_2H_6, and whose structural formula is $H-\overset{\overset{\displaystyle H}{|}}{\underset{\underset{\displaystyle H}{|}}{C}}-\overset{\overset{\displaystyle H}{|}}{\underset{\underset{\displaystyle H}{|}}{C}}-H.$ Ethane (see figure 21) is also a gas present in natural gas, although in a much lower percentage than methane. The difference in the molecular formulas of methane and ethane appears to be one carbon and two hydrogen atoms. When we examine the structural formulas of each, we can see that the difference is indeed $H-\overset{\displaystyle |}{\underset{\displaystyle |}{C}}-H.$

Propane is the next hydrocarbon in this series, and its molecular formula is C_3H_8 which is one carbon and two hydrogen atoms different from ethane. Propane's structural formula is $H-\overset{\overset{\displaystyle H}{|}}{\underset{\underset{\displaystyle H}{|}}{C}}-\overset{\overset{\displaystyle H}{|}}{\underset{\underset{\displaystyle H}{|}}{C}}-\overset{\overset{\displaystyle H}{|}}{\underset{\underset{\displaystyle H}{|}}{C}}-H,$ which is one $H-\overset{\displaystyle |}{\underset{\displaystyle |}{C}}-H$ "unit" bigger than ethane. Propane (see figure 22) is an easily liquified gas which is used as a fuel.

The next hydrocarbon in the series is **butane**, another rather easily liquified gas used as a fuel. Together, butane and propane are known as

**Figure 21
Structural Diagram of
Ethane.**

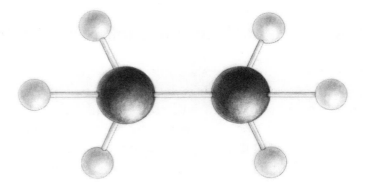

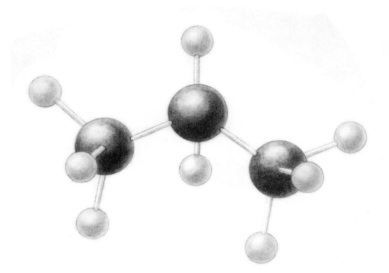

**Figure 22
Structural Diagram of
Propane.**

the LP (liquified petroleum) gases. Butane's molecular formula is $C_4H_{10,}$ which is CH_2 bigger than propane, and the structural formula

H—C—C—C—C—H   contains this   H—C—H   "unit"

difference.

As you can see, up to this point, the series begins with a one-carbon-atom compound, methane, and proceeds to add one carbon atom to the chain for each succeeding compound. Since carbon will form four convalent bonds, it must also add two hydrogen atoms to satisfy those two unpaired electrons and allow carbon to satisfy the octet rule, thus achieving eight electrons in the outer ring. In every hydrocarbon, whether saturated or unsaturated, *all* atoms must reach stability. There are only two elements involved in a hydrocarbon, hydrogen and carbon; hydrogen *must* have two electrons in the outer ring, and carbon *must* have eight electrons in the outer ring. Since the carbon-hydrogen bond is always single, the rest of the bonds must be carbon-carbon, and these bonds must be single, double, or triple, depending on the compound.

Continuing in the alkane series (which, by the way, is also called the *paraffin* series because the first solid hydrocarbon in the series is paraffin, or candle wax), the next compound is pentane. This name is derived from the Greek word *penta*, for five. As its name implies, it has five carbon atoms; its molecular formula is $C_5H_{12,}$ and its structural formula is

$$H-\underset{\underset{H}{|}}{\overset{\overset{H}{|}}{C}}-\underset{\underset{H}{|}}{\overset{\overset{H}{|}}{C}}-\underset{\underset{H}{|}}{\overset{\overset{H}{|}}{C}}-\underset{\underset{H}{|}}{\overset{\overset{H}{|}}{C}}-\underset{\underset{H}{|}}{\overset{\overset{H}{|}}{C}}-H,$$

which again confirms the

$$H-\underset{|}{\overset{|}{C}}-H$$

"unit" difference.

From pentane on, the Greek prefix for the numbers five, six, seven, eight, nine, ten, and so on are used to name the alkanes, the Greek prefix corresponding to the number of carbon atoms in the molecule. The first four members of the alkane series do not use the Greek prefix method of naming, simply because their *common* names are so universally accepted: thus the names methane, ethane, propane, and butane.

The next six alkanes are named pentane, hexane, heptane, octane (a very familiar name), nonane, and decane. Their molecular formulas are C_5H_{12}, C_6H_{14}, C_7H_{16}, C_8H_{18}, C_9H_{20}, and $C_{10}H_{22}$, respectively, and the structural formulas are shown in figure 23 for the series up to ten carbon atoms. The alkanes do not stop at the ten-carbon chain, however. Since these first ten represent flammable gases and liquids, and most of the *derivatives* of these compounds comprise the vast majority of hazardous materials that you will encounter, we have no need to go any further in the series. Nevertheless, be advised that there are formulas and names for the longest carbon chain you can think of, and for which the Greeks had a numerical prefix. As a matter of fact, you can write the formula for *any* alkane, if you know the number of carbon atoms in the molecule, using the *general formula* for the alkanes: C_nH_{2n+2}. The letter *n* stands for the number of carbon atoms in the molecule. The number of hydrogen atoms then becomes two more than twice the number of carbon atoms. Since there is more than one analogous series of hydrocarbons, you must remember that each series is unique; the alkanes are defined as the analogous series of saturated hydrocarbons with the general formula C_nH_{2n+2}.

Isomers

Although we are not going to investigate further any alkanes whose carbon content is more than ten atoms, you must be aware that within each analogous series of hydrocarbons there exist *isomers* of the compounds within that series. An isomer is defined as a compound with the same molecular formula as another compound but with a

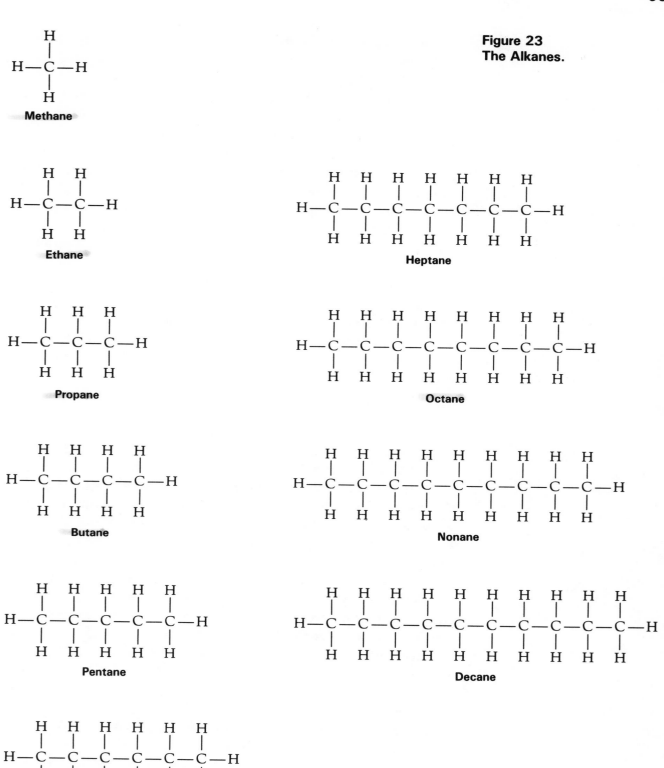

**Figure 23
The Alkanes.**

Methane

Ethane

Propane

Butane

Pentane

Hexane

Heptane

Octane

Nonane

Decane

different structural formula. In other words, if there is a different way in which the carbon atoms can align themselves in the molecule, a different compound with different properties will exist.

In figure 23, we show the structural formulas of the first ten compounds in the analogous series known as the alkanes. If we look at the first compound, methane, we can see that there is only one way to arrange the four hydrogen atoms around the one carbon atom. Similarly, there is only one way to arrange the two carbon atoms and the surrounding six hydrogen atoms in the second compound, ethane. In propane, the third compound, it appears as though, if you were to draw the third carbon atom above the second carbon atom (as if it were making a left turn) the result would look like a different arrangement of the carbon atoms. The length of a chain of carbon atoms, however, is determined by counting the total number of carbon atoms connected to each other in a continuous chain, no matter if the carbon atoms are arranged in a straight line, or if they are arranged in a zigzag. Counting the carbon atoms in this way, the chain length is three, so the compound is still propane. Since the first three alkanes can each be drawn in only one way, they have no isomers.

Beginning with the fourth alkane, butane, we find we can draw a structural formula of a compound with four atoms and ten hydrogen atoms in two ways; the first is as the *normal* butane exists in figure 23, and the second is as shown is figure 24, with the name *isobutane*. With isobutane, you will notice that no matter how you count the carbon atoms in the longest chain, you will always end with three. Notice that the structural formula *is* different—one carbon atom attached to three other carbon atoms—while in butane (also called *normal butane*), the largest number of carbon atoms another carbon atom can be attached to is two. This fact *does* make a difference in certain properties of the compounds. Look at the list of the first ten alkanes and the isomers of two of them in table 4. Notice that the *molecular formulas* of butane and isobutane are the same and, therefore, so are the molecular weights. But consider the 38-degree difference in melting points, the 20- degree difference in boiling points, and the 310-degree difference in ignition temperatures. The structure of the molecule clearly plays a part in the properties of the compounds. Incidentally, there are only two different ways to draw the structural formula of the four-carbon alkane, butane. Try to see if you can draw more.

Moving to the five-carbon alkane, pentane, there are three ways to draw the structural formula of this compound with five carbon atoms and twelve hydrogen atoms. The isomers of normal pentane are called *isopentane* and *neopentane*. The structural formulas of these three

67

compounds are shown in figure 24, while the properties are shown in table 4. Again you can see the three identical molecular formulas and three identical molecular weights, but significantly different melting, boiling, and flash points and different ignition temperatures. We are not going any further in discussing isomers and their different properties yet identical molecular formulas; be advised, however, that isomers *do* exist, and their properties *are* different. For your information, the next alkane, hexane, has four isomers, heptane has eight, octane has seventeen, and so on. A favorite exercise that chemistry professors love to assign a class that has not been doing its homework is to start with butane and go through decane, drawing all possible isomers. This separates the chemistry majors from the rest.

What we have just seen is the *structure effect* and how it produces differences in the properties of compounds with the same molecular formulas. This particular structure effect is called the *branching* effect, and the isomers of all the straight-chain hydrocarbons are called *branched hydrocarbons*. There is another structural effect; it is produced simply by the length of the chain formed by consecutively

Table 4 / Properties of Alkanes

Compound	Formula	Atomic Weight (°F.)	Melting Point (°F.)	Boiling Point (°F.)	Flash Point (°F.)	Ignition Temperature (°F.)
Methane	CH_4	16	−296.5	−259	gas	999
Ethane	C_2H_6	30	−298	−127	gas	882
Propane	C_3H_8	44	−306	−44	gas	842
Butane	C_4H_{10}	58	−217	31	gas	550
Pentane	C_5H_{12}	72	−201.5	97	< −40	500
Hexane	C_6H_{14}	86	−139.5	156	−7	437
Heptane	C_7H_{16}	100	−131.1	209	25	399
Octane	C_8H_{18}	114	−70.2	258	56	403
Nonane	C_9H_{20}	128	−64.5	303	88	401
Decane	$C_{10}H_{22}$	142	−21.5	345	115	410
Butane	C_4H_{10}	58	−217	31	gas	550
Isobutane	C_4H_{10}	58	−255	11	gas	860
Pentane	C_5H_{12}	72	−201.5	97	< −40	500
Isopentane	C_5H_{12}	72	−256	82	< −60	788
Neopentane	C_5H_{12}	72	2	49	< −20	842

Figure 24
Isomers of Butane and
Pentane.

Compound	Molecular Formula	Structural Formula

Butane — C_4H_{10}

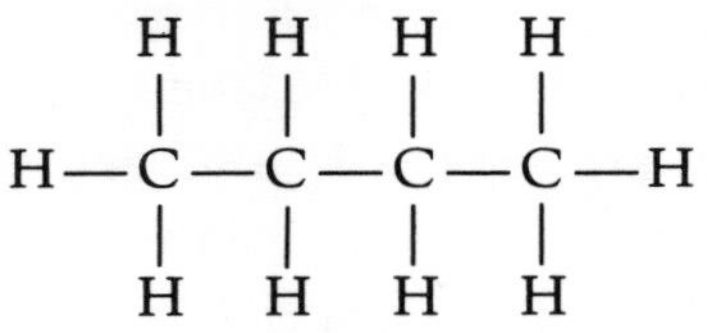

$$H-\overset{\overset{\displaystyle H}{|}}{\underset{\underset{\displaystyle H}{|}}{C}}-\overset{\overset{\displaystyle H}{|}}{\underset{\underset{\displaystyle H}{|}}{C}}-\overset{\overset{\displaystyle H}{|}}{\underset{\underset{\displaystyle H}{|}}{C}}-\overset{\overset{\displaystyle H}{|}}{\underset{\underset{\displaystyle H}{|}}{C}}-H$$

Isobutane — C_4H_{10}

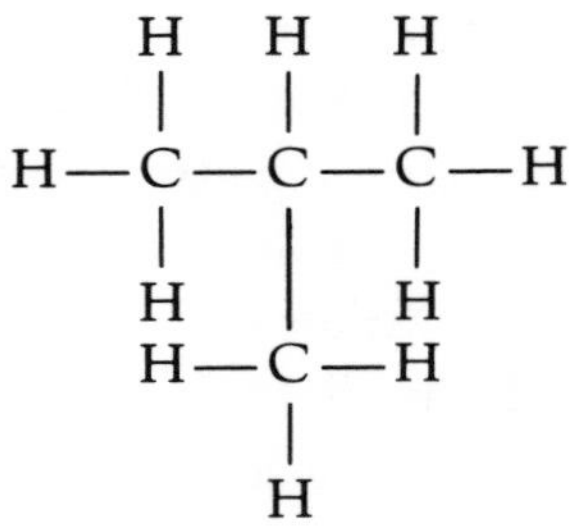

Pentane — C_5H_{12}

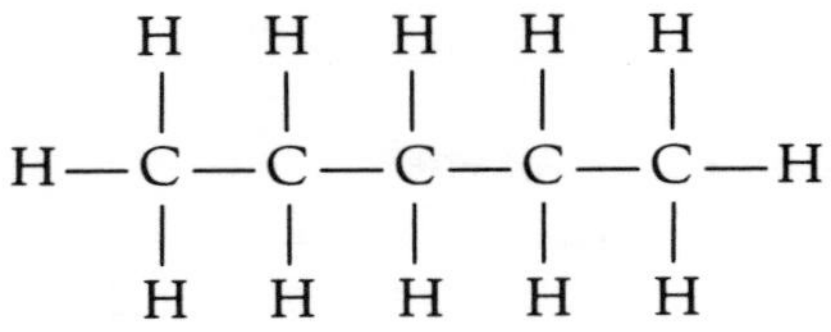

Isopentane — C_5H_{12}

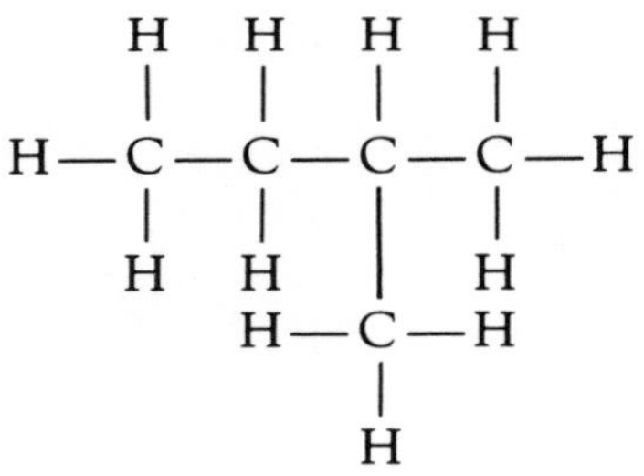

Neopentane — C_5H_{12}

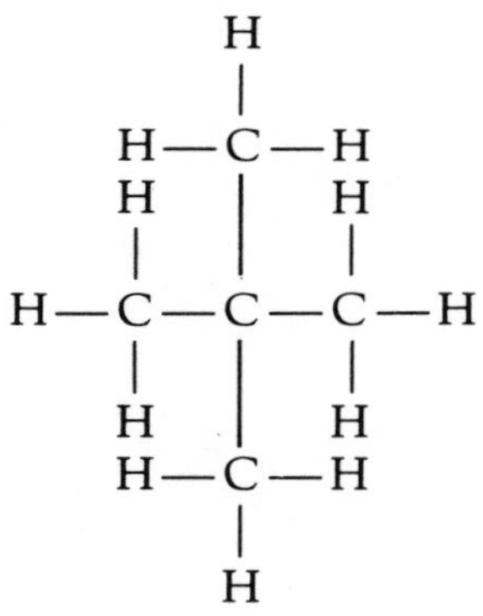

attached carbon atoms. If you will refer again to figure 23, you will note the increasing length of the carbon chain from methane through decane. The difference in each succeeding alkane is that "unit" made up of one carbon atom and two hydrogen atoms; that "unit," which, of course, is not a chemical compound itself, has a molecular weight of fourteen. Therefore each succeeding alkane in the analogous series weighs fourteen atomic mass units more than the one before it and fourteen less than the one after it. This *weight* effect is the reason for the increasing melting and boiling points, the increasing flash points, and the decreasing ignition temperatures. The increasing weights of the compounds also account for the changes from the gaseous state of the first four alkanes, to the liquid state of the next thirteen alkanes, and finally to the solid state of the alkanes, starting with the 17-carbon atom alkane, heptadecane. Using the general formula of the alkanes, C_nH_{2n+2}, what is the molecular formula for heptadecane?

This is a general rule of chemistry: the larger a molecule (that is, the greater the molecular weight), the greater affinity each molecule will have for each other molecule, therefore slowing down the movement. The molecules, duly slowed from their frantic movement as gases, become liquids, and, as the molecules continue to get larger, they are further slowed from their still rapid movement as liquids and become solids.

You will be responsible for knowing the molecular formulas and structural formulas of the first ten alkanes, and of isobutane, isopentane, and neopentane. More important, you must know the change in properties caused by the structure effect, and be able to deduce which compounds in the alkanes are more hazardous than others in the series, thereby becoming able to rank them by increasing hazard to you and to the public.

Other Straight-Chain Hydrocarbons

The straight-chain hydrocarbons we looked at in the first section, plus their isomers in the second section, represent just one group of straight-chain hydrocarbons, the saturated hydrocarbons known as the alkanes. There are other series of hydrocarbons that are *unsaturated*; one of those is important in the study of hazardous materials. Additionally, the first hydrocarbon in another series is the only hydrocarbon important in that series. Let us look at those hydrocarbon series.

Alkenes

The series of unsaturated hydrocarbons that contain just one double bond in the structural formula of each of its members is the analogous series known as the *alkenes*. Notice that the name of this analogous series is similar to the analogous series of *saturated* hydrocarbons known as the *alkanes*, but the structural formula is significantly different. Remembering that the definition of a saturated hydrocarbon is a hydrocarbon with nothing but single bonds in the structural formula and that an unsaturated hydrocarbon is a hydrocarbon with *at least one* multiple bond in the structural formula, we would expect to find a multiple bond in the structural formulas of the alkenes, and we do. The names of all the hydrocarbons in this analogous series end in *-ene*. The corresponding names for this series of hydrocarbons is similar to the alkanes, with the only difference being the above-mentioned ending. Thus, in the alkene series ethane becomes ethene, propane is propene, butane's counterpart is butene, the five-carbon straight-chain hydrocarbon in the alkene series is pentene, as opposed to pentane in the alkane series, and so on. The compound-to-compound comparison between the two analogous series is shown in figure 25, along with the structural formulas of the compounds.

You will notice that these compounds are indeed all hydrocarbons; that is, they are covalently bonded compounds containing only hydrogen and carbon. The differences in their structural formulas are apparent; the alkanes have only single bonds in their structural formulas, while the alkenes have one (and only one) double bond in *their* structural formulas. You will notice, however, that there are different numbers of hydrogen atoms in the two analogous series. This difference is due to the octet rule that carbon must satisfy. Since one pair of carbon atoms shares a double bond, this fact reduces the number of electrons the carbons need (collectively) by two, so there are two fewer hydrogen atoms in the alkene than in the corresponding alkane.

In any hydrocarbon compound, carbon will form four covalent bonds. In a saturated hydrocarbon, the four bonds will all be *single bonds*. The definition of an unsaturated hydrocarbon, however, is a hydrocarbon with at least one multiple bond, and the alkenes are an analogous series of unsaturated hydrocarbons containing just one double bond (which, of course, is a multiple bond). The double bond must be formed with another carbon atom since hydrogen atoms can form *only* single bonds and, in a hydrocarbon compound, there are no other elements but hydrogen and carbon. Therefore, in forming a

Methane CH_4

Figure 25
Alkanes and Alkenes.

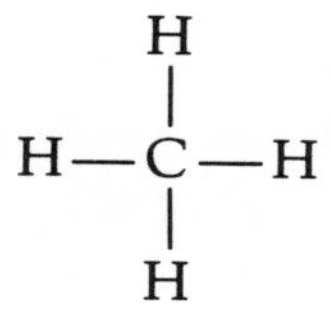

Ethane C_2H_6

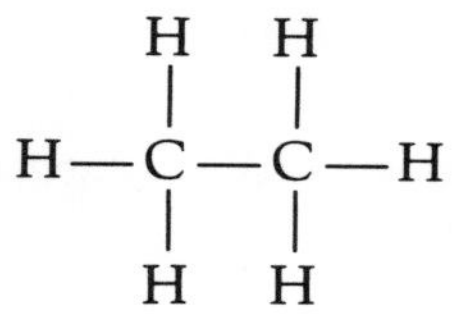

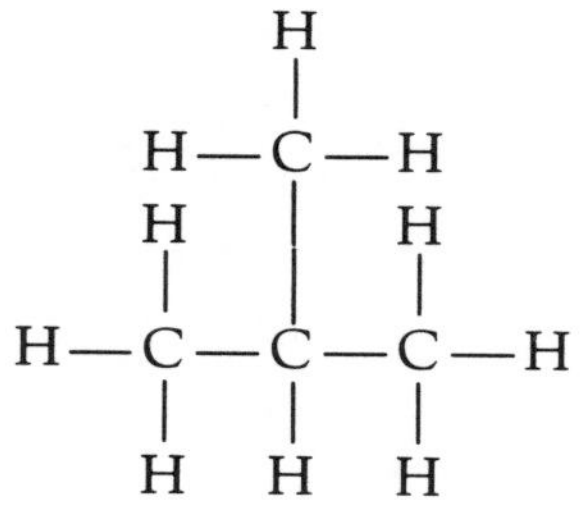

Ethene C_2H_4

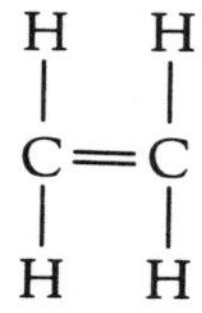

Propane C_3H_8

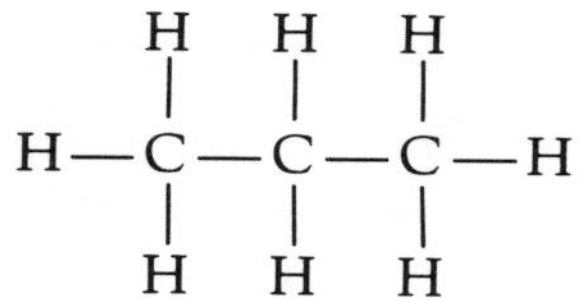

Propene C_3H_6

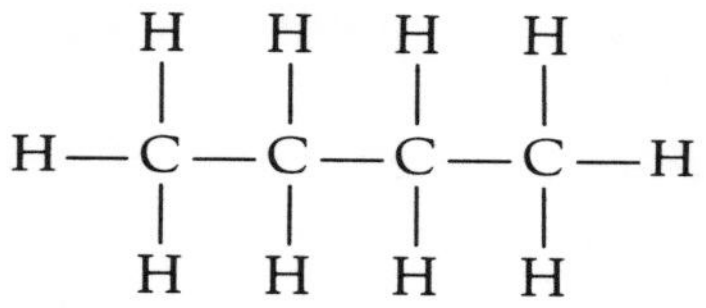

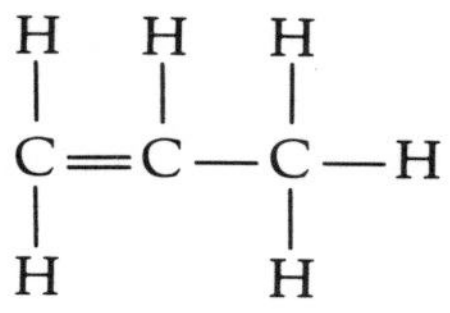

Butane C_4H_{10}

1-Butene C_4H_8

Isobutane C_4H_{10}

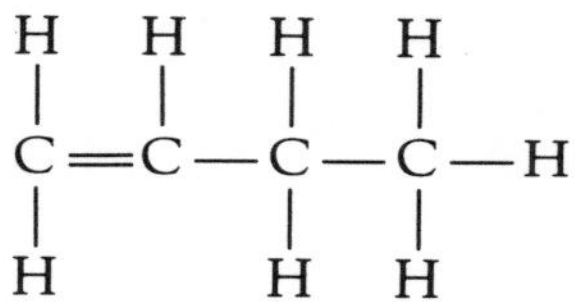

2-Butene C_4H_8

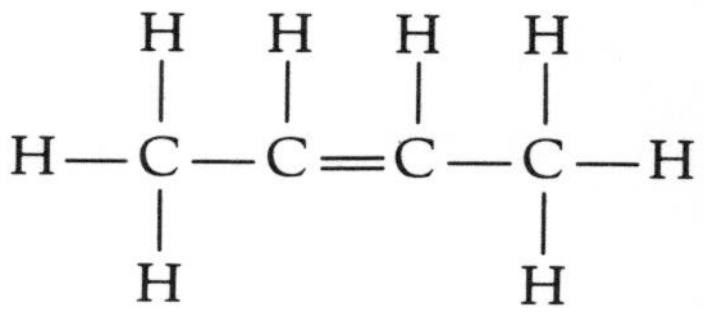

2-methyl propene C_4H_8

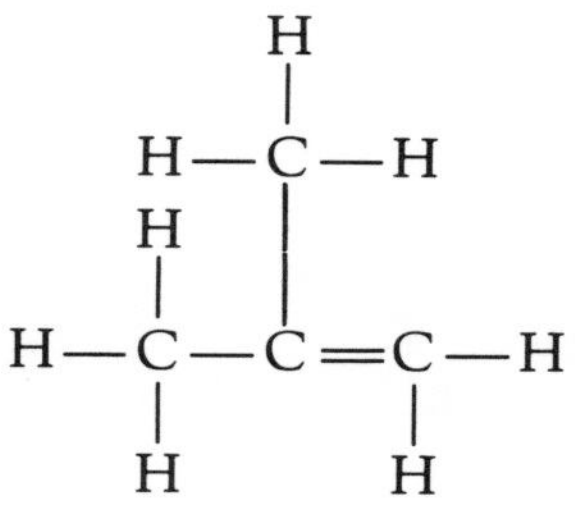

double bond with another carbon atom and to satisfy the octet rule, the alkene must form fewer bonds with hydrogen, resulting in less hydrogen in the structural formula of each alkene than in the corresponding alkane. Indeed, there are two fewer hydrogen atoms in each of the alkenes than in the alkane with the same number of carbon atoms. This is also shown by the general molecular formula of the alkenes, C_nH_{2n}, as opposed to the general molecular formula of the alkanes, which is C_nH_{2n+2}.

You will also note that there is no one-carbon alkene corresponding to methane, the one-carbon alkane; surely you recognize the logical reason for it. If the alkenes are defined as an analogous series of hydrocarbons containing one double bond (which they are), where would the double bond be situated in the structural formula of a hydrocarbon with one carbon? It cannot be between the carbon and any of the hydrogens, since hydrogen can *never* form more than one covalent bond, and there is no other carbon atom in the structural formula. Therefore, the first compound in the alkene series is ethene, while the corresponding two-carbon compound in the alkane series, ethane, is the second compound in the series, with methane the first.

Although the naming of the alkenes is the same as the alkanes, with only the ending changed from -ane to -ene, there is a problem with the names of the first three alkenes. The systematic names of hydrocarbons came a long while after the simplest (that is, the shortest chain) of the compounds in each series was known and named. You will recall that, in naming the alkanes, the system of using the Greek names for numbers as prefixes begins with pentane, rather than with methane. That situation occurred because methane, ethane, propane, and butane were known and named long before it was known that there was an almost infinite length to the chain that carbon could form and that a systematic naming procedure would be needed. Before the new system was adopted, the *common* names for the shortest-chain compounds had become so entrenched that those names survived unchanged; they do so to this day. Therefore, not only are the first four compounds in the alkane series named differently from the rest of the series, the corresponding two-, three-, and four-carbon compounds are *not* generally known as ethene, propene, and butene. Their common names are ethylene, propylene, and butylene. The problem is that younger chemists know them as ethene, propene, and butene. So, for the foreseeable future, you may see either name used to represent these compounds; you must recognize them as the same materials, no matter which name is used. The saving grace is, however, that no matter *what* name is used, the molecular formulas and the structural formulas will always be the same. Remember that more than one compound may have the same molecular formula (isomers), but that a structural formula is unique to one compound. In addition, you will find many chemicals each of which possesses more than one chemical name, for the same reason we mentioned above. There may never come a time when each chemical has one and only one name by which it will be universally recognized, so your job of memorization of chemical names *seems* to be an impossible one; however, the most valuable, and therefore the most common, organic chemicals are those that have the shortest carbon chains. This fact is also true of their *derivatives*, which we will study later in Chapter Five. Your job will be simplified if you become familiar with the compounds that *usually* contain five or fewer carbon atoms.

As you have no doubt guessed by now, the inclusion of a double bond in the structural formula will have a profound effect on the properties of a compound. Table 5 illustrates those differences. Just be advised that the presence of a double bond (and, indeed, a triple bond) between two carbon atoms in a hydrocarbon increases the chemical activity of the compound tremendously over its corresponding saturated hydrocarbon. The smaller the molecule (that is, the shorter the chain), the more pronounced this activity is. A case in point is the unsaturated hydrocarbon ethylene. Disregarding for the present the differences in combustion properties between it and ethane, ethylene is so chemically active that, under the proper conditions, instead of

Table 5 / Properties of Alkenes

Compound	Formula	Molecular Weight	Melting Point (°F.)	Boiling Point (°F.)	Flash Point (°F.)	Ignition Temperature (°F.)
Ethylene	C_2H_4	28	−272.2	−155.0	gas	1,009
Propylene	C_3H_6	42	−301.4	−53.9	gas	927
1-Butene	C_4H_8	56	−300.0	21.7	gas	700
2-Butene	C_4H_8	56	−218.2	38.7	gas	615
1-Pentene	C_5H_{10}	70	−265.0	86.0	32	523
2-Pentene	C_5H_{10}	70	−292.0	98.6	32	NA
1-Hexene	C_6H_{12}	84	−219.6	146.4	−15	487
2-Hexene	C_6H_{12}	84	−230.8	154.4	−5	473
1-Heptene	C_7H_{14}	98	−119.2	199.9	28	500
1-Octene	C_8H_{16}	112	−152.3	250.3	70	446

NA = Not Applicable

burning, it will undergo a rather specialized chemical reaction called polymerization, which, if it is uncontrolled, is a much more violent reaction than combustion. This tendency to polymerize is due to the presence of the double bond. The tendency to polymerize decreases as the molecule gets bigger (the chain is longer). Much will be said in later chapters about polymerization and its products, the polymers.

Although you will be responsible for knowing the names and formulas for the first nine straight-chain hydrocarbons in the series, only the first four or five are important in the study of hazardous materials. Few, if any, of the isomers of the alkenes are common; therefore, aside from knowing what they are, they will not pose any great danger to you—unless, of course, they are involved in a spill, leak, or other type of incident.

There are other hydrocarbon compounds that contain multiple bonds, but we will restrict our discussion to those compounds containing just one multiple bond in their molecules. This is because the compounds containing just one multiple bond are the most valuable commercially and therefore the most common. There is, however, a simple way to recognize when you are dealing with a compound that may contain *two* double bonds; that is a name in which the Greek prefix "di-" is used. As example would be the compound butadiene. You should recognize from the first part of the name ("buta-") that there are four carbon atoms in the chain, and that there is a double bond present (the ending "-ene"); however, just before the -ene ending is the prefix "di-," meaning two. Therefore, you should be able to recognize that you are dealing with a four-carbon hydrocarbon with two double bonds. This idea may be beyond the scope of this textbook, but it provides a good example of how, if you concentrate on just a few principles, you can tell an enormous amount about a chemical simply by looking at its name. At least, it is a beginning in the recognition that this study has as its goal.

Isomers of the Alkenes

As in the alkanes, it is possible for carbon atoms to align themselves in different orders to form isomers, which, you will recall, are compounds with the same molecular formulas but different structural formulas. A listing of the first few isomers is shown in figure 25. You will notice that not only is is possible for the carbon atoms to form branches which produce isomers, but it is also possible for the double bond to be situated between different carbon atoms in different compounds. This different position of the double bond also results in different structural formulas, which, of course, are isomers. Just as in the alkanes, isomers of the alkenes have different properties. These are shown in table 5.

We do not spend a great deal of time on hydrocarbons other than the alkanes, because it is the alkanes and their derivatives that are the most common in commerce; therefore they are the hazardous materials that you have the greatest probability of encountering. You still need to know how to recognize when you are dealing with one of them, however, simply because the unsaturated hydrocarbons and their derivatives are *more active chemically* than the saturated hydrocarbons and *their* derivatives, and more active chemically may be translated as meaning *more hazardous*.

The Alkynes

There exists another analogous series of unsaturated hydrocarbons that contain just one multiple bond, but, instead of being a double bond, it is a *triple* bond. This analogous series is called the *alkynes* and the names of all the compounds end in *-yne*, rather than -ane as in the alkanes, or -ene as in the alkenes. The only compound in this series that is at all common happens to be an extremely hazardous material. It is a highly unstable (to heat, shock, and pressure), highly flammable gas that is the first compound in the series. What we have learned about naming straight-chain hydrocarbons tells us this two-carbon unsaturated hydrocarbon with a triple bond between its two carbon atoms is called ethyne, and indeed this is its *proper* name. It is, however, and will probably *always* be, known by its common name, acetylene; the -ene ending could be confusing, so you will have to memorize the fact that acetylene is an alkyne rather than an alkene. Its molecular formula is C_2H_2, and its structural formula is $H-C \equiv C-H$. The fact that it contains this triple bond makes it extremely active chemically; that is what is meant by its instability to heat, shock, and pressure. You remember that it takes energy to start a chemical reaction, and heat, shock, and pressure are forms of energy. The fact that the triple bond contains so much energy tied up in the structure means that it will release this energy at the slightest urging, which is the input of some slight amount of external energy. When this input energy strikes the molecule of acetylene, the triple bond breaks, releasing the internal energy of the bonds. This produces either great amounts of heat or an explosion, depending on the way in which the external energy was applied. We will have more to say about bond energies shortly, but acetylene is loaded with them.

There are no other alkynes that are commercially important, and so acetylene will be the only member of this series that will concern us for the purposes of this book. There *are* other alkynes, however, along with hydrocarbons that might have one double bond *and* a triple bond present in the molecule; again, you must recognize the rare occasions when this situation occurs. Therefore, we will have no isomers of the alkynes to concern ourselves with, either.

Naming Straight-Chain Hydrocarbons

As we mentioned previously, there is a system for naming the straight-chain hydrocarbons, based on an agreed-upon method of retaining the first three or four common names, then using Greek prefixes that indicate the number of carbon atoms in the chain. For isomers, the same system is used, always using the name of the compound that is attached to the chain *and* the name of the chain. This, of course, is all confusing to you now, but it will, we hope, become clearer shortly.

Go back to the first analogous series of hydrocarbons we studied, in the first section. These were the alkanes, a series of saturated hydrocarbons, all ending in -ane. For these hydrocarbons and other hydrocarbons to react, a place on the hydrocarbon chain must exist for the reaction to take place. Since all the bonds from carbon to hydrogen are already used, an "opening" on one of the carbon atoms must exist for it to be able to react with something else. This "opening" occurs when one of the hydrogen atoms is removed from its bond with a carbon atom, thus causing that carbon to revert back to a condition of instability, with seven electrons in its outer ring, or, as we now state, with one unpaired electron. This one unpaired electron (or half of a covalent bond, or "dangling" bond) wants to react with something, and it will, as soon as another particle which is ready to react is brought near. This chain of carbon atoms (from one carbon to another to another, and so on) with a hydrogen atom missing is a particle that was once a compound, and its name is a *radical*. You may have heard them referred to as "free radicals," which, indeed, they are, for ever so short a period of time. (Remember nature and the impossibility of instability?) Since only in a very small context will they be referred to as free radicals, for the sake of simplicity we will refer to them as radicals. Radicals are created by energy being applied to them in a chemical reaction or in a fire, but no matter how they come about, radicals are important to us. Remember that a hydrocarbon compound with at least one hydrogen atom removed *is no longer a compound*, but a chemical particle known as a radical.

The radicals have names of their own; they are derived from the name of the alkane. When a hydrogen atom is removed from the alkane hydrocarbon, the name is changed from -ane to -*yl*. Therefore, when a hydrogen is removed from the compound methane, the *methyl* radical is formed. When a hydrogen atom is removed from the compound ethane, the ethyl radical is formed. In the same manner, the propyl radical comes from propane, the butyl radical comes from butane, and so on. Similarily, isobutane will produce the isobutyl radical, and isopentane will produce the isopentyl radical. A list of hydrocarbons and the radicals produced from them when a hydrogen is removed is shown is table 6. You will note that there are only a few radicals from compounds other than the alkanes which are important, so you will not have to memorize a list of radicals that originated as alkenes or alkynes.

These radicals are known as hydrocarbon "backbones," and will become increasingly important when we get to hydrocarbon derivatives. Just now, it is useful to know them in order to be able to name other hydrocarbons. For instance, isobutane is more properly named methyl propane. If you refer to your list of isomers in figure 24, you may be able to reason this out. Another isomer with a different proper name is isopentane, more properly called methyl butane. Neopentane is also named 2,2-dimethyl propane. There are only a few rules you must know to be able to do this naming properly, and they are presented in the list below. As we stated in the introduction to the book, however, it is not our intention to make chemists out of anyone studying to learn about hazardous materials. You will not be expected

Table 6 / Radicals

Methane	CH_4	Methyl	$-CH_3$
Ethane	C_2H_6	Ethyl	$-C_2H_5$
Propane	C_3H_8	n-Propyl	$-C_3H_7$
		Isopropyl	$-C_3H_7$
Butane	C_4H_{10}	n-Butyl	$-C_4H_9$
Isobutane	C_4H_{10}	Isobutyl	$-C_4H_9$
		sec-Butyl	$-C_4H_9$
		tert-Butyl	$-C_4H_9$
Ethylene	C_2H_4	Vinyl	$-C_2H_3$
Benzene	C_6H_6	Phenyl	$-C_6H_5$

to know the proper names of the isomers and their derivatives. These rules are presented for your information only; you can refer to them if you run across a long-named compound, to see if you can figure out its structure.

1. Find the longest continuous chain and name it as if it were an alkane.
2. Name the side branches in the same manner.
3. Identify the number of the carbon atom on the longest chain to which the branch is attached by counting from the end of the chain nearest to the branch.
4. If it is possible that there could be *any* confusion as to which carbon atom is meant, put the number in front of the name of the compound, followed by a dash.
5. If there is more than one branch, you *must* use the numbers to identify the carbon atom to which they are attached.
6. If the branches are identical, use the prefixes di- for two, tri- for three, tetra- for four, and so on.

In this manner, the four isomers of hexane are named 2-methyl pentane, 3-methyl pentane, 2,2-dimethyl butane, and 2,3-dimethyl butane.

In any case, you will not be expected to be able to write the molecular formula for every hydrocarbon and all of the isomers possible. What you *will* be expected to do is to *recognize* what class of hazardous materials is involved, to be able to make some sort of judgment as to the danger posed by the substance in question in its present condition, *and* if conditions change.

Aromatic Hydrocarbons

Up to this point, we have been discussing straight-chain hydrocarbons, both saturated and unsaturated, with the unsaturated hydrocarbons containing only one multiple bond. The unsaturated hydrocarbons studied were the alkenes with one double bond and the alkynes (really only one compound, acetylene) with one triple bond. There are other straight-chain hydrocarbons that are unsaturated containing *more than one* multiple bond, some with more than one double bond, and some with a mixture of double bonds and triple bonds. The combinations and permutations are endless, but there are very few of these highly unstable materials around. We will mention them only as they occur in the course of our studies.

There is a large body of hydrocarbons that is very important in chemistry, in the stream of commerce, and therefore important to first responders to a hazardous-materials incident. These hydrocarbons are different in that they are not straight-chain hydrocarbons but have a structural formula that can only be called *cyclical*. The most common and most important hydrocarbon in this group is called *benzene*. It is the first and simplest of the six-carbon cyclical hydrocarbons called *aromatic* hydrocarbons.

Benzene has been widely used in many industries for a long time; from the first time it was synthesized, its structural formula baffled chemists. Its molecular formula was known to be C_6H_6, but it did not behave like hexane, hexene, or any of their isomers. You would expect it to be similar to these other six-carbon hydrocarbons in its properties. Using hexane and 1-hexene as comparisons, look at table 7, the list of comparative properties.

As you can see, there are some major differences between benzene and the straight-chain hydrocarbons of the same carbon content. Hexene's ignition temperature is very near to hexane's, but certainly there must be some explanation for such a wide variation in other temperatures. The flash point difference is not great, but look at the melting point differences. These major differences should not occur unless there were another factor controlling them, and there is. The only explanation for these differences is structure; since the molecular formula of benzene was known to be correct, the only way this

Table 7 / Differences Between Properties of Benzene and of Straight-Chain Hydrocarbons

Compound	Formula	Melting Point (°F.)	Boiling Point (°F.)	Flash Point (°F.)	Ignition Temperature (°F.)	Molecular Weight
Hexane	C_6H_{14}	−139.5	156.0	−7	500	86
1-Hexane	C_6H_{12}	−219.6	146.4	⟨−20	487	84
Benzene	C_6H_6	41.9	176.2	12	1,044	78

molecular formula could exist, with benzene's properties, is to exist in a cyclical form, shown here:

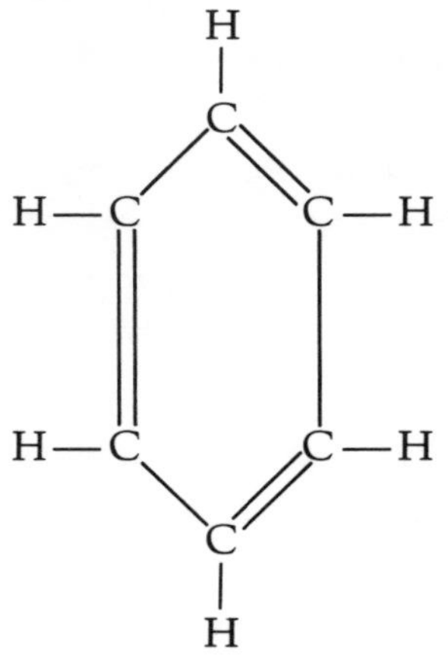

```
              H
              |
              C
            ⫽   ⟍
     H — C       C — H
          ‖       |
          |       ‖
     H — C       C — H
            ⟍   ⫽
              C
              |
              H
```

The alternating double bonds should impart very different properties to benzene, however, and the fact is that they do not. The only possible way for the benzene molecule to exist is in this manner:

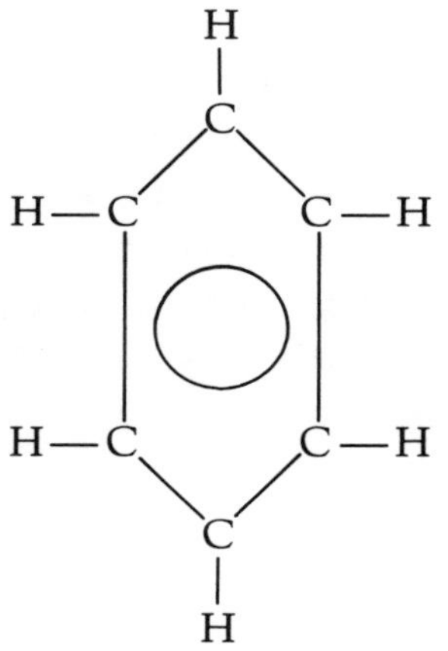

```
              H
              |
              C
            ⟋   ⟍
     H — C       C — H
        |   ( O )   |
     H — C       C — H
            ⟍   ⟋
              C
              |
              H
```

The circle is drawn within the hexagonal structural to show that the electrons that should form a series of alternating double bonds are really spread among all six carbon atoms. It is the only structure possible that would explain the unique properties of benzene. This structural formula suggests *resonance*; that is, the possibility that the electrons represented by the circle are actually alternating back and forth between and among the six carbon atoms.

This particular hexagonal structure is found throughout nature in many forms, almost always in a more complicated way, usually connected to many other "benzene rings" to form many exotic compounds, most of which are not really understood. We will, however, be concerned with just benzene and a few of its *derivatives*.

Those hydrocarbon compounds include toluene and xylene, whose structural formulas are shown in figure 26 and compared with benzene.

The properties given in figure 26 are for reference and not for memorization. They are presented to show the differences caused in these properties by molecular weight and structural formulas. These compounds are the most valuable and therefore most common of the aromatic hydrocarbons, and they are all flammable liquids. We will look at these structures again when we reach hydrocarbon derivatives.

There are other cyclical hydrocarbons, but they do not have the structural formulas of the aromatics, unless they are benzene-based. They will be discussed in future chapters only as each is mentioned as a hazardous material. These cyclical hydrocarbons may have three, four, five, or seven carbons in the cyclical structure, in addition to the six-carbon ring of the aromatics. None of them has the stability or the chemical properties of the aromatics.

**Figure 26
Comparison of Benzene and
Some of Its Derivatives.**

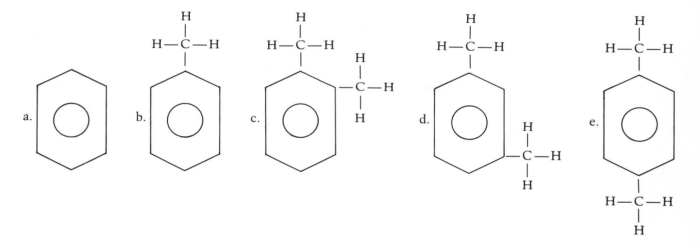

Compound	Formula	Melting Point (°F.)	Boiling Point (°F.)	Flash Point (°F.)	Ignition Temperature (°F.)	Molecular Weight
a. Benzene	C_6H_6	41.9	176.2	12	1,044	78
b. Toluene	C_7H_8	−138.1	231.3	40	997	92
c. o-xylene	C_8H_{10}	−13.0	291.2	90	867	106
d. m-xylene	C_8H_{10}	−53.3	281.9	81	982	106
e. p-xylene	C_8H_{10}	55.8	281.3	81	984	106

Glossary

Alkanes: An analogous series of saturated hydrocarbons with the general formula C_nH_{2n+2}.

Alkenes: An analogous series of unsaturated hydrocarbons with the general formula C_nH_{2n}; the alkenes all contain just one double bond between carbon atoms.

Alkynes: An analogous series of unsaturated hydrocarbons with the general formula C_nH_{2n-2}; the alkynes all contain just one triple bond between carbon atoms.

Analogous Series: A series of compounds that have differences of the same type between compounds. That is, in an analogous series of hydrocarbon compounds, each compound is different from the one succeeding it (one more carbon atom in the chain) or the one preceding it (one less carbon atom in the chain) by a specific "unit," or part of the compound, such as one carbon atom and two hydrogen atoms.

Aromatic: The name originally given to cyclical compounds containing the benzene "ring," because the first benzene-type compounds isolated smelled "good."

Branching: A configuration in which a carbon atom attaches itself to another carbon atom that has two or three other carbon atoms attached to it, forming a *branch*, or side chain. When the carbon attaches to another carbon that has only one other carbon attached to it, a straight chain is formed, rather than a branched chain.

Chain: The way carbon atoms react with each other, producing covalent bonds between them, resembling a chain with carbon atoms as the links.

Common Name: The name originally given to a compound upon its discovery, prior to the adoption of an organized system of assigning proper names.

Cyclical: The structure of certain molecules where there is no end to the carbon chain; the molecule is a closed structure resembling a ring, where what would be the "last" carbon in the chain is bonded to the "first" carbon in the chain. There are cyclical compounds in which the closed structure contains the atoms of other elements in addition to carbon.

Derivative: A compound made from a hydrocarbon by substituting another atom or group of atoms for one of the hydrogen atoms in the compound.

Diatomic: Two atoms; as in a *diatomic* molecule, which contains two atoms bound covalently to each other.

Free Radical: An atom or group of atoms bound together chemically with at least one unpaired electron. A free radical is formed by the introduction of energy to a covalently bonded molecule, when that molecule is broken apart by the energy. It cannot exist free in nature and therefore must react quickly with other free radicals present.

General Formula: The general molecular formula for an analogous series of compounds that will give the actual molecular formula for any member of the series as long as the number of carbon atoms in the compound is known. This number is substituted for the letter "*n*" in the formula.

Hydrocarbon: A covalent compound containing *only* hydrogen and carbon.

"Iso-": The prefix (meaning the same) given to a compound having the same number and kind of atoms as another compound, as in *iso*mer.

Isomer: A compound with a molecular formula identical to another compound but with a different structural formula. That is, a compound may possess exactly the same elements, and exactly the same number of atoms of those elements as another compound, but those atoms are arranged in a different order from the first compound.

Molecular Formula: A method of representing a molecule by a written formula, listing which atoms and how many of them are in the molecule, without showing how they are bonded to each other.

"Neo": A prefix given to an isomer of another compound. It exists in compounds that were named long ago and is used only when the compound it best known by its common name.

"Normal": The designation given to a straight-chain compound that has isomers. The designation in the molecular formula is an "n-" in front of the formula.

Olefins: A synonym for the alkene series.

Paraffin Series: An older name given to the alkanes.

Polymerize: The chemical reaction whereby a compound reacts with *itself* to form a polymer.

Proper Name: An agreed-upon system of naming organic compounds according the the longest carbon chain in the compound.

Radical: An atom or group of atoms bound together chemically that has one or more unpaired electrons; it cannot exist in nature in that form, so it reacts very fast with another radical present, to form a new compound; also known as a "free" radical.

Resonance: A phenomenon whereby a structure, to satisfy the rules of covalent bonding, should be fluctuating (resonating) back and forth between two alternate molecular structures, both of which are "correct" for the molecule. It is a way of explaining what cannot be explained using only the rules of covalent bonding.

Saturated: A hydrocarbon possessing only single covalent bonds between carbon atoms.

Straight Chain: The configuration of the molecule of a hydrocarbon when a carbon atom attaches itself to another carbon atom that has only one other carbon atom already attached to it.

Structural Effect: The effect upon certain properties of an analogous series of compounds by *branching*. Properties such as boiling point, flash point, ignition temperature, and others change as branches are added to compounds, including isomers.

Structural Formula: A drawing of the molecule, showing all the atoms of the molecule and how they are bonded to each other atom.

Unsaturated: A hydrocarbon with at least one multiple bond between two carbon atoms somewhere in the molecule.

Weight Effect: The change produced in certain properties, including flash point, boiling point, and water solubility, as the molecular weight (calculated by adding the atomic weights of all the atoms in the molecule) of compounds in an analogous series is increased or decreased.

5

Hydrocarbon Derivatives

Introduction

All the hydrocarbons discussed in the last chapter, with the exception of the aromatic hydrocarbons, have one thing in common: they are all *—(ALL ARE USED TO BURN)* used for their combustion properties. The first two compounds in the alkane series, methane and ethane, are very efficient, clean-burning fuels. Methane is the principal ingredient in natural gas, and ethane is a distant second, but it too is present in natural gas. Propane and butane are the LP (liquid petroleum) gases; they too are used extensively as heating fuels. The next six compounds, pentane, hexane, heptane, octane, nonane, and decane, all flammable liquids, are the materials that make up gasoline, along with a few other additives. The next four to six compounds, in different ratios, make up kerosene, diesel fuel, and the other fuel oils. The longer-chained alkanes are the solid hydrocarbons which are useful in many other ways. Of course, as we consider the members of the alkanes or any other analogous series, it is understood that the isomers of these compounds are included in the mentioned uses. In some cases, the isomers are more desirable than the straight-chain hydrocarbon because of their different properties.

The short-chain alkenes are most valued for their ability to polymerize and, as such, are the backbone of the rubber and plastics industries. They have other uses, of course, but polymerization is the main reason for their popularity. The longer-chain alkenes have very specialized uses, and they are rare when compared to the alkanes. The only alkyne that is commercially important is acetylene; its major use is in cutting and welding metals, because of its high flame temperature. For these reasons, we spend very little, if any, time on long-chain unsaturated hydrocarbons. They are not commercially valuable and are therefore not much used. Since they are not much used, they are not made in any significant quantity and therefore not shipped in any large quantities. If they are neither shipped nor used very much, it follows logically that you, as a first responder to a hazardous-materials incident, will not run across them very often. This is particularly true of the alkynes; there are a few commercially valuable alkynes in addition to acetylene (ethyne), but they are rare.

The aromatic hydrocarbons are used mainly as solvents and as feedstock chemicals for chemical processes that produce other valuable chemicals. As far as cyclical hydrocarbons are concerned, the aromatic hydrocarbons are the only compounds we discussed. These compounds all have the six-carbon benzene ring as a base, but be advised that there are three-, four-, five-, and seven-carbon rings. These materials will be considered as we examine their occurrence as hazardous materials. You will find that, after the alkanes, the aromatics are the next most common chemicals shipped and used in commerce. The short-chain olefins (alkenes) such as ethylene and propylene may be shipped in larger quantities because of their use as monomers, but for sheer numbers of different compounds, the aromatics will surpass even the alkanes in number, although not in volume.

Quite a long time ago, man discovered how to alter these hydrocarbon compounds, to produce other compounds that had desirable properties. These early chemists were encouraged to *synthesize* many of the naturally occurring compounds that were wanted, for whatever reason. These compounds included spices, fragrances, dyes, and those with many other uses, including medicines. As the Industrial Age developed, requirements for chemicals with special properties increased to the point that literally thousands of new chemicals were developed annually; most of them entered the stream of commerce, as the products made from them or because of them came more and more into demand by society. Since the hydrocarbons were so readily available in this country, much of this industrial development was centered around *hydrocarbon derivatives*. A hydrocarbon derivative is a compound with a *hydrocarbon backbone* and a *functional group* attached to it chemically. A hydrocarbon backbone is defined as a molecular fragment that began as a hydrocarbon compound and has had at least one hydrogen atom removed from the molecule. Such a fragment is also known as a *radical*; you may be more familiar with them as free radicals. A functional group is defined as an atom or a group of atoms, bound together chemically, which impart specific chemical properties to a molecule; they may also be referred to as radicals. A hydrocarbon derivative then is really a compound made up of two specific parts; the first part comes from a hydrocarbon, and the second may have many different origins (which includes coming from a hydrocarbon), depending on the chemical makeup of the functional group. The only reason for making a hydrocarbon derivative is because someone is willing to buy it or to buy a product made from it, or because it facilitates the manufacture of a product or a service for which someone is willing to pay.

The hydrocarbon backbone may come from an alkane, an alkene, an alkyne (indeed, any saturated or unsaturated hydrocarbon), or from an aromatic hydrocarbon or other cyclical hydrocarbon. That is to say, *any* hydrocarbon compound may form the hydrocarbon backbone portion of the hydrocarbon derivative, as long as it has been converted to a radical, by removal of one or more hydrogens, in preparation for the reaction (see figure 27). The functional group may have many origins, with chemists using as reactants any chemical compound that will produce the desired functional group. The functional groups include the halogens (fluorine, chlorine, bromine, and iodine, represented by

an "X"), the hydroxyl radical (–OH), the carbonyl group $\left(-\overset{\overset{\displaystyle O}{\displaystyle \|}}{C}-\right)$,

oxygen (–O–), the carboxyl group $\left(-\overset{\overset{\displaystyle O}{\displaystyle \|}}{C}-OH\right)$, the peroxide radical

(–O–O–), the amine radical $\left(-\overset{\overset{\displaystyle H}{\displaystyle |}}{N}-H\right)$, and even other hydrocarbon radicals (– R). When these functional groups are chemically attached to hydrocarbon backbones, they form compounds called hydrocarbon derivatives, and each functional group imparts a separate set of chemical and physical properties to the molecule formed by this chemical attachment.

Just as the alkanes and alkenes had general formulas, the hydrocarbon derivatives all have general formulas. The hydrocarbon backbone will provide a portion of the general formula, and the functional group will provide the other part. In each case, the hydrocarbon derivative will be represented by the formula R–, and the hydrocarbon backbone will have its own specific formula. Let's begin by looking at just what sort of compounds these *substituted hydrocarbons*

Figure 27
Hydrocarbon Radicals

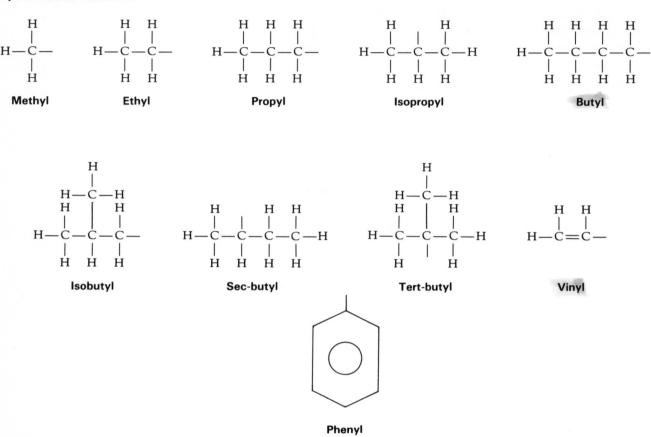

are. Substituted hydrocarbon is simply another name for hydrocarbon derivative, because the functional group is substituted for one or more hydrogen atoms in the chemical reaction.

Halogenated Hydrocarbons

A *halogenated hydrocarbon* is defined as a derivative of a hydrocarbon in which a hydrogen atom is replaced by a halogen atom. Since all of the halogens react similarly, and the number of hydrocarbons (including all saturated hydrocarbons, unsaturated hydrocarbons, aromatic hydrocarbons, other cyclical hydrocarbons, and all the isomers of these hydrocarbons) is very large, you can see that the number of halogenated hydrocarbons can also be very large. Luckily for the student of hazardous materials, the most common hydrocarbon derivatives are those of the first four alkanes and the first three alkenes (and, of course, the isomers of these hydrocarbons). There are some aromatic hydrocarbon derivatives, but, again, they are of the simplest structure. This narrowing down of the number of compounds to a manageable few saves the hazardous-materials student from an impossible task of memorization. Whatever the hydrocarbon backbone turns out to be, it is represented in the general formula by its formula, which is R –. Therefore, the halogenated hydrocarbons will have formulas such as R – F, R – Cl, R – Br, and R – I for the respective substitution of fluorine, chlorine, bromine, and iodine on to the hydrocarbon backbone. As a rule, the general formula can be written R – X, with the R as the hydrocarbon backbone, the X standing for the *halide* (any of the halogens), and the "–" the covalent bond between the hydrocarbon backbone and the halogen. R – X is read as "alkyl halide."

Remember, whenever a hydrocarbon backbone (a hydrocarbon radical) is formed from an alkane, the name is changed from an ending of -ane to -yl. Methane thus becomes the methyl radical when a hydrogen atom is removed from the molecule, ethane becomes ethyl, propane becomes propyl or isopropyl (why an isopropyl radical when there is no such compound as isopropane?), butane becomes butyl or secondary butyl, and isobutane becomes isobutyl or tertiary butyl. In general, these radicals of the alkanes are referred to as alkyl redicals. There are two other important radicals; they are the *vinyl* radical, which is produced when a hydrogen atom is removed from ethylene, and the *phenyl* radical, which results when a hydrogen atom is removed from benzene. Any other radical that arises in our studies will be explained as we come to it. Refer to table 5 for the molecular formulas for these hydrocarbons backbones. The structural formulas for these radicals are shown in figure 27. There is a corresponding radical for every hydrocarbon that exists, as long as you start with the hydrocarbon, and remove at least one hydrogen. The radicals of the longer-chain hydrocarbons, however, are not important, so we will not spend any time on them.

The common question that is constantly asked is: why do these chemicals exist? (Why does man make these materials?) The answer has already been given. Society (you and I and everyone else) wants certain goods and services and are willing to pay for them. In other words, money can be made by providing people with these products and services, and the manufacturing processes demand these chemicals as intermediates or as raw materials to make the product. Sometimes these chemicals are themselves the end products. Whatever the reasons, society insists on *new* and *better* products, so the manufacturing segment of society employs many people to invent and design chemicals and processes in order to satisfy its customers (again, you and me). We will not consider the origins and uses of all the hydrocarbon derivatives, but we will examine specific compounds to show their structure; perhaps this will help you to understand their chemistry.

To form the simplest halogenated hydrocarbon, you start with the simplest hydrocarbon, which is methane. Remember, halogenated means that a halogen atom has been substituted for a hydrogen atom in a hydrocarbon molecule. The most common halogenated hydrocarbons are the chlorinated hydrocarbons, perhaps because of the abundance of chlorine existing in the earth's crust and the oceans as the chloride ion. Therefore the simplest chlorinated hydrocarbon is methyl chloride, whose molecular formula is CH_3Cl. The structural formula for methyl chloride is:

$$\begin{array}{c} H \\ | \\ H-C-Cl \\ | \\ H \end{array}$$

You can see that indeed one chlorine atom has been substituted for one hydrogen atom and has taken its place in the structural formula. Methyl chloride has many uses, such as an herbicide, as a topical anesthetic, extractant, and low-temperature solvent, and as a catalyst carrier in low-temperature polymerization. It is a colorless gas that is easily liquified and is flammable; it is also toxic in high concentrations. Methyl chloride is the common name for this compound, while its *proper* name is chloromethane. Proper names, you will recall, are determined by the longest carbon chain in the molecule, and the corresponding hydrocarbon's name is used as the "last" name of the compound. Any substituted groups are named first, and a number is used to designate the carbon atom that the functional group is attached to, if applicable.

It is possible to substitute more than one chlorine atom for a hydrogen atom on a hydrocarbon molecule; such substitution is done *only* when the resulting compound is commercially valuable or is valuable in another chemical process. An example is methylene chloride (the common name for dichloromethane), which is made by substituting two chlorine atoms for two hydrogen atoms on the methane molecule. Its molecular formula is CH_2Cl_2, and its structural formula is:

$$\begin{array}{c} H \\ | \\ Cl-C-Cl \\ | \\ H \end{array}$$

Methylene chloride is a colorless, volatile liquid with a sharp, ether-like odor. It is listed as a non-flammable liquid, but it *will* ignite at 1,224°F.; it is narcotic at high concentrations. It is most commonly used as a "stripper" of paints and other finishes, so you will find it in almost all furniture-restoring operations. It is also a good degreaser and solvent extractor and is used in some plastics processing.

Since we mentioned that it is possible to substitute more than one halogen on the hydrocarbon molecule (that is, up to the total number of hydrogen atoms in the molecule), we will continue with the halogenation of methane, since the resulting compounds are commercially valuable. Substituting a third chlorine on the methane molecule results in the compound whose proper name is trichloromethane (tri- for three; chloro- for chlorine; and methane, the hydrocarbon's name for the one-carbon chain). It is more commonly known as chloroform. Its molecular formula is $CHCl_3$, and its structural formula is:

$$Cl - \underset{\underset{\displaystyle Cl}{|}}{\overset{\overset{\displaystyle H}{|}}{C}} - Cl$$

*(MOST POISON GASES ARE HIGHLY FLAM. ALSO)

Chloroform is a heavy, colorless, volatile liquid with a sweet taste and characteristic odor. It is classified as non-flammable, but it will burn if exposed to high temperatures for long periods of time. It is narcotic by inhalation and toxic in high concentrations. It is an insecticide and a fumigant and is very useful in the manufacture of refrigerants.

As long as we have gone this far (we will *not* do this for the rest of the hydrocarbon derivatives), we might as well show the compound formed by the total chlorination of methane. Its proper name would be tetrachloromethane (tetra- for four); its common name is carbon tetrachloride. You all know "carbon tet" as a fire-extinguishing agent that is no longer used since it has been classified as a carcinogen. It is still present, though, and its uses include refrigerants, metal degreasing, and chlorination of organic compounds. Its molecular formula is CCl_4, and its structural formula is:

$$Cl - \underset{\underset{\displaystyle Cl}{|}}{\overset{\overset{\displaystyle Cl}{|}}{C}} - Cl$$

The stepwise chlorination of methane was shown simply because all four chlorinated hydrocarbon compounds produced from methane are commercially useful. You must be aware that since all the halogens react similarly, it is possible to form *analogues* of methyl chloride (methyl fluoride, methyl bromide, methyl iodide), methylene chloride (substitute fluoride, bromide, and iodide in this name also), chloroform (fluoroform, bromoform, and iodoform), and carbon tetrachloride (tetrafluoride, tetrabromide, and tetraiodide). Each of these halogenated hydrocarbons has some commercial value, and you will encounter them at some time. It will behoove you to add the properties of these chemicals to your storehouse of knowledge. You *may not* run across them under their common names but by their proper names. You should become comfortable with shifting back and forth between these naming systems, since you may find either name (but seldom both) on a label.

What was true for one hydrocarbon compound is true for most hydrocarbon compounds, particularly straight-chain hydrocarbons; that is, you may subsitute a functional group at each of the bonds where a hydrogen atom is now connected to the carbon atom, remembering that you do this only if you have a need for the resulting product. Where four hydrogen atoms exist in methane, there are six hydrogen atoms in ethane; you recall that the difference in make-up from one compound to the next in an analogous series is the "unit" made up of one carbon and two hydrogens. Therefore it is *possible* to substitute six functional groups on to the ethane molecule, although it would be highly unlikely that there would be a great demand for the resulting compound. You should also be aware that the functional groups that would be substituted for the hydrogens need not be the same; that is, you may substitute chlorine at one bond, fluorine at another, the hydroxyl radical at a third, an amine radical at a fourth, and so on. Obviously there would have to be a real need for a product such as this for a chemist to go to all that trouble (many times there is such a need).

Substituting one chlorine atom for a hydrogen atom in ethane produces ethyl chloride, a colorless, easily liquifiable gas with an ether-like odor and a burning taste, which is highly flammable and moderately toxic in high concentrations. It is used to make tetraethyl lead and other organic chemicals. Ethyl chloride is an excellent solvent and analytical reagent, as well as an anesthetic. Its molecular formula is C_2H_5Cl, and its structural formula is:

$$\begin{array}{ccc} & H & H \\ & | & | \\ H - & C - & C - Cl \\ & | & | \\ & H & H \end{array}$$

Remember, we are using chlorine as the functional group, but it may be any of the other halogens. In addition, we are giving the common names, while the proper names *may* be used on the labels and shipping papers. Ethyl chloride's proper name is chloroethane.

Substituting another chlorine produces ethylene dichloride (proper name 1,2-dichloroethane). In this case, an isomer is possible, which would be the chlorinated hydrocarbon where both chlorines attached themselves to the same carbon atom, whereby 1, 1-dichloroethane is formed. The chemist and chemical engineer control which compound is made, since these compounds have slightly different properties and different demands in the marketplace. You can see why the numbering system is used to designate which compound we mean. You can also see that, as further chlorination of ethane occurs, we would have to use the proper name to designate which compound is being made. One of the analogues of ethylene dichloride is ethylene dibromide, a toxic material that is most efficient and popular as a grain fumigant, but that has been known to be a carcinogen in test animals for some time.

We could continue to proceed through the halogenation of the alkanes, but it would just be repetition of what we have already done. There are many uses for the halogenated hydrocarbons, and you will meet them quite often. None of them are good for you if you contact them in high concentrations for long periods of time, nor are they good for the people who work with them. Many of them are flammable; most are combustible. Some halogenated hydrocarbons are classified as neither, and a few are excellent fire-extinguishing agents (the Halons®), but they will all decompose into smaller, more harmful molecular fragments when exposed to high temperatures for long periods of time.

The Alcohols

The compounds formed when a hydroxyl group (–OH) is substituted for a hydrogen are called alcohols. They have the general formula R–OH. The hydroxyl radical looks exactly like the *hydroxide* ion, but it is definitely not an ion. Where the hydroxide ion fits the definition of a complex ion—a chemical combination of two or more atoms that have collectively lost or (as in this case) gained one or more electrons—the hydroxide radical is a molecular fragment produced by separating the – OH from another compound, and it has no electrical charge. It does have an unpaired electron waiting to pair up with another particle having its own unpaired electron. The alcohols, as a group, are flammable liquids in the short-chain range, combustible liquids as the chain grows longer, and finally solids that will burn if exposed to high temperatures, as the chain continues to become longer. As in the case of the halogenated hydrocarbons, the most useful, and therefore the most valuable, alcohol compounds are of the short-carbon-chain variety. Just as in the case of the halogenated hydrocarbons, the simplest alcohol is made from the simplest hydrocarbon, methane. Its name is methyl alcohol, its molecular formua is CH_3OH, and its structural formual is:

$$
\begin{array}{c}
\text{H} \\
| \\
\text{H} - \text{C} - \text{OH} \\
| \\
\text{H}
\end{array}
$$

When we say that methyl alcohol may be made by substituting the hydroxyl radical for one of methane's hydrogen atoms, the statement is true, but it may not be precisely how methyl alcohol is made commercially. (The statement is made for illustrative purposes only.) Nature produces a tremendous amount of methyl alcohol, simply by the fermentation of wood, grass, and other materials made to some degree of cellulose. As a matter of fact, methyl alcohol is known as wood alcohol, along with names such as wood spirits and methanol (its proper name; the proper names of all alcohols end in -ol). Methyl alcohol is a clear, colorless liquid with a characteristic alcohol odor. In volume, it is the twenty-first highest chemical produced in the United States, has a flash point of 54°F., and is highly toxic. It has too many commercial uses to list here, but among them are as a denaturant for ethyl alcohol (the addition of the toxic chemical methyl alcohol to ethyl alcohol in order to form denatured alcohol, to discourage you from drinking it or kill you if you do), antifreezes, gasoline additives, and solvents. No further substitution of hydroxyl radicals is performed on methyl alcohol.

The most widely known alcohol is ethyl alcohol, simply because it is the alcohol in alcoholic drinks. It is also known as grain alcohol, or by its proper name, ethanol. Ethyl alcohol is a colorless, volatile liquid with a charactistic odor and a pungent taste. It has a flash point of 55°F., is classified as a depressant drug, and is toxic when ingested in large quantities. Its molecular formula is C_2H_5OH, and its structural formula is:

$$
\begin{array}{cc}
\text{H} & \text{H} \\
| & | \\
\text{H} - \text{C} - \text{C} - \text{OH} \\
| & | \\
\text{H} & \text{H}
\end{array}
$$

In addition to its presence in alcoholic beverages, ethyl alcohol has many industrial and medical uses, such as a solvent in many manufacturing processes, as antifreeze, antiseptics, and cosmetics.

The substitution of one hydroxyl radical for a hydrogen atom in propane produces propyl alcohol, or propanol, which has several industrial uses. Its molecular formula is C_3H_7OH, and its structural formula is:

$$
\begin{array}{ccccccc}
 & H & & H & & H & \\
 & | & & | & & | & \\
H - & C & - & C & - & C & - OH \\
 & | & & | & & | & \\
 & H & & H & & H &
\end{array}
$$

Propyl alcohol has a flash point of 77°F. and, like all the alcohols, burns with a pale blue flame. More commonly known is the isomer of propyl alcohol, isopropyl alcohol. Since it is an isomer, it has the same molecular formula as propyl alcohol, but the following structural formula:

$$
\begin{array}{ccccccc}
 & H & & OH & & H & \\
 & | & & | & & | & \\
H - & C & - & C & - & C & - H \\
 & | & & | & & | & \\
 & H & & H & & H &
\end{array}
$$

Isopropyl alcohol has a flash point of 53°F., so you can see the structure effect caused by the different position of the hydroxyl radical. Its ignition temperature is 850°F., while propyl alcohol's ignition temperature is 700°F., another effect of the different structure. Isopropyl alcohol, or 2-propanol (its proper name) is used in the manufacture of many different chemicals, but you probably know it best as rubbing alcohol. Perhaps now you can answer the question posed earlier as to how can there be an isopropyl radical when there is no such compound as isopropane!

It is not necessary to proceed further, as the above-mentioned alcohols are by far the most common, and they are the ones with which you will come in contact most. For example, butyl alcohol is not as commonly used as the first four in the series, but it *is* used. Secondary butyl alcohol and tertiary butyl alcohol (so named because of the type of carbon atom in the molecule to which the hydroxyl radical is attached) must be mentioned because they are flammable liquids, while isobutyl alcohol has a flash point of 100°F. All of the alcohols of the first four carbon atoms in the alkanes, therefore, are extremely hazardous because of their combustion characteristics.

It was stated earlier that there was no need to mention any further hydroxyl substitution of methane because no chemical useful to society was produced (aside from its impossibility), but the same is not true for the further substitution of ethane. Whenever a hydrocarbon backbone has two hydroxyl radicals attached to it, it becomes a special type of alcohol known as a *glycol*. The simplest of the glycols, and the most important, is ethylene glycol, whose molecular formula is $C_2H_4(OH)_2$, and whose structural formula is:

$$HO-\overset{\overset{\displaystyle H}{|}}{\underset{\underset{\displaystyle H}{|}}{C}}-\overset{\overset{\displaystyle H}{|}}{\underset{\underset{\displaystyle H}{|}}{C}}-OH$$

The molecular formula can also be written CH_2OHCH_2OH and may be printed as such on some labels. Ethylene glycol is a colorless, thick liquid with a sweet taste, is toxic by ingestion and by inhalation, and among its many uses is the one most familiar to you, that of a permanent antifreeze and coolant for your automobile. It is a combustible liquid with a flash point of 240°F.

The only other glycol that is fairly common is propylene glycol, which has a molecular formula of $C_3H_6(OH)_2$. It is a combustible liquid with a flash point of 210°F., and its major use is in organic synthesis, particularly of polyester resins and cellophane.

The last group of substituted hydrocarbons produced by adding hydroxyl radicals to the hydrocarbon backbone are the compounds made when *three* hydroxyl radicals are substituted; these are known as glycerols. The name of the simplest of this type of compound *is* just glycerol. Its molecular forumla is $C_3H_5(OH)_3$, and its structural formula is:

$$H-\overset{\overset{\displaystyle H}{|}}{\underset{\underset{\displaystyle OH}{|}}{C}}-\overset{\overset{\displaystyle H}{|}}{\underset{\underset{\displaystyle OH}{|}}{C}}-\overset{\overset{\displaystyle H}{|}}{\underset{\underset{\displaystyle OH}{|}}{C}}-H$$

Glycerol is a colorless, thick, syrupy liquid with a sweet taste, has a flash point of 320°F., and is used to make such diverse products as candy and explosives, plus many more. Other glycerols are made, but most of them are not classified as hazardous materials.

The Ethers

The ethers are a group of compounds with the general formula R–O–R'. The R, of course stands for any hydrocarbon backbone, and the R' also stands for any hydrocarbon backbone, but the designation R' is used to indicate that the second hydrocarbon backbone may be different from the first. In other words, both the hydrocarbon backbones in the formula may be the same, but the " ' " is used to indicate that it may also be different. R–O–R as the general formula for the ethers is also correct. The fact that there are two hydrocarbon backbones on either side of an oxygen atom means that there will be two hydrocarbon names used.

The simplest of the ethers would be ether that has the simplest hydrocarbon backbones attached; those backbones are the radicals of the simplest hydrocarbon, methane. Therefore, the simplest of the ethers is *dimethyl ether*, whose formula is CH_3OCH_3, and whose structural formula is:

$$
\begin{array}{ccccc}
 & H & & H & \\
 & | & & | & \\
H- & C & -O- & C & -H \\
 & | & & | & \\
 & H & & H &
\end{array}
$$

Dimethyl is used because there are two methyl radicals, and "di-" is the prefix for two. This compound could also be called methyl methyl ether, or just plain methyl ether, but it is better known as dimethyl ether. It is an easily liquified gas that is extremely flammable, has a relatively low ignition temperature of 662°F., and is used as a solvent, a refrigerant, a propellant for sprays, and a polymerization stabilizer.

The next simplest ether is the ether with the simplest alkane as one of the hydrocarbon backbones and the next alkane, which is methyl ethyl ether. Its molecular formula is $CH_3OC_2H_5$, and its structural formula is:

$$
\begin{array}{ccccccc}
 & H & & H & H & \\
 & | & & | & | & \\
H- & C & -O- & C & - & C & -H \\
 & | & & | & | & \\
 & H & & H & H &
\end{array}
$$

It is a colorless gas with the characteristic ether odor. It has a flash point of 31°F., and an ignition temperature of only 374°F. This property, of course, makes it an extreme fire and explosion hazard.

The next simplest ether is actually the one most commonly referred to as "ether." It is diethyl ether, whose molecular formula is

$C_2H_5OC_2H_5$, sometimes written as $(C_2H_5)_2O$, and whose structural formula is:

$$H-\underset{\underset{H}{|}}{\overset{\overset{H}{|}}{C}}-\underset{\underset{H}{|}}{\overset{\overset{H}{|}}{C}}-O-\underset{\underset{H}{|}}{\overset{\overset{H}{|}}{C}}-\underset{\underset{H}{|}}{\overset{\overset{H}{|}}{C}}-H$$

This ether is the compound that was widely used as an anesthetic in many hospitals and may still be found stored in Civil Defense hospital storage units. One of the hazards of all ethers, and particularly diethyl ether because of its widespread use, is that once ethers have been exposed to air, they possess the unique capability of adding an oxygen atom to their structure and converting to a dangerously unstable and explosive organic peroxide. The peroxide-forming hazard aside, diethyl ether has a flash point of $-56°F$. and ignition temperature of $356°F.$; it is a colorless, volatile liquid with the characteristic ether odor. In addition to its use as an anesthetic, it is useful in the synthesis of many other chemicals. Of course, it is an extremely hazardous material.

Another important ether is vinyl ether, a colorless liquid with the characteristic ether odor. Its molecular formula is $C_2H_3OC_2H_3$ and its structural formula is:

$$H-\overset{\overset{H}{|}}{C}=\overset{\overset{H}{|}}{C}-O-\overset{\overset{H}{|}}{C}=\overset{\overset{H}{|}}{C}-H$$

Vinyl ether has a flash point of $-22°F$. and an ignition temperature of $680°F$. It is highly toxic by inhalation and is used in medicine and in the polymerization of certain plastics.

The Ketones

The ketones are a group of compounds with the general formula $R-\overset{\overset{O}{||}}{C}-R'$. The $-\overset{\overset{O}{||}}{C}-$ functional group is known as the carbonyl group or carbonyl radical; it appears in many different classes of hydrocarbon derivatives. There are only a few important ketones, and they are all extremely hazardous. The first is the simplest, again with two methyl radicals, one on either side of the carbonyl group. Its molecular formula is CH_3COCH_3, and its structural formula is:

$$
\begin{array}{ccccc}
 & H & O & H & \\
 & | & || & | & \\
H- & C & -C & -C & -H \\
 & | & & | & \\
 & H & & H &
\end{array}
$$

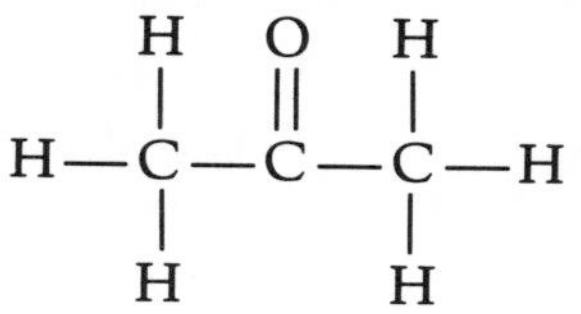

Its proper name is propanone (propa- because of the relationship to the three-carbon alkane, propane, and -one because it is a ketone); it could logically be called dimethyl ketone, but it is universally known by its common name, acetone. Acetone is a colorless, volatile liquid with a sweet odor, has a flash point of 15°F. and an ignition temperature of 1,000°F., is narcotic in high concentrations, and could be fatal by inhalation or ingestion. It is widely used in manufacturing many chemicals and is extremely popular as a solvent.

The next most common ketone is methyl ethyl ketone, commonly referred to as MEK. Its molecular formula is $CH_3COC_2H_5$, and its structural formula is:

$$
\begin{array}{ccccccc}
 & H & O & H & H & \\
 & | & || & | & | & \\
H- & C & -C & -C & -C & -H \\
 & | & & | & | & \\
 & H & & H & H &
\end{array}
$$

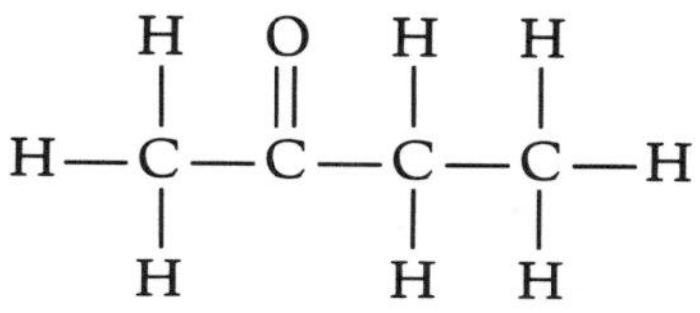

MEK has a flash point of 24°F. and an ignition temperature of 960°F. It is a colorless liquid with a characteristic ketone odor. It is as widely used as acetone and is almost as hazardous..

The Aldehydes

The aldehydes are a group of compounds with the general formula R-CHO. The aldehyde functional group is always written -CHO, even though this does not represent the aldehyde's structural formula. It is written in this way so that the aldehydes will not be confused with R-OH, the general formula of the alcohols.

The simplest of the aldehydes is formaldehyde, whose molecular formula is HCHO, and whose structural formula is:

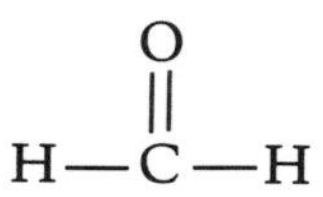

$$
\begin{array}{ccc}
 & O & \\
 & || & \\
H- & C & -H
\end{array}
$$

You can see that the second hydrocarbon backbone of the ketone has been replaced by a hydrogen atom. Formaldehyde is a gas that is extremely soluble in water; it is often sold commerically as a 50 percent solution of the gas in water. The gas itself is flammable, has an ignition temperature of 806°F. and a strong, pungent odor, and is toxic by inhalation. Inhalation at low concentrations over long periods of time has produced illness in many people. Beside its use as an embalming fluid, formaldehyde is used in the production of many plastics and in the production of numerous other chemicals.

The next aldehyde is acetaldehyde, a colorless liquid with a pungent taste and a fruity odor. Its molecular formula is CH_3CHO, and its structural formula is:

$$H-\overset{\displaystyle \overset{H}{|}}{\underset{\displaystyle \underset{H}{|}}{C}}-\overset{\displaystyle \overset{O}{||}}{C}-H$$

It has a flash point of $-40°F.$, an ignition temperature of 340°F., and is toxic by inhalation. Acetaldehyde is used in the manufacture of many other chemicals.

Other important aldehydes are propionaldehyde, butyraldehyde, and acrolein. We will spend more time on these compounds when we reach Chapter Seven, in the material on flammable liquids.

The Peroxides

The peroxides are a group of compounds with the general formula $R-O-O-R'$. This chemical combination of two oxygen atoms bonded covalently between two hydrocarbon backbones is so unstable and so hazardous that we will merely mention them here but will devote a large part of Chapter Fourteen, on unstable hazardous materials, to them. *All* peroxides are hazardous materials, but the organic peroxides may be the most hazardous of all.

The Esters

The esters are a group of compounds with the general formula $R-O-O-R'$. They are not generally classified as hazardous materials, except for the acrylates, which are monomers and highly flammable. Few of the rest of the class are flammable, so no further time will be spent on them here. There are some esters that are hazardous, and we will consider them in Chapter Seven, in the material on flammable liquids.

The Amines

The amines are a group of compounds with the general formula $R-NH_2$, and all the common amines are hazardous. As a class the amines pose more than one hazard, being flammable, toxic, and, in some cases, corrosive. The amines are an analogous series of compounds and follow the naming pattern of the alkyl halides and the alcohols; that is, the simplest amine is methyl amine, with the molecular formula of CH_3NH_2, and the structural formula:

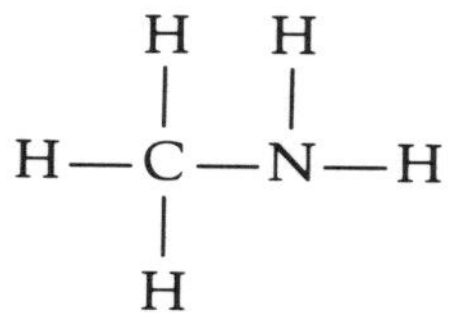

$$
\begin{array}{cccc}
& H & H & \\
& | & | & \\
H- & C- & N- & H \\
& | & & \\
& H & &
\end{array}
$$

Methyl amine is a colorless gas with an ammonia-like odor and an ignition temperature of 806°F. It is a tissue irritant and toxic, and it is used as an intermediate in the manufacture of many chemicals.

Ethyl amine is next in the series, followed by propyl amine, isopropyl amine, butyl amine and its isomers, and so on. We will devote more time to the amines in Chapter Seven.

Other Hydrocarbon Derivatives

There are more hydrocarbon derivatives, but we will not cover them here. Some are more important commercially than others, so their incidence of exposure will be directly proportional to that importance. The organic acids are important too, and, although they are not very corrosive, they will be covered in that same chapter, Chapter Thirteen.

Glossary

Alcohol: The hydrocarbon derivative in which a hydroxyl radical (–OH) is substituted for a hydrogen atom and which has the general formula $R-OH$.

Aldehyde: A hydrocarbon derivative with the general formula $R-CHO$.

Alkyl: The general name for a radical of an alkane; an alkyl halide is a halogenated hydrocarbon whose hydrocarbon backbone originated as an alkane.

Amine: The hydrocarbon derivative in which an amine group $(-NH_2)$ is substituted for a hydrogen atom and which has the general formula $R-NH_2$.

Analogue: A compound in one analogous series that has a property in common with a compound in another analogous series; for example, methyl chloride is an analogue of methyl fluoride.

Aromatic: A name given to cyclical hydrocarbons based on the benzene ring.

Carbonyl: The functional group with the structural formula

$$-\overset{\overset{\textstyle O}{\|}}{C}-.$$

Carcinogen: A cancer-causing agent.

"Di—": The prefix that means two.

Ester: The hydrocarbon derivative with the general formula

$$R-\overset{\overset{\textstyle O}{\|}}{C}-O-R'$$

Ether: A hydrocarbon derivative with the general formula $R-O-R'$.

Functional Group: An atom or group of atoms, bound together chemically, that has an unpaired electron, which when it attaches itself to the hydrocarbon backbone, imparts special properties to the new compound thus formed.

Glycerol: A series of substituted hydrocarbons with three hydroxyl radicals substituted for hydrogen atoms.

Glycol: A hydrocarbon derivative with two hydroxyl radicals substituted for two hydrogen atoms.

Halide: A halogenated compound.

Halogenated: A compound that has had a halogen atom substituted for another hydrogen atom. A halogenated hydrocarbon is a hydrocarbon that has had at least one hydrogen atom removed and replaced by a halogen.

Halogenation: The chemical reaction whereby a halogen is substituted for another atom, usually a hydrogen atom.

Halogens: The elements of group VIIA: fluorine, chlorine, bromine, iodine, and astatine.

Hydrocarbon Backbone: The molecular fragment that remains after a hydrogen atom is removed from a hydrocarbon; the hydrocarbon portion of a hydrocarbon derivative.

Hydrocarbon Derivative: A compound that began as a hydrocarbon, had a hydrogen atom removed from the chain somewhere, and had a functional group attached to replace the hydrogen atom.

Hydroxyl: The functional group of the alcohols; the structural formula is $-O-H$, usually written $-OH$.

Ketone: A hydrocarbon derivative with the general formula

$$R—\overset{\overset{\textstyle O}{\|}}{C}—R'$$

"Mono-": The prefix that means one.

Monomer: A simple, small molecule that has the special capability of reacting with *itself* to form a giant molecule called a polymer.

Peroxide: The hydrocarbon derivative with the general formula $R-O-O-R'$; also the name of the peroxide radical which has the structural formula $-O-O-$.

Phenyl: The general name for the radical of benzene.

Polymerization: The chemical reaction in which a special compound, called a monomer, combines with itself to form a long-chain molecule called a polymer.

Radical: An atom or group of atoms bound together chemically that has one or more unpaired electrons; it cannot exist in nature in that form, so it reacts very fast with another radical present to form a new compound; also known as a "free" radical.

Substituted: A compound that has hand one or more of its atoms removed and replaced by atoms of other elements in the molecule. A substituted hydrocarbon is a compound that has had a hydrogen atom removed and another atom substituted for it.

Synthesize: To make a molecule to duplicate a molecule made in nature.

Unit: A molecular fragment that repeats itself in a series.

"Tetra-": The prefix that means four.

"Tri-": The prefix that means three.

Vinyl: The general name for the radical of ethylene.

6

Fire and Pyrolysis

Introduction

As we have seen in the previous five chapters, it is necessary for the serious student of hazardous materials to have a rudimentary background in chemistry in order to be able to understand the materials with which he will be dealing. Before we discuss the first actual class of hazardous materials, we need to be sure that, since we now have a background in the composition and molecular structure of many materials, we will be able to understand the chemical reactions in which these materials participate. Therefore, since the first several classes of hazardous materials we study are hazardous materials that will burn, we need to look a little more closely at the chemical reaction that we call combustion to make sure we understand just what is happening while material is burning. Understanding the combustion process and the clearly definable parts into which it is divisible gives us knowledge of how to interrupt that process, and, of course this interruption of the combustion process is called extinguishment.

Theories of Fire

The chemical reaction that we call fire, or combustion, is really an *oxidation* reaction. Oxidation is defined as the chemical combination of oxygen with any substance (this definition is incomplete, but for our current purposes, it will suffice). In other words, whenever oxygen (and, as will be explained later, some other materials) combines chemically with a substance, that substance is said to have been *oxidized*. Rust is an example of oxidized iron. In this case, the chemical reaction is very slow. Slow is a relative term; if it is your automobile that has rusted, *you* may consider it a very fast reaction. The very rapid oxidation of a substance is called *combustion*, or fire. The important thing for you to remember is that a chemical reaction *is* occurring, and it may be your job to stop it. Knowing the various theories of fire will give you insight into the various ways of interrupting the reaction.

There are currently three theories of fire that are generally accepted by the fire services. They are: the fire triangle, the tetrahedron of fire, and the life cycle of fire. Of the three, the first is the oldest and best known, the second is accepted as more fully explaining the chemistry of fire, while the third and least known is a more detailed version of the fire triangle, going a little more into the physical aspects of fire. Let us now begin a quick review of these three theories.

The Fire Triangle

The fire triangle (see figure 28) is by far the oldest (and, it is hoped, the best understood) of the three theories; so far as it goes, it is still valid. It is quite simplistic in nature, gives a basic understanding of the three entities that are necessary for a fire, and, coincidentally, that are common to all three theories. Simply put, the fire triangle theory states that there are three things necessary to have a fire: fuel, oxygen, and heat. It likens these three things to the three sides of a triangle, stating that as long as the triangle is not complete, that is, the legs are not touching each other to form the closed or completed triangle (i.e.,

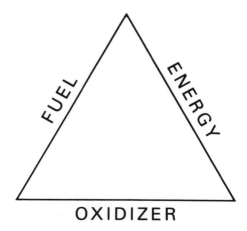

**Figure 28
The Fire Triangle.**

one or more of the three necessary essential ingredients is missing), it is impossible to have a fire. The theory, as stated, is still correct. Without fuel to burn, there can be no fire. If there is no oxygen present, there can be no fire (technically, this is *not* correct, but we can make the fire triangle theory technically correct by changing the oxygen leg to an *oxidizer* leg). Finally, without heat, there can be no fire. This last statement must also be brought up to date. The fact is that heat is just one form of energy; it is really *energy* that is necessary to start a fire. This difference is mentioned because there are some instances where light or some other form of energy may be what is needed to start the combustion reaction. It is best to change the "heat" leg of the fire triangle to the "energy" leg. Therefore, our "updated" fire triangle now has three sides representing fuel, *oxidizer*, and *energy*. Let us look a little more closely at each of the three components.

A *fuel* may be defined as anything that will burn. It is important for you to grasp this definition, because most firefighters consider only flammable gases and liquids as fuels. Many others include wood and coal as fuels, because we all recognize that they will burn, but we also tend to forget the *metals*, which under many circumstances are more hazardous than almost any other type of fuel. By defining fuels as *anything* that will burn, we are sure to include everything.

Fuels may be categorized into the following classes:

1. Elements (which include the metals, and some non-metals such as carbon, sulphur, and phosphorus)
2. Hydrocarbons
3. Carbohydrates (including mixtures that are made up partially of cellulose, like wood and paper)
4. Many covalently bonded gases (including carbon monoxide, ammonia, and hydrogen cyanide)
5. All other organic compounds

As you can see, this list of materials that burn is quite long, and you must not forget that the list includes not only the *pure substances* such as the elements and compounds that make up the list, but *mixtures* of those elements and compounds. Examples of mixtures would include natural gas, which you remember is a mixture of methane (principally), ethane, and a few other compounds, and gasoline, which, you also recall, is a mixture of the first six liquid alkanes (pentane, hexane, heptane, octane, nonane, and decane), plus a few other compounds. Wood (another mixture), and wood-related products, like paper, are excellent fuels, as are many polymers such as rubber, plastics, wool, silk, and the above-mentioned cellulose, which makes wood and paper the excellent fuels that they are.

The second leg of the fire triangle is oxygen or, in our updated version, the oxidizer leg. We changed this because oxygen, although it is the most common oxidizing agent you will encouter, is not the *only* oxidizer. We will cover this in greater detail in Chapter Eleven, on oxidizing agents. Another problem with calling this second leg the oxygen leg is that most firefighters consider only oxygen from the atmosphere when they think of oxygen, and do not consider other sources (this, again, will be covered later). Since the greatest source of oxygen *is* the atmosphere, however, this has to be considered the source that must be eliminated as one of the ways to control a fire. Whatever the source, you must realize that *oxygen does not burn.*

The third leg of the fire triangle is what was once called the heat leg, but we have updated to call it the energy leg, so as to consider *all* forms capable of providing the source of energy needed to start the combustion process. This energy can be provided in one or more of four ways. The energy can be generated chemically by the combustion of some other fuel, or it can be generated by some other *exothermic* chemical reaction. Exothermic is defined as the emission or liberation of heat (or energy). This is the opposite of *endothermic*, which is defined as the taking-in or absorption of heat (or energy).

Energy may also be generated by mechanical action; that is, the application of physical force by one body upon another. Examples of this are the energy created by the friction of one matter upon another or the compression of a gas (the increased incidence of collision of gas molecules against each other as more and more of them are packed ever more tightly in the same confined space). The force of friction in one case may produce energy that manifests itself as heat, while friction in the other case may result in a discharge of static electricity. Static electricity is created whenever molecules move over and past other molecules. This happens whether the moving molecules are in the form of a gas, a liquid, *or* a solid. (This is the reason why leaking natural gas under high pressure will ignite. This is also the reason why two containers must be *bonded*—connected by an electrical conductor—when you are pouring flammable liquids from one container to another. In any case, the amount of energy present and/or released could be more than enough to start the combustion reaction.)

A third method of generation of energy is electrical—much like the discharge of static electricity. This method may manifest itself as heat, as produced in an electrical heater, as arcing in an electrical motor or in a "short" circuit, or as the tremendous amount of energy released as lightning.

The fourth method of generation of energy is nuclear. Nuclear energy may be generated by the fission (splitting) of the atoms of certain elements and by the fusion (or joining together) of the nuclei of certain elements.

Once the energy—in many cases, heat—is generated, it must be transmitted to the fuel (the "touching" of the fuel and energy legs). This process is accomplished in three ways: conduction (the transfer of heat *through* a medium, such as a pan on a stove's heating element), convection (the transfer of heat *with* a medium, such as the heated air in a hot-air furnace), and radiation (the transfer of heat which is not dependent on *any* medium).

These three entities, fuel, oxidizer, an energy, make up the three legs of the fire triangle. It is a physical fact, a law of nature that cannot be repealed, that when fuel, oxidizers, and energy are brought together in the proper amounts, a fire *will* occur. If the three are brought together slowly, and over a long period of time, the oxidation will occur slowly, as in the rusting of iron. If the three are of a particular combination, the resulting oxidation reaction might even be an explosion. Whatever form the final release of energy takes, the thing that cannot be changed is that the chemical reaction *will* occur.

The Tetrahedron of Fire

The second popular theory of fire is the tetrahedron of fire theory (see figure 29). This theory encompasses the three concepts in the fire triangle theory but adds a fourth "side" to the triangle, making it a pyramid, or tetrahedron; this fourth side is called the "chain reaction of burning." This theory states that when energy is applied to a fuel like a hydrocarbon, some of the carbon-to-carbon bonds break, leaving an unpaired electron attached to one of the molecular fragments

Figure 29
The Fire Tetrahedron.

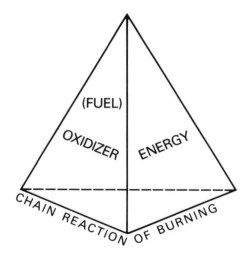

caused by the cleavage of the bond, thus creating a *free radical*. This molecular fragment with the unpaired electron, or "dangling" bond, is highly reactive, because its very existence is against the law of nature; it will seek out some other material to react with, to satisfy the octet rule. The same energy source that provided the necessary energy to break the carbon-to-carbon bond may have also broken some carbon-to-hydrogen bonds, creating more free radicals, and also broken some oxygen-to-oxygen bonds, creating oxide radicals. This mass breaking of bonds creates the free radicals in a particular space, and in a number large enough to be near each other, so as to facilitate the recombining of these free radicals with whatever other radicals or functional groups may be nearby. The breaking of these bonds releases the energy stored in them, so that this subsequent release of energy becomes the energy source for still more bond breakage, which in turn releases *more* energy. Thus the fire "feeds" upon itself by continuously creating and releasing more and more energy (the chain reaction), until one of several things happens: either the fuel is consumed, the oxygen is depleted, the energy is absorbed by something other than the fuel, or this chain reaction is broken. Thus a fire usually begins as a very small amount of bond breakage by a relatively small energy (ignition) source and builds *itself* up higher and higher, until it becomes a raging inferno, limited only by the fuel present (a fuel-regulated fire) or the influx of oxygen (an oxygen-regulated fire). The earlier in the process that the reaction can be interrupted, the easier the extinguishment of the fire will be.

The tetrahedron of fire theory is considerably more complex than what has just been presented, encompassing the actual mechanism of bond breakage, and the stepwise formation of all sorts of free radicals, but you now have read enough of this theory to grasp it. The theory claims that the propagation of all hydrocarbon fires (or fires involving hydrocarbon derivatives) depends upon the formation of the hydroxyl (−OH) radical, which is found in great quantities in all such fires. If this theory is correct, it would appear that the hydroxyl radical may be the key to the extinguishment of most fires.

The Life Cycle of Fire

The third theory of fire is the life cycle of fire theory (see figure 30). This particular theory breaks the combustion process down into six parts, rather than the three of the fire triangle or the four of the tetrahedron of fire theory, but three of the steps in this theory are the same as the only three steps in the fire triangle theory.

In the life cycle of fire theory, the first step is the input heat, which

Figure 30
The Life Cycle of Fire.

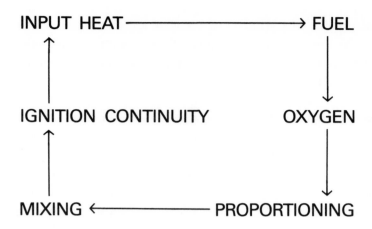

is defined as the amount of heat required to produce the evolution of vapors from the solid or liquid. The input heat will also be the ignition source and must be high enough to reach the ignition temperature of the fuel; it must be continuing and self-generating and must heat enough of the fuel to produce the vapors necessary to form an ignitable mixture with the air near the source of the fuel.

The second part of the life cycle of fire theory is the fuel, essentially the same as the fuel in the tetrahedron of fire and the fire triangle. It was assumed without so stating in the fire triangle theory, and is true in all three theories, that the fuel must be in the proper form to burn; that is, it must have vaporized, or, in the case of a metal, almost the entire piece must be raised to the proper temperature before it will begin to burn.

The third part is oxygen, and Dawson Powell, who developed the life cycle of fire theory, in *The Mechanics of Fire* (Providence: Grinnell Corporation, 1955) concerns himself only with atmospheric oxygen, because his theory centers around the *diffusion flame*, which is the flame produced by a spontaneous mixture (as opposed to a pre-mixed mixture) of fuel gases or vapors and air. This theory concerns itself with air-regulated fires, so airflow is crucial to the theory; this is why only atmospheric oxygen is discussed. Ignoring oxygen and the halogens that are generated from oxidizing agents is viewed by many as a flaw in this theory. The author of the theory, however, is very specific on the conditions that must exist for fire to be explained by his theory.

The fourth part of the theory is proportioning, or the occurrence of intermolecular collisions between oxygen and the hydrocarbon molecule (the "touching" together of the oxidizer leg and the fuel leg of the fire triangle). The speed of the molecules and the number of collisions depend on the heat of the mixture of oxygen and fuel; the hotter the mixture, the higher the speed. A rule of thumb is used in chemistry that states the speed of any chemical reaction doubles for roughly every 18°F. (10°C.) rise in temperature (remember the kinetic molecular theory?).

The fifth step is mixing; that is, the ratio of fuel to oxygen must be right before ignition can occur (flammable range). Proper mixing *after* heat has been applied to the fuel to produce the vapors needed to burn is the reason for the "backdraft" explosion that occurs when a fresh supply of air is admitted to a room where a fire has been smoldering.

The sixth step is ignition continuity, which is provided by the heat being radiated from the flame back to the surface of the fuel; this heat must be high enough to act as the input heat for the continuing cycle of fire. In a fire, chemical energy is converted to heat; if this heat is converted at a rate faster than the rate of heat loss from the fire, the heat of the fire increases; therefore the reaction will proceed faster, producing more heat faster than it can be carried away from the fire, thus increasing the rate of reaction even more. When the rate of conversion of chemical energy falls below the rate of dissipation, the fire goes out. That is to say, the sixth step, ignition continuity, is also the first step of the next cycle, the input heat. If the rate of generation of heat is such that there is not enough energy to raise or maintain the heat of the reaction, the cycle will be broken, and the fire will go out.

As stated above, the life cycle of fire theory expands the fire triangle. The fire triangle simply states that when three essential ingredients (fuel, oxidizer, and energy) are brought together in the *proper* amounts, there *will be* a fire. The life cycle of fire theory adds the concepts of flash point and ignition point (heat input) and flammable range (mixing).

Theories of Fire Extinguishment

Remembering that the theories of fire were presented so that you could see the components or steps of the combustion process, it must follow logically that if you can interrupt the reaction at any particular step, the reaction *should* stop—and it will. In the fire triangle, if you can prevent the three components from coming together, you will *prevent* a fire. Therefore it is equally logical that if, while a fire is in progress, you can remove one or more of the "legs," the fire will go out. Your job, then, is either to remove the fuel, prevent the energy from reaching the fuel-oxidizer mixture (removing the heat so the temperature drops below the ignition temperature), prevent the oxidizer from reaching the fuel, or a combination of these three actions.

In the tetrahedron of fire theory, the only difference is the fourth side, the chain reaction of burning, which is caused by the formation and subsequent reaction of free radicals, particularly the hydroxyl radical. Again, it is logical to assume that if the formation and reaction of free radicals is occurring in a fire, the prevention of the formation and/or the reaction of these radicals should extinguish the fire. The tetrahedron of fire theory calls for the application of "free-radical quenchants" to the fire, which will either prevent their formation or react with them after they are formed, to keep them from reacting with more fuel or oxygen radicals. Indeed, whatever you can do that will prevent either the formation of free radicals, or the interception of those already formed so that they will not react to cause the formation of more free radicals will extinguish the fire. The dry chemical fire-extinguishing agents work in this way.

In the life cycle of fire theory, the interruption of or the prevention of any of the six steps will result in the extinguishment of the fire. In this theory, an interesting option appears to be the interruption of the ignition continuity, such as using dynamite to "blow out" the flame of an oil-well fire, so that the fire is extinguished. The use of your breath to "blow out" the flame from a burning match or candle is a more common way of interrupting ignition continuity. Your breath will redirect the heat away from the fuel (and the fuel away from the heat), preventing the radiated heat from contacting the surface and causing the generation of flammable vapors by pyrolysis. Your breath has removed the energy faster than it was being generated.

No matter how sophisticated an approach a theory presents, all fires are extinguished in the same way, by interruption of some vital step in their sequence of events. You can probably develop a theory of fire on your own; the chances are that the more steps into which you break the reaction down, the more opportunities you will provide for yourself to extinguish the fire.

How Water Acts as a Fire Extinguisher

We all recognize the fact that fire is an exothermic (heat-liberating) reaction; as we learned in the preceding discussion of the various theories of fire, there must be a continuous feedback of energy (heat) to keep the reaction going. We also know that heat is dissipated from the fire by one or more of the methods of transferring heat: conduction, convection, and radiation. Heat energy is also fed back to the fire by radiation from the flame, and this source of heat keeps the fire going. If we could devise a way to interrupt that feedback of heat to the fuel, the continuity of the fire would be broken, and the fire would go out. What we need is a fire-extinguishing agent that would somehow "siphon" heat energy away from the fire, reduce the temperature of the material burning, and cool the surroundings below the ignition temperature of the fuel, so that there would not be a re-ignition of flammable vapors once the fire was extinguished. Not only would we need such an extinguishing agent, we would need a way to deliver it to the seat of the fire.

As you already know, water is the extinguishing agent that performs this task; we also have many ways to deliver it, not only to the seat of the fire, but also above it and around it, if the situation so dictates. Water really has a lot of drawbacks, however, and after reviewing them you may wonder why it is used. The answer is really very simple; we will come to it in due course.

Some of the drawbacks to the use of water as an extinguishing agent include its propensity to conduct electricity (which, of course, is deadly if the water is applied incorrectly), its low viscosity (which allows it to run off a wall instead of sticking there), and a high surface tension (which prevents it from penetrating tightly arranged materials). Water also allows heat to be radiated through it, freezes at a *relatively* high temperature, splashes about, and displaces many flammable liquids, causing them to spread rapidly, while burning all the time! This list of problems also includes the fact that water itself will violently react with many of the hazardous materials it is supposed to control. With all these problems, why even think about using the stuff to extinguish fires?

In addition to the fact that water is relatively inexpensive and is *usually* available in large quantities, there are two specific properties of water that make it invaluable. Those properties are its *latent heat of vaporization* and its *specific heat*. The latent heat of vaporization of a substance is defined as the amount of heat a material must absorb when it changes from a liquid to a vapor or gas. The specific heat of a substance is defined as the *ratio* between the amount of heat necessary to raise the temperature of a substance and the amount of heat necessary to raise the same weight of water by the same number of degrees.

The specific heat of water is important because it is so high in relation to the specific heat of other materials; this fact means that it takes more energy to raise the temperature of water than just about any other material. Therefore the temperature of the materials to which water has been applied will drop faster than the temperature of water will rise. The specific heat may be reported as the number of calories needed to raise the temperature of one gram of the material 1°C., or the number of British Thermal Units (BTUs) needed to raise one pound of the material, 1°F. This latter measurement for water is more familiar to firefighters; the value is 1.0 BTU per pound; it is also 1.0 calorie per gram.

Therefore, when water is applied to a fire, it begins absorbing heat from the fire, thereby cooling the fire down while the water heats up. For every BTU absorbed, the temperature of the water will rise 1°F. per pound of water involved. The important thing to remember here is that the rise in temperature of the water is caused by heat energy absorbed *from the fire*. The water is "siphoning" the heat away from the burning material. The temperature of the water will continue to rise, as long as the fire is producing heat, until it reaches its boiling point of 212°F. At this time the latent heat of vaporization of water comes into play. At 212°F. the water is still a liquid and will remain a liquid unless more energy is received from the fire. At this time, there is a phase change from liquid to vapor, *with no increase in temperature*; that is, water as a liquid at 212°F. converts to water *vapor* at 212°F. It is at this phase change that the latent heat of vaporization of water does its work, for while water will absorb 1 BTU per pound for every increase of 1°F., up to 212°F., *at* 212°F. when the phase change occurs, *970* BTUs are absorbed per pound. That sudden, rapid, and massive withdrawal of heat energy from the fire at this time is what gives water its tremendous fire-extinguishing capabilities, which are so valuable as to overcome the previously mentioned disadvantages. Heat is withdrawn from the burning material so rapidly, and in such large quantities, that the temperature of the burning fuel drops dramatically, usually well below its ignition temperature. When this happens, of course, the fire goes out. The latent heat of vaporization also explains why steam (which is invisible) at 212°F. is *hotter* than boiling water at 212°F. The live steam has 970 BTU's of energy more than the boiling water.

This latent heat of vaporization also explains why materials wet with water are difficult, and sometimes impossible, to ignite. If a combustible substance has absorbed enough water to be considered wet, or just damp, this water will act as a barrier to ignition by its evaporation as it is heated. As heat is applied to the wet substance, the water begins to evaporate (go through the phase change from a liquid to a vapor). To make this phase change, the water *must* absorb 1 BTU for every pound of water present for every 1°F. it rises until it reaches 212°F., whereupon it must absorb 970 BTUs for every pound of water present. Before *any* combustible material that has been wet with water can burn, the water (which has preferentially been absorbing the applied heat and thus keeping the combustible material itself from heating to its ignition temperature) must be driven off. If in the process of driving off the water enough heat energy from the potential ignition source has been used up so that there is not enough left (for example a burnt-out match) to raise the combustible to its ignition temperature, there will be no fire.

Pyrolysis

The word pyrolysis comes from two Greek words meaning "fire" and "breakdown." Pyrolysis, therefore, is defined as breakdown by heat. The "breaking down" is actually the cleavage of covalent bonds in hydrocarbon compounds and in hydrocarbon derivative compounds. This breaking of the bonds between carbon atoms is also called "cracking," and, indeed, it is the same reaction that takes place in a catalytic cracking tower ("cat cracker") at an oil refinery, used to pyrolytically "crack" long-chain hydrocarbons from petroleum into short-chain hydrocarbons (pentane, hexane, heptane, octane, and so on) to use as fuels (gasoline, kerosene, fuel oils, and the like). It is different from combustion in that pyrolysis takes place in the absence of air, or at least where there is not enough oxygen or other oxidizer to support combustion. Pyrolysis *does* occur in air, simply because near the surface of the material that is undergoing pyrolysis, the fuel is too rich to ignite. Once the fuel enters the flammable range, and the heat energy is sufficient to reach the ignition temperature of the gases, combustion will occur.

Wood, you recall, is not a compound but instead a mixture of compounds, the principal compound being cellulose, a naturally occurring polymer that contains covalently bonded carbon, hydrogen, and oxygen (see figure 31). A polymer is a "giant" molecule, which literally means that it is huge when compared to other molecules. The repeating unit in the polymer is the cellulose molecule, which means that this six-carbon-chain molecule is connected to another cellulose molecule, which is connected to another, and so on, and so on, until we have a new compound: this giant molecule called the cellulose polymer. Visualize these thousands and thousands of carbon atoms strung together, in a long chain (we will be discussing polymers later in

Figure 31
Pyrolysis of a Log of Wood.

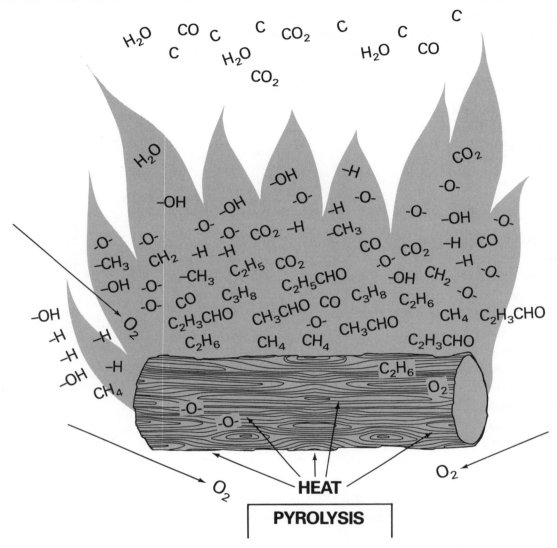

Chapter Twelve). When energy is applied to the molecule in the form of heat, the molecule, which had been vibrating rather slowly (remember, the kinetic molecular theory states that all molecules are in constant motion), begins to vibrate rather rapidly, until this vigorous motion causes one of the carbon-to-carbon bonds to break (cleave). As more and more heat is applied, the molecule moves faster and faster, causing more bonds to break, on both the remaining long-chain molecules and the shorter-chain molecules which resulted from earlier bond cleavage. As all the compounds in the vicinity of the heat source absorb more and more heat, more and more bonds break, and the resulting compounds are shorter and shorter chains. The end result of breaking covalent bonds between carbon atoms in hydrocarbon and hydrocarbon derivative compounds (in other words, organic compounds) is the formation of molecules of the simplest hydrocarbon, methane—similar to the process that takes place within your body, as digestion takes place. Methane, as you know, is a highly flammable gas; it is the methane and some of the other short-chain hydrocarbons (ethane, propane, and so on) that actually burn, when mixed with the right amount of oxygen and in contact with the proper ignition source.

The key to pyrolysis, of course, is heat. The original source of heat is the ignition source; before it can raise the fuel to ignition temperature, it must first produce the fuel by breaking down the solid cellulose. Once the methane and other short-chain hydrocarbons have been created and have begun to rise from the solid material, the continuing energy from the ignition source produces combustion. As the gases are consumed, the energy needed to continue the processes of pyrolysis and combustion comes from the heat of the flame radiated back to the surface of the fuel. If the ignition source is constant, and oxygen is excluded from the process, pyrolysis alone will occur; there will be a large buildup of flammable gases, many or most of them heated to energy levels above their ignition temperatures, just waiting for that third leg of the fire triangle to snap shut and complete the triangle. When the air comes rushing into the superheated gases, the dreaded "backdraft" explosion will occur.

Within the same room, as flammable vapors are formed pyrolytically but are heated to temperatures below their ignition temperatures and rise toward the ceiling—some gases are lighter than air, while others rise because they are hot—they will gather there, being further heated by heat radiating from the ignition source or a fire burning on or near the floor. The gases will also radiate heat back to unburned portions of the room. When the gases, or at least one of them, are heated to the ignition temperature, all the gases (assuming they are in the flammable range) will ignite, producing a "flashover."

Pyrolysis explains all flaming combustion, by producing the fuels that burn in the flame. Some experts have claimed that solids, like liquids, do not burn, that it is only the flammable gases produced by the breakdown of solid hydrocarbons and hydrocarbon derivatives that burn. While it is true that the flame consists of a diffusion of simple gases—not *always* hydrocarbon gases, since carbon monoxide and hydrogen cyanide may have been formed by the combustion reaction—and oxygen in a combustion reaction, while it is true that liquids do not burn, the fuel that burns is the gases that are volatilized and sometimes pyrolyzed from the liquid. While it is also true that gases are indeed liberated and "cracked" from solid organic fuels, mix in the flame over the solid, and burn, it is nevertheless true that wood, coal, and other solid organic fuels *will* burn on their surfaces. This is called glowing, or surface burning, as opposed to flaming, or space burning, which is the process that occurs as a result of pyrolytic breakdown of solid organic fuel. This glowing, or surface burning, is quite obvious to anyone who has enjoyed a fire in the fireplace; after all the gases are driven off in the flaming, or space burning, what remains is a solid char, or charcoal, if you prefer, that produces the pleasant glowing fire long after the flames are gone. Of course, glowing, or surface burning, is how charcoal burns, and how soft or hard coal will burn, once the volatile gases are driven off.

Another example of surface burning is the manner in which metals burn. There is no space burning because there are no gases generated (pyrolyzed) by the heated metal. The glowing combustion, or surface burning of metals, is the same phenomenon as the glowing combustion of charcoal. It is direct combination of the metal with oxygen at the surface of the metal.

The flame itself is explained by pyrolysis. All flames appear above the fuel, as the gases, which are hot, rise with the thermal column. Many of the gases are not fully "cracked," that is, broken down completely to methane; as they rise (including the methane and ethane), they will be too concentrated (too rich) right near the surface of the fuel and will not begin to burn until the proper amount of oxygen is mixed with them. Once the proper mixture is achieved, the gases begin to burn, radiating more heat to the surface, producing more flammable gases. The space between the fuel surface and the bottom of the flame is the area where the gases are too rich to burn; once they do burn, they produce even more heat. This contributes to the turbidity of the thermal column causing the flame to quiver and "dance."

The end products of pyrolysis, the simplest hydrocarbons such as methane, are still not the materials that burn. Further input of energy is necessary to break a carbon-to-hydrogen covalent bond on the molecules of these hydrocarbons, in order to produce a free radical. It is the reaction of these free radicals with the oxygen radicals formed from the bond cleavage of O_2 that is actually the combustion process. The breaking of the carbon-hydrogen bonds (from the simple hydrocarbons formed) and the oxygen-oxygen bonds (from atmospheric oxygen) produces carbon monoxide (CO), carbon dioxide (CO_2), and the hydroxyl radical (–OH). The hydroxyl radical then reacts with more hydrogen radicals (H–) to form water (H–O–H or H_2O). All this is usually accompanied by the diffusion flame in which these reactions take place and by the evolution of heat energy.

Pyrolysis explains the "crown" fires in a burning forest, as the heat from the approaching fire radiates out in all directions, pyrolyzing the grass, producing flammable gases that are further heated, by radiation, to their ignition temperatures, and, when properly mixed with oxygen, produce a rolling flame across the top of the field. As trees are heated by the radiated energy of an approaching fire, not only will the water trapped in the trunk cause a cracking and splitting of the tree, but the ignition of pyrolyzed gases will cause the tree to appear to explode. All fires produced at a distance by radiated heat cause pyrolysis first, followed by the explosive breaking into flame of the gases produced.

This process explains how fires are spread from one building to the interior of another, by radiated heat through the windows, producing pyrolysis of the combustibles there, which generates tremendous quantities of flammable gases. Further radiated heat raises these gases to their ignition temperatures—and a fire has jumped from one building to another, through a closed window! This is also how fires, which have broken *out* of a window below, start a fire on the floor above, by radiating heat through the intact window and then proceeding to leapfrog to the top of the building.

Pyrolysis of liquids is possible but only in liquids of very high molecular weight, such as asphalt, tar, and heavy oils, simply because the heating of a lighter liquid may produce evaporation first. Once these vapors are in the air, combustion will occur almost instantaneously. Pyrolysis of plastics is another matter. Theoretically, since plastics are polymers just as cellulose is, you would expect them to be subject to pyrolysis just as cellulose is. Plastics, however, are different in their properties; while it *is* possible for plastics to pyrolyze, they usually melt and flow away before pyrolysis occurs.

Bond Energies

Bond energy is defined as the energy required to break a covalent bond *or* the energy released when a bond is formed. When materials burn, there is a tremendous amount of energy released, which means many *other* bonds are being formed. These bonds are those formed when the final or the intermediate combustion products are formed. The final combustion products of any hydrocarbon are water, carbon monoxide, and carbon dioxide. Carbon monoxide may also be considered an intermediate combustion product, since it is flammable and will be further oxidized to carbon dioxide, which *is* a final combustion product. The amount of energy released depends, of course, on the amount and type of materials that burn. The amount of energy contained in each bond depends on the atoms between which the bond is formed. The various bond energies possessed by the different bonds are listed in table 8. The values are listed in kilocalories (units of a thousand calories). You can calculate the energy liberated by the combustion of any substance if you know what the composition of that material is and how much of it is burning.

Notice that the energy contained in the triple bond of acetylene is considerably higher than the double bond of an ethylene molecule, and the double bond has more energy than the single carbon-to-carbon bond. This fact accounts for the tremendous amount of energy released when acetylene burns, producing an extremely high flame temperature. This amount of energy, packed into the space between two carbon atoms, is what makes the acetylene molecule so unstable. You can calculate the heat of combustion of any material by adding up the

Table 8 / Chemical Bonds and Bond Energies

Chemical Bond	Average Bond Energy (kcal)	Chemical Bond	Average Bond Energy (kcal)
C – H	98.2	C – C	80.1 (isobutane)
C – F	116	C – C	84.9 (solid carbon)
C – Cl	81	C = C	140.0 (ethylene)
C – Br	68	C = C	123.8 (benzene)
C – I	52	C ≡ C	193.3
H – H	104.2	C – N	72.8
H – F	135	C = N	147
H – Cl	102.1	C ≡ N	212.6
H – Br	86.7	C – O	85.5
H – I	70.6	C = O	176 (aldehydes)
F – F	37	C = O	179 (ketones)
Cl – Cl	57.9	C – P	138
Br – Br	45.4	C – S	65
I – I	35.6	C = S	165
C – C	77.7 (ethane)	C – Si	63
C – C	79 (propane)	C – Si	78 (silicones)
C – C	79.6 (normal butane)		

bond energies of all the covalent bonds in the compound. The bigger the organic molecule, that is, the more bonds that are present and that will be broken, the higher the caloric value of the fuel. There are more calculated bond energies than we have presented, but these others are beyond the scope of this textbook.

Glossary _____

Backdraft: The term given to a type of explosion caused by the sudden influx of air into a mixture of gases, which have been heated to above the ignition temperature of at least one of them.

BTU: British Thermal Unit: The amount of energy required to raise one pound of water 1°F.

Calorie: The amount of energy required to raise one gram of water 1°C.

Combustion: A rapid chemical combination of a substance with oxygen, usually accompanied by the liberation of heat and light.

✳**Conduction:** The transfer of heat through a medium.

✳**Convection:** The transfer of heat with a medium.

"Cracking": The breaking of covalent bonds, usually between carbon atoms.

Diffusion Flame: The flame produced by the spontaneous mixture of fuel vapors or gases and air.

Endothermic: The absorption of heat.

Evaporization: The changing of a liquid to a vapor.

Exothermic: The liberation of heat.

Flashover: A situation in which heat radiated back to the floor by superheated gases near the ceiling causes the entire room to burst into flame, almost instantaneously.

Free Radical: An atom or group of atoms bound together chemically with at least one unpaired electron. A free radical is formed by the introduction of energy to a covalently bonded molecule, when that molecule is broken apart by the energy. It cannot exist free in nature and therefore must react quickly with other free radicals present.

Fuel: Anything that will burn.

Heat: A form of energy; the total amount of vibration in a group of molecules.

Ignition Continuity: The continuation of burning caused by the radiated heat of the flame.

Input Heat: The amount of heat required to produce the evolution of vapors from a solid or liquid.

Latent Heat of Vaporization: The amount of heat a substance must absorb when it changes from a liquid to a vapor or gas.

Oxidation: The chemical combination of any substance with oxygen.

Proportioning: The occurrence of intermolecular collisions between oxygen and hydrocarbon molecules.

Pyrolysis: The breakdown of a molecule by heat.

Quenchant: Any substance that absorbs another.

Radiation: The transfer of heat with no medium.

Specific Heat: The ratio between the amount of heat necessary to raise the temperature of a substance and the amount of heat necessary to raise the same weight of water the same number of degrees.

Volatilization: The changing of a liquid to a vapor.

7

Flammable and Combustible Liquids

Introduction

Before we begin the study of flammable liquids, we must get clear in our own minds what certain key terms mean, so that later, when we are in the midst of a hazardous-materials incident involving flammable liquids, there will be no misunderstanding of the properties and principles concerned in controlling such an incident. These terms, properties, and definitions are crucial to understanding not only flammable liquids, but flammable gases as well. The definitions will be presented in a logical, orderly manner; when a new term is used in one definition, it will be explained in the next one or two definitions.

Starting with the most basic terms, before we define flammable liquid, we must define the word "liquid." A liquid is defined as a fluid having a vapor pressure no higher than 40 psia (pounds per square inch absolute, which means that the atmospheric pressure of 14.7 psi is included) at 100°F. *Vapor pressure* is defined as the pressure exerted by the escaping vapor against the sides of its container at equilibrium, the state at which the vapor pressure has stabilized and is no longer rising or falling. All liquids have a vapor pressure, which is caused by the evaporation of the liquid. *Evaporation* is nothing more than the escape of the molecules of the liquid through the surface of the liquid. The kinetic molecular theory states that all molecules are in constant motion. In a liquid, the molecules are sliding over each other in all directions; that is, the molecules of the liquid are moving in all directions, including toward the surface of the liquid. Most of the molecules are held within the liquid by the surface of the liquid and by the intermolecular forces that attract molecules to each other. To escape the surface of the liquid, a molecule must possess enough energy to penetrate the surface. The energy possessed by the individual molecules depends upon the temperature of the liquid. Since some evaporation is occurring all the time, all the molecules probably possess enough energy to penetrate the surface and pass into the space above the liquid but are prevented from doing so by the constant collisions between molecules. Those molecules that do escape the surface are those whose paths have caused them to avoid an intermolecular collision while near the surface.

If the liquid is in a closed container, the molecules that have evaporated will continue to move in straight lines in all directions until they collide with other molecules or with the walls of the container. When they collide with the walls of the container, they provide us with an opportunity to "count" such collisions; it is this "counting" of the collisions with a sensitive instrument such as a pressure gauge that tells us how many molecules are colliding with the sides of the container at any given time. This value is shown on the face of the gauge as a certain amount of pressure that we call vapor pressure. If heat is applied to the liquid in the container, the kinetic molecular theory states that the molecules will absorb this energy and begin to move faster. As they move faster, the number of molecules that escape from the liquid increases, and the number of collisions against the sides of the container increases; because of the increased number of molecules now in the space above the liquid, *and* because all those molecules are moving faster as a result of the energy absorbed. This increased number of collisions will cause the pressure gauge to show a higher pressure than before.

Since the molecules are moving in *all* directions, there will be a certain number that are moving back toward the surface of the liquid; these will re-enter the liquid and move about in the liquid as the other liquid molecules are doing. When the number of molecules re-entering the liquid just equals the number of molecules escaping the liquid, *equilibrium* is reached. At this point the pressure gauge stops rising, and at this point the vapor pressure is determined. It is important to realize that vapor pressure is temperature-dependent, so that when a vapor pressure value is given, the temperature at which the vapor pressure is recorded must also be reported. The definition for a liquid clearly spells out at what temperature the vapor pressure must be measured. This temperature and pressure are arbitrary; that is, they have been selected and agreed upon for the sake of uniformity. Therefore, any fluid that has a vapor pressure higher than 40 psia at 100°F. is, by definition, a gas, while any fluid with a vapor pressure of 40 psia or less when measured at 100°F. is a liquid.

A *flammable liquid* is defined as a liquid (that is, a fluid having a vapor pressure not exceeding 40 psia at 100°F.) that has a flash point below 100°F. A *combustible liquid* is defined as a liquid with a flash point of 100°F. or higher. The dividing point is very important to the shippers of these hazardous materials. This is not to say that a liquid with a flash point of 100°F. (a combustible liquid) is very much less hazardous than a liquid with a flash point of 99°F. (a flammable liquid). That one-degree difference is insignificant to the firefighter who has to contend with a spill. The difference is important to the people who do not want to have their product classified as a flammable liquid. Therefore, groups of manufacturers, such as those who make kerosene, have banded together to have the dividing line set at 100°F. (another arbitrary decision), then very carefully formulate their material, by properly blending the right alkanes, to give kerosene a flash point of 100°F., so that it will have a flash point low enough to be useful (or valuable), and high enough to be classified as a combustible liquid rather than a flammable liquid.

Now that we have drawn the distinction between flammable liquids and combustible liquids, we must define that property which provides the distinction, flash point. *Flash point* is defined as the *minimum* temperature of the liquid at which it gives off vapors sufficient to form an ignitable mixture with the air near the surface of the liquid or the container. It is extremely important to understand the concept of flash point, since it is *by far* the most important property of flammable liquids with which you will be concerned. This is not to say that the other properties we will be defining are not important; rather, more than any *other* property, flash point will determine the degree of hazard that is created by an incident involving flammable liquids. The concept of flash point is not at all complicated, as long as you realize that it is the *temperature* of the liquid that determines the amount of vapor that will be released into the air, to be mixed with oxygen, and then to wait for a suitable ignition source. Remember our discussion about the kinetic molecular theory and the absorption of energy by the molecules? As the molecules move faster when they absorb this energy, more and more molecules escape the surface of the liquid to the space above it. At the exact *temperature* that this ratio is "right" for ignition with the proper ignition source, you have reached the flash point for this particular liquid.

Flash points are determined experimentally, using both open and closed containers for the determination. It is not hard to figure out that the open container (open cup) will have more meaning to first-response personnel than the "closed-cup" method. The temperature of the liquid being heated is monitored very carefully as the heat is applied, an ignition source providing the proper ignition temperature is placed near the surface of the liquid or near the container, and at the instant a "flash" occurs across the surface of the liquid, the temperature is noted, and is reported as the flash point of the liquid tested. You must be very careful in using reference sources to determine the flash point of any given liquid, since many are reported with significantly different values in different books. In order to be on the safe side, you should consult several reference sources and select the lowest flash point listed for that particular liquid. If you find differences of a few degrees, this disagreement is insignificant. But where the flash point may be listed in one book as 56°F., and another has it at 75°F. (above normal room temperature), you can see how you might be severely injured by the difference on a spring day when the ambient temperature is 62°F.!

In some reference books, a temperature is given for the fire point, which technically is the temperature of the liquid at which there will be generated vapors sufficient to cause *self-sustained combustion*, once ignition occurs. This is usually a few degrees higher than the flash point, but it is not to be considered an important property of the liquid. The difference is so small that it matters *only* in the laboratory where the experiments are being performed. The most important property of flammable liquids for you is the flash point, because you are *not* in a laboratory performing experiments with the hazardous materials under proper control, but out in the real world, where you can rest assured that if the flash point of the liquid is reached and there is a suitable ignition source, there *will* be a fire, the fire point notwithstanding.

Let us be sure the principle behind flash point is understood. If you have a flammable liquid with a flash point of 65°F., and the ambient temperature (the temperature of the surrounding air) is 63°F., you *cannot* tell if you have reached the flash point of the liquid unless you know what the temperature of the *liquid* is. The liquid may have been heated by the rays of the sun, or by the presence nearby of some other heat source radiating energy that has been absorbed by the liquid or by its container, which will pass it along to the liquid. If the ambient temperature is 68°F., you can be sure that the temperature of the liquid will eventually rise to that level, unless it is insulated. In other words, the controlling factor in flash point is the temperature of the liquid itself. If this temperature never reaches the flash point, you *will not* have a fire, no matter what the ignition source is. Once the flash point is reached, however, the vapors necessary to break into flame (a rather tame way of saying explode) are present and *will* ignite with the proper ignition source. Once the flash point has been reached, the vapors are ready to burn; below the flash point, there are not enough vapors present to ignite.

If you were to warm two containers of liquid, one of gasoline and the other of kerosene, to 70°F., and then place a burning match near the surface of both liquids, what would happen? With the kerosene, nothing, but the gasoline will ignite immediately. Why? Simply because we were at (or above) the flash point for gasoline, but below the flash point for kerosene. The vapors of gasoline were there, in the proper mixture with oxygen, waiting for the ignition source; while there were *some* kerosene vapors present, there were not enough to be ignited (the mixture was too lean). You must be careful with the ignition source in conducting this experiment, because while trying to prove that there are insufficient kerosene vapors present since the kerosene was below its flash point, you may be radiating enough heat to the surface of the kerosene to raise it to its flash point, thereby causing an ignition of vapors you were trying to prove were *not* there.

We have used the term *ignitable mixture* several times; we now define it as a mixture of fuel and air within the flammable range that is capable of the propagation of flame away from the ignition source when ignited. *Flammable range* is defined as the proportion of gas or vapor in air that is between the upper and lower flammable limits. *Lower flammable limit* is defined as the minimum concentration of vapor or gas in air below which it is not possible to ignite the vapors with a proper ignition source (the mixture is too lean). *Upper flammable limit* is defined as the maximum proportion of vapors or gas in the air above which it is not possible to cause ignition with a proper ignition source (the mixture is too rich). Lower flammable limit and upper flammable limit may also be referred to as the lower explosive limit (lel), and the upper explosive limit (uel), because an explosion usually takes place when gases or vapors within the flammable range are ignited.

The next property to consider is *ignition temperature*, which is defined as the minimum temperature to which a material must be raised to initiate or cause self-sustained combustion. The terms ignition temperature and auto-ignition temperature mean exactly the same thing. What is being said is that if the flammable liquid has produced enough vapors to be within the flammable range, the vapors *will* ignite if the ignition temperature is reached. The important implication here is that there need be no open flame or spark, nor anything else that a layperson might consider as an ignition source. All that is needed is for the temperature of the vapors to be raised to their ignition temperature, assuming, of course, that the vapors *are* in the flammable range.

Another important property of flammable liquids is *specific gravity*, which is defined as the weight of a solid or liquid, as compared to the weight of an equal volume of water. This is an important property only if the flammable liquid or combustible liquid is not soluble in water. The specific gravity of water is 1.0, by definition, and any substance that does not dissolve in water and has a specific gravity higher than 1.0 will sink, while any substance that does not dissolve in water and has a specific gravity of 1.0 or less will float. It is very important for you to know the specific gravity of the liquid in question, since the application of water may spread it about. Of course, if the liquid in question is soluble in water, its specific gravity becomes unimportant.

The *vapor density of the liquid* is also a highly important property. The vapor density is defined as the relative density of a vapor or gas (with no air present) as compared to air. The vapor density of dry air is 1.0 (again, by definition). If a vapor or gas has a vapor density of less than 1.0, the vapor or gas will rise in air, and the "lighter" the gas or vapor (that is, the lower the vapor density), the more rapidly the gas or vapor will disperse in air. If the vapor density is higher than 1.0, the gas or vapor will sink to the ground and then seek out low spots. The "heavier" a gas or vapor, that is, the higher the vapor density, the more the gas or vapor will "hang together," the farther it will flow along the ground downhill, much like a liquid, and the longer the time it will take to disperse. Vapor densities are to gases and vapors as specific gravities are to liquids and solids. It is not necessarily correct to speak of the specific gravity of a gas, although some reference books do so. The correct term for gases and vapors is vapor density.

If you know the molecular formula for the liquid or gas in question, it is very simple to determine its vapor density, simply by calculating the molecular weight of the material. Molecular weight is calculated by adding up all the atomic weights of all the atoms of all the elements in the molecule and comparing that to the *average* molecular weight of air, which is 29. If the molecular weight of the substance in question (the average molecular weight if the substance is a mixture, and you know its composition) is 29, it will have a vapor density of 1.0 (that is, 29/29 = 1.0). If it has a molecular weight of 44, as does propane, its vapor density will be about 1.5 (44/29 = 1.517). Butane has a molecular weight of 58, so its vapor density is 2.0. Natural gas is principally methane, so its vapor density is 0.55; natural gas is actually a mixture, but since so much of it is methane, it is acceptable to substitute methane's properties for it. By knowing the composition of a liquid or gas, you can quickly determine whether the vapors or gases will rise and diffuse or hang together and flow along the ground, seeking out an ignition source.

The *boiling point* of a substance is defined as the temperature at which the vapor pressure of the liquid just equals atmospheric pressure. It is important to grasp the fact that the boiling point of any substance is pressure-sensitive; that is, although you may report the boiling point of a substance as a particular temperature, this temperature is recorded at a specific vapor pressure, which means that particular temperature at which its vapor pressure just equals atmospheric pressure, whatever that may be at that place on Earth at that time. The boiling points of liquids are usually recorded at sea level, where atmospheric pressure equals 14.7 psia. It is at its boiling point that a liquid will liberate the most vapors; that is, the liquid evaporates at its highest rate at its boiling point.

It is impossible to elevate the temperature of a liquid above its boiling point, except when the liquid is under pressure. Since a liquid will evaporate at its fastest rate at its boiling point, if the liquid is under pressure and heated above its boiling point, anything that will cause this pressure to drop will cause tremendous amounts of vapor to be generated in a remarkably short time. It is this phenomenon you witness when there is a tiny (or large) leak in the cooling system of your car's engine: the coolant in the system is above its boiling point, when the pressure is suddenly dropped to that of the surrounding atmosphere by the leak. The coolant is caught in a situation that is contrary to the laws of physics, and, in an attempt to come into instantaneous compliance with those laws, gives up massive amounts of vapor.

The *evaporation rate* of a liquid is defined as the rate of change from a liquid to a vapor at temperatures below the boiling point. Since all liquids (and some solids) have vapor pressures, they will be evaporating at all times and temperatures, except when they are under pressure.

The *melting point* of a substance is that definite temperature at which a solid changes to a liquid; the reverse, of course, is the definition of the *freezing point* of a substance. An important part of the definition is *that definite temperature*, which means that at precisely that temperature a material melts or freezes, not that it begins to melt or freeze just before it reaches that temperature.

Heat is defined as the total amount of vibration in a group of molecules (the kinetic molecular theory again), while *temperature* is defined as a measure of how fast the molecules are vibrating. When we ask whether one material is "hotter" than another, do we mean which of the two materials has a higher temperature, or do we mean which material has much more vibration in the total of its molecules? Ordinarily, the former is usually meant, while technically it should be the latter. Heat is a form of energy, so when a material is said to be "hot," it would be correct to say that that material possessed more energy in total, because of the number of molecules and the amount of vibration they possessed. Therefore, a gallon of water at 100°F. would be "hotter" (that is, possess more energy) than one ounce of water at 200°F. Of course, the water at 200°F. would "feel" hotter, since it has a higher temperature, but the energy in the gallon of water would be much greater. Water is much more dramatic at 212°F. liquid, as compared with water at 212°F. vapor (steam). The liquid water at 212°F. has the same temperature as the steam at 212°F., but the liquid possesses only the energy absorbed at the rate of one BTU per pound of water for every degree Fahrenheit that it had been raised, while the steam possesses the same amount of energy *plus* the 970 BTUs absorbed for every pound of water that vaporized to steam. That is why so much more tissue damage is done by live steam than by boiling water at 212°F.

Of all the definitions presented to you in this rather long section, far and away the most important of all, as it pertains to liquids that produce vapors which burn, is *flash point*. That is because whenever the liquid reaches or exceeds this temperature, the liquid has thus produced the fuel to burn, and all that is needed for a fire to occur is a suitable ignition source. The definition of flash point includes the fact that the vapors will be in the flammable range; since this "flash" will occur out in the real world, rather than in some well-controlled laboratory, the "flash" will produce a real fire. This situation is not to belittle the importance of flammable range and ignition temperature, since these properties are also extremely important, but it is the flash point that announces the fuel is ready to burn, and *will*, when the ignition temperature of the vapor has been reached.

Types of Flammable Liquids

Hydrocarbons

Most common

The first type of flammable liquid we will consider is the rather large class of compounds known as hydrocarbons; in this group we will include *mixtures* of hydrocarbons, as well as the pure compounds. The first group of hydrocarbons are the alkanes, which you will remember as an analogous series of unsaturated hydrocarbons, the first four of which are gases. The first liquid in the series is pentane, followed by hexane, heptane, octane, nonane, and decane, plus the myriad isomers of these compounds. These first six liquids in the series, in various combinations of themselves plus a few other compounds, make up the most common of all the flammable liquids, the mixture known as gasoline (see figure 32). Gasoline is a typical hydrocarbon, even though it is a mixture. Like all hydrocarbons, whether they are compounds or mixtures, gasoline is insoluble in water, simply because water is a polar substance, while all the hydrocarbons are non-polar. In chemistry, there is an axiom which states "like dissolves like," so for a liquid to dissolve in another liquid, they would both have to be either polar or non-polar.

**Figure 32
A Gasoline Fire.**

The alkenes are another group of hydrocarbons which include flammable liquids as members of their analogous series. The liquids start with the compound 1-pentene (the 1- stands for the position of the double bond in the chain; the presence of a multiple bond means the hydrocarbon is unsaturated) and continues through hexene, heptene, octene, and nonene. The numbers are omitted for simplicity; you should assume that all the isomers of these compounds are flammable, even if the flash point is at 100°F. or slightly higher.

The next group of hydrocarbons are the alkynes, but since we have decided that the only important member is acetylene (a gas), we will not list here the members of the series that are flammable liquids. Be advised, however, that the first few liquids in the family are flammable, and watch out for them.

The other hydrocarbon family that contains flammable liquids is the aromatic family, which starts with benzene and contains, among others, toluene and the xylenes. There are *cyclical* compounds that contain other than six carbons in the structure; they include cyclopentane, cyclohexane, cycloheptane, and the corresponding cycloalkenes.

There are, as mentioned before, many mixtures made from the above compounds, too many to list here. But you should be aware that many refiners of petroleum products will purify their products only so far; when it no longer becomes profitable for them to separate the various compounds from the mixtures produced, they will sell the mixtures under several names. Those names include naphtha, Stoddard Solvent, lighter fluid, petroleum spirits, and many others. Treat these mixtures as flammable liquids, no matter what you determine their flash points to be (see figure 33).

Table 9 is a list of *selected* flammable hydrocarbon liquids and some of the most important properties (temperatures are given in degrees Fahrenheit and LEL and UEL are given as percentages of vapors in air).

Figure 33
Placards for Flammable and/or Combustible Liquids.

Table 9 / Some Flammable Hydrocarbon Liquids and Their Properties

Compound	Flash Point (°F.)	Ignition Temperature (°F.)	LEL (%)	UEL (%)
pentane	< −40	500	1.5	7.8
isopentane	< −60	788	1.4	7.6
neopentane	< −20	842	1.4	7.5
hexane	−7	437	1.1	7.5
heptane	25	399	1.1	6.7
octane	56	403	1.0	6.5
nonane	88	401	0.8	2.9
1-pentene	0	523	1.5	8.7
1-hexene	< −20	487	NA	NA
1-heptene	28	500	NA	NA
1-octene	70	446	NA	NA
1-nonene	78	NA	NA	NA
benzene	12	1,044	1.3	7.1
toluene	40	896	1.2	7.1
o-xylene	90	867	1.0	6.0
m-xylene	81	982	1.1	7.0
p-xylene	81	984	1.1	7.0
cyclopentane	< −20	682	0.7	2.4
cyclohexane	−4	473	1.3	8.0
cycloheptane	69	NA	NA	NA

NA = Not Applicable

LEL   Lower explosive limit

UEL   Upper explosive limit

Halogenated Hydrocarbons

Halogenated hydrocarbons include more compounds than just the alkyl halides. They include any compound that started as a hydrocarbon and which has had one or more halogen atoms substituted for hydrogen atoms. This means that a certain halogenated hydrocarbon may have as its hydrocarbon backbone a radical formed from an alkane, an alkene, a diene (an unsaturated hydrocarbon with two double bonds), an aromatic hydrocarbon, or indeed, any other hydrocarbon imaginable. Halogenated hydrocarbons run the gamut from flammable liquid to combustible liquid to liquids that are almost impossible to ignite, to fire-extinguishing agents. No attempt will be made here to list their flash points, ignition temperatures, or flammable ranges, but the most common of them appear at the end of this chapter in a general list of the most common flammable liquids. They are not generally soluble in water, although some compounds may be slightly soluble.

Alcohols

Alcohols are a group of compounds characterized by the presence of the hydroxyl radical attached to a hydrocarbon backbone, as in their *general formula* R–OH. As a group, alcohols have generally lower heats of combustion than some other classes of flammable liquids because they are already partially oxidized (that is, there is some oxygen in the molecule). Alcohols are polar materials, so they are generally soluble in water. Like any hydrocarbon derivative, however, the longer the chain gets, the more the compound begins to have the same properties of the hydrocarbon from which it was derived. The alcohol series has no gases; even the lightest, methyl alcohol, is a liquid. Table 10 is a list of the most common alcohols and their flammability properties. Temperatures are given in degrees Fahrenheit and LEL and UEL are given as percentages of vapors in air.

Ethers

The ethers, as a class of flammable liquids, are generally more dangerous than the other classes of flammable liquids because their flash points are usually lower, and the hazard of the anesthetic

Table 10 / Some Flammable Alcohols and Their Properties

Compound	Flash Point (°F.)	Ignition Temperature (°F.)	LEL (%)	UEL (%)
methyl alcohol	52	725	6.0	36.0
ethyl alcohol	55	685	3.3	19.0
propyl alcohol	74	775	NA	13.7
isopropyl alcohol	53	750	2.0	12.7
butyl alcohol	84	650	1.4	11.2
isobutyl alcohol	82	780	1.7	10.6
sec-butyl alcohol	75	761	1.7	9.8
tert-butyl alcohol	52	892	2.4	8.0
allyl alcohol	70	713	NA	NA

NA = Not Applicable

properties poses an additional danger. Solubility in water varies and, generally, the shorter the hydrocarbon chain, the more soluble the ether will be. The ether molecule is characterized by the presence of *two* hydrocarbon backbones or radicals, one on either side of the oxygen atom; they have the general formula $R-O-R'$, with the possibility that both hydrocarbon radicals might be the same or might be different. An additional hazard of the ethers is that the molecule can be oxidized in an unusual and unexpected way: an additional oxygen atom may enter the ether molecule to form the corresponding peroxide, $R-O-O-R'$, which is potentially explosive. The problem is especially severe with diethyl ether, more commonly known as ethyl ether, or just plain ether. This material was for some time the most common anesthetic used in hospitals, and much of it may still be around. The problem occurs whenever a container of ether has been opened and thereby exposed to the oxygen in the atmosphere. The general rule of safety with ethers is that once a container has been opened, it should be used completely or disposed of but never saved. Table 11 is a list of some of the most common ethers and their flammability properties. Temperatures are given in degrees Fahrenheit and LEL and UEL are given as percentages of vapors in air.

Table 11 / Some Flammable Ethers and Their Properties

Compound	Flash Point (°F.)	Ignition Temperature (°F.)	LEL (%)	UEL (%)
methyl ethyl ether	−35	374	2.0	10.1
diethyl ether	−49	320	1.9	36.0
methyl propyl ether	⟨−4	NA	NA	NA
ethyl propyl ether	⟨−4	NA	1.7	9.0
ethyl butyl ether	40	NA	NA	NA
isopropyl ether	−18	830	1.4	7.9
dibutyl ether	77	382	1.5	7.6
divinyl ether	⟨−22	680	1.7	27.0
ethylene oxide	⟨0	804	3.6	100

NA = Not Applicable

Ketones

The ketones are a class of flammable liquids characterized by the presence of the carbonyl group, $-\overset{\overset{\displaystyle O}{\|}}{C}-$, with hydrocarbon radicals on both sides to form the general formulam $R-\overset{\overset{\displaystyle O}{\|}}{C}-R'$. The shortest-chain ketones are soluble in water, and their flash points are generally higher than the ethers. The ketones are more widely used as solvents and as such will be found in the open at many manufacturing plants which need them as degreasers and other cleaners. The ketones have many other uses and are often found in the home as a solvent for many adhesives. Table 12 lists the flammability properties of some of the most common ketones. Temperatures are given in degrees Fahrenheit and LEL and UEL are given as percentages of vapors in air.

Aldehydes

The aldehydes are a group of hydrocarbon derivatives characterized by the presence of the carbonyl group, as in the ketones, but with a hydrogen attached to the carbon rather than to the second hydrocarbon radical, giving as a result the general formula R–CHO. The aldehydes as a class have flash points similar to the ketones, if just slightly lower, and, again, the shortest-chain aldehydes are soluble in water. Formaldehyde, the simplest, is a gas, although it is usually transported dissolved in water. In addition, the aldehydes are all irritants, acrolein being the major irritant in smoke given off by burning wood and wood products, along with other Class A combustibles.

Table 12 / Some Flammable Ketones and Their Properties

Compound	Flash Point (°F.)	Ignition Temperature (°F.)	LEL (%)	UEL (%)
acetone	−4	869	2.1	13.0
methyl ethyl ketone	16	759	1.7	11.4
methyl butyl ketone	77	795	NA	NA
methyl isobutyl ketone	64	840	1.2	8.0
methyl propyl ketone	45	846	1.5	8.2
mesityl oxide	87	652	1.4	7.2

NA = Not Applicable

Table 13 / Some Flammable Aldehydes and Their Properties

Compound	Flash Point (°F.)	Ignition Temperature (°F.)	LEL (%)	UEL (%)
acetaldehyde	−38	347	4.0	60.0
propionaldehyde	−22	405	2.6	17.0
butyraldehyde	−8	425	1.9	12.5
paraldehyde	96	460	1.3	NA
acrolein	−15	428	2.8	31.0

NA = Not Applicable

Table 13 shows some of the aldehydes' flammability properties. Temperatures are given in degrees Fahrenheit and LEL and UEL are given as percentages of vapors in air.

Amines

Amines are hydrocarbon derivatives that have the general formula $R-NH_2$, but they may also be considered to be derivatives of ammonia (NH_3), with a hydrocarbon radical replacing one of the hydrogen atoms of the ammonia. When only one of the hydrogens is replaced on the ammonia by a hydrocarbon radical, the result is a

Table 14 / Some Flammable Amines and Their Properties

Compound	Flash Point (°F.)	Ignition Temperature (°F.)	LEL (%)	UEL (%)
ethyl amine	⟨0	725	3.5	14.0
diethyl amine	−9	594	1.8	10.1
triethyl amine	20	NA	1.2	8.0
propyl amine	−35	604	2.0	10.4
dipropyl amine	63	570	NA	NA
isopropyl amine	−35	756	NA	NA
diisopropyl amine	30	600	1.0	7.1
butyl amine	10	594	1.7	9.8
sec-butyl amine	16	NA	NA	NA
allyl amine	−20	705	2.2	22.0

NA = Not Applicable

primary amine; when two hydrogens are replaced, it is a *secondary* amine; and when all three hydrogens are replaced, it is a *tertiary* amine. Most common amines are of all three types. Most amines are foul-smelling materials, they are all toxic to some degree, and they have, as a class, relatively low flash points. The smell of rotting flesh is due to the formation of amines, along with the putrification or digestion of other nitrogen-containing organic materials. The amines are water-soluble. Table 14 lists the flammability properties of the most common amines. Temperatures are given in degrees Fahrenheit and LEL and UEL are given as percentages of vapors in air.

Esters

The esters are the flavors and fragrances of nature. They have the general formula

$$R - \overset{\overset{\displaystyle O}{\|}}{C} - O - R',$$

and most have distinctive, pleasant odors. They are very slightly soluble in water. Table 15 gives the flammability properties of some of the more common esters. Temperatures are given in degrees Fahrenheit and LEL and UEL are given as percentages of vapors in air.

Table 15 / Some Flammable Esters and Their Properties

Compound	Flash Point (°F.)	Ignition Temperature (°F.)	LEL (%)	UEL (%)
methyl formate	− 2	840	4.5	23.0
methyl acetate	14	850	3.1	16.0
methyl acrylate	27	875	2.8	25.0
methyl propionate	28	876	2.5	13.0
methyl butyrate	57	NA	NA	NA
ethyl formate	− 4	851	2.8	16.0
ethyl acetate	24	800	2.0	11.5
ethyl acrylate	50	702	1.4	14.0
ethyl propionate	54	824	1.9	11.0
ethyl butyrate	75	865	NA	NA
vinyl acetate	18	756	2.6	13.4

NA = Not Applicable

Combustible Liquids

Remember that the dividing line for flammable liquids and combustible liquids is 100°F., with everything *below* 100° being flammable, and all liquids *at 100°* or higher being combustible. This distinction is not to say that a flammable liquid whose flash point is 99°F. is hazardous and a combustible liquid whose flash point is 100°F. is not hazardous; remember ignition temperatures, and the part they play in starting a fire. You may encounter a combustible liquid whose ignition temperature is considerably *lower* than that of a flammable liquid, without a very large difference in their flash points, which could make the combustible liquid more hazardous than the flammable liquid. Even though a strict definition of combustible liquids might not include liquids with flash points above 200°F., beware of *all* organic liquids. Just because the flash point is high does not take the liquid out of the range of hazardous materials. The heat of an approaching fire could turn a container of even the heaviest oils into a raging inferno!

The animal and vegetable oils are categorized as non-hazardous materials because of high flash points, but the presence of one or more double bonds in their structures makes them candidates for sponaneous ignition. Moreover, they will burn fiercely, once ignited. Under no circumstances should you consider combustible liquids non-hazardous. Kerosene with a flash point of 100°F., formulated precisely so it would not be classified as a flammable liquid, can be just as dangerous as a flammable liquid with a flash point a few degrees lower.

Other Flammable Liquids

Table 16 is a list of flammable liquids. There may be some disagreement among references as to the exact flash points of these liquids, but you should consider them all as flammable. No attempt has been made to list *all* the flammable liquids in existence. The materials already listed in the text are *not* repeated here. Table 16 is presented as a quick checklist only. You should prepare Hazardous-Materials Data Sheets on these and all other substances presented as hazardous materials.

Table 16 / Other Flammable Liquids

acetal	butyronitrile	2,5-dimethylfuran
acetonitrile	carbon disulfide	2,3-dimethylhexane
acetyl chloride	chlorobenzene	2,4-dimethylhexane
acrylonitrile	2-chloro-1,3-butadiene	1,1-dimethylhydrazine
allyl acetate	2-chlorobutene-2	dimethylpentaldehyde
allyl bromide	1-chlorohexane	dimethylpentane
allyl chloride	1-chloropropylene	dimethyl sulfide
allyl chloroformate	2-chloropropylene	dioxane
allyl ether	crotonaldehyde	dioxolane
allyl trichlorosilane	crotanyl alcohol	divinyl acetylene
amyl acetate	cumene	epichlorohydrin
sec-amyl acetate	cyclohexene	ethoxyacetylene
amyl alcohol	cyclohexanone	ethylbenzene
sec-amyl alcohol	cyclohexylamine	ethyl borate
amyl amine	cyclohexyl chloride	ethylbutyl amine
sec-amyl amine	1,5-cyclooctadiene	2-ethyl-1-butene
amyl bromide	cyclopentene	ethyl butyl ether
amyl chloride	denatured alcohol	2-ethylbutyraldehyde
amylene	di-sec-butylamine	ethyl butyrate
amyl formate	dibutyl ether	ethyl chloride
amyl mercaptan	1,3-dichlorobutene	ethyl chloroformate
benzotrifluoride	dichloroethylene	ethyl crotonate
butadiene monoxide	dichloropropene	ethylcyclobutane
2,3-butanedione	dicyclopentadiene	ethylcyclohexane
1-butanethiol	diethyl carbonate	N-ethylcyclohexylamine
2-butanethiol	diethyl glycol	ethylcyclopentane
butyl acetate	diethyl ketone	ethyl dichlorosilane
sec-butyl acetate	dihydropyran	ethylenediamine
butyl bromide	diisobutylamine	ethylene dichloride
butyl chloride	diisobutylene	ethylenelmine
sec-butyl chloride	diisopropylamine	ethylidene dichloride
tert-butyl chloride	diketene	ethyl isobutyrate
butylene oxide	2,2-dimethylbutane	ethyl mercaptan
1,2-butylene oxide	2,3-dimethylbutane	ethyl methacrylate
butyl formate	2,3-dimethyl-1-butene	4-ethylmorpholine
tert-butyl hydroperoxide	2,3-dimethyl-2-butene	ethyl nitrate
N-butyl isocyanate	1,3-dimethylbutylamine	ethyl nitrite
butyl nitrate	1,3-dimethylcyclohexane	ethyl propenyl ether
tert-butyl peracetate	1,4-dimethylcyclohexane	ethyl propionate
butyl propionate	dimethyldichlorosilane	ethyltrichlorosilane
2-butyne	dimethyl dioxane	fluorobenzene

Table 16 / Other Flammable Liquids (cont.)

furan	2-methyl-1 butene	methylpyrrolidine
furfurylamine	2-methyl-2-butene	methyltetahydrofuran
gasoline	3-methyl-1-butene	methyltrichlorosilane
1,4-hexadiene	N-methylbutylamine	2-methylvaleraldehyde
hexanal	methyl butyl ketone	methyl vinyl ketone
hexanone	3-methyl butynol	naphtha
hexene	2-methyl butyraldehyde	nickel carbonyl
hexylamine	methyl butyrate	nitroethane
hydrocyanic acid	methyl carbonate	nitromethane
iron carbonyl	methylcyclohexane	1-nitropropane
isoamyl acetate	4-methylcyclohexene	2-nitropropane
isoamyl chloride	methylcyclopentane	nonene
isobutyl acetate	methyldichlorosilane	2,5-norbornadiene
isobutyl alcohol	2-methyl-4-ethylhexane	tert-octylamine
isobutyl amine	3-methyl-4-ethylhexane	1,3-pentadiene
isobutyl chloride	2-methyl-3-ethylpentane	pentamethylene oxide
isobutyl formate	2-methylfuran	2,4-pentanedione
isobutyraldehyde	2-methylhexane	1-pentyne
isobutyrnitrile	3-methylhexane	petroleum ether
isoheptane	methylhydrazine	pinene
isohexane	methyl isoamyl ketone	piperidene
isooctane	methyl isobutyl ketone	1,3-propanediamine
isooctene	methyl isocyanate	propargyl alcohol
isopentaldehyde	methyl methacrylate	propargyl bromide
isoprene	4-methylmorpholine	propenyl ethyl ether
isopropenyl acetate	2-methyl-1,3-pentadiene	propionic nitrile
isopropenyl acetylene	4-methyl-1,3-pentadiene	propionyl chloride
isopropyl acetate	methylpentaldehyde	propyl acetate
isopropyl amine	2-methyl pentane	propyl benzene
isopropyl chloride	3-methyl pentane	n-propyl butyrate
isopropyl formate	2-methyl-1-pentene	propyl chloride
jet fuel B	4-methyl-1-pentene	propylenediamine
jet fuel JP-4	2-methyl-2-pentene	propylene dichloride
metaldehyde	4-methyl-2-pentene	propylene oxide
methallyl alcohol	2-methyl-2-propanethiol	n-propyl ether
methallyl chloride	2-methylpropanal	propyl formate
3-methoxypropylamine	methyl propionate	propyl nitrate
methylal	methylpropyl acetylene	propyltrichlorosilane
methyl borate	methyl n-propyl ether	pyridine
3-methyl-2-butanethiol	methyl propyl ketone	styrene
2-methyl-2-butanol	methylpyrrole	tetrahydrofuran

Table 16 / Other Flammable Liquids (cont.)

2,2,3,3-tetramethyl pentane	2,3,3-trimethyl pentane	4-vinyl cyclohexene
2,2,3,4-tetramethyl pentane	2,3,4-trimethyl-1-pentene	vinyl ethyl ether
thiophene	2,4,4-trimethyl-1-pentene	vinylidene chloride
toluol	2,4,4-trimethyl-2-pentene	vinylisobutyl ether
trichloroethylene	3,4,4-trimethyl-2-pentene	vinylisopropyl ether
trichlorosilane	tripropylene	vinyl propionate
triethylamine	valeraldehyde	vinyltrichlorosilane
2,2,3-trimethyl-1-butene	vinyl allyl ether	m-xylene
trimethylchlorosilane	vinyl butyl ether	o-xylene
2,2,5-trimethyl hexane	vinyl butyrate	p-xylene
2,2,3-trimethyl pentane	vinyl-2-chloroethyl ether	
2,2,4-trimethyl pentane	vinyl crotonate	

Glossary

Boiling Point: The temperature at which the vapor pressure of a liquid just equals atmospheric pressure.

Combustible Liquid: A liquid with a flash point at 100°F. or higher.

Evaporation: The process by which molecules of a liquid escape through the surface of the liquid into the air space above.

Flammable Liquid: A liquid with a flash point below 100°F.

Flammable Range: The proportion of gas or vapor in air between the upper and lower flammable limits.

Flash Point: The minimum temperature of a liquid at which it gives off vapors sufficient to form an ignitable mixture with air.

Freezing Point: The temperature at which a liquid changes to a solid.

Gas: A state of matter defined as a fluid with a vapor pressure exceeding 40 psia at 100°F.

Heat: A form of energy; the total amount of vibration in a group of molecules.

Ignition Temperature: The minimum temperature to which a substance must be raised before it will ignite.

Kinetic Molecular Theory: A theory that states all molecules are in constant motion at all temperatures above absolute zero; molecules will move (or vibrate) faster at higher temperatures because of the energy absorbed.

Liquid: A fluid with a vapor pressure no higher than 40 psia.

Lower Flammable Limit: The minimum concentration of gas or vapor in air below which it is not possible to ignite the vapors.

Melting Point: The temperature at which a solid changes to a liquid.

Specific Gravity: The weight of a solid or liquid as compared to the weight of an equal volume of water.

Temperature: A measure of how fast a group of molecules are moving.

Upper Flammable Limit: The maximum concentration of gas or vapor in air above which it is not possible to ignite the vapors.

Vapor Density: The relative density of a vapor or gas as compared to air.

Vapor Pressure: The pressure exerted by vapor molecules on the sides of a container, at equilibrium.

8

Compressed Gases

Introduction

Gases are one of the three states of matter, the other two being solids and liquids. Gases are the least dense of the three states, having the least intermolecular attraction (that is, the molecules of gases have the least attraction for each other), and, consequently, gases have much more freedom to move around than matter in either of the other two states. Gases will always fill the containers in which they are placed; their molecules will move randomly and endlessly in all directions (another manifestation of the kinetic molecular theory).

Gases are separated from liquids, although both are fluids, by their vapor pressures, which have been set arbitrarily. Gases are defined as fluids with vapor pressures higher than 40 psia (pounds per square inch absolute) at 100°F.; this definition includes the vapor pressure of air: 14.7 psi. This definition with which we must deal may seem to be unnecessary, but it becomes extremely important when the hazardous material with which we are confronted is in liquified form. If the material is indeed a gas in its natural form (that is, it is a gas at room temperature and atmospheric pressure), we must realize that it is actually in the container at a temperature above its boiling point. This fact means that if there were a sudden drop in pressure such as would occur if there were a leak in the container, the liquified gas would attempt to convert to the gaseous state as rapidly as possible, following the natural law that a material can never reach a temperature higher than its boiling point, except when it is under pressure. Liquified gases will remain in the liquid state for some time as they are boiling away, because, as they boil, the liquid's latent heat of vaporization causes tremendous amounts of energy (in the form of heat) to be absorbed as the liquid vaporizes. Cryogenic gases, which we will cover later, in Chapter Ten, seem to boil away much more slowly than liquified gases, and in fact they do. The latent heat of vaporization is operating here also; remember, however, that the heat energy absorbed as the liquid converts to a gas must come from somewhere for evaporation to occur; it usually comes from the liquid itself, as well as from the surrounding environment. In the case of cryogenics, the material is so cold that it has very little energy to contribute to the evaporation process.

Physical Forms

Compressed gases come in two forms, pressurized and liquified. Pressurized gases are those gases whose boiling points are extremely low, around −150°F., or those gases that the seller does not wish to liquify because the buyer is satisfied with the gas in its pressurized form (that is, the gas in the container is still in gaseous form, except that it is now under a pressure higher than atmospheric pressure). Liquified gases are those gases whose boiling points are "higher," from −150°F., up to about 32°F. Pressurized gases are shipped in cylinders, while liquified gases are shipped in tanks of various sizes (see figures 34 and 35). Gases are liquifed to save transportation space,

**Figure 34
Placards for
Compressed Gases.**

**Figure 35
Placards for Flammable
Compressed Gases.**

since the vapor-to-liquid ratio for gases is very high. Propane has an expansion ratio of 270 to 1; that is, one cubic foot of liquid will vaporize to 270 cubic feet of gas. The expansion ratios of the cryogenic gases are even higher. It is important to remember that all gases are compressible.

Containers for compressed gases come with various safety relief devices. Most such containers have a spring-loaded valve, which is pre-set to activate at some elevated temperature, but one which is still well below the design strength of the container. For instance, the container may be built to withstand pressures of 1,000 psi, but the spring-loaded valve will be set to activate at 250 psi. As the spring-loaded valve opens, the gas is vented until the pressure drops below 250 psi, when the valve re-seats itself, and the container is sealed once again. A spring-loaded valve is the only safety relief device that can reset itself after the pressure drops, to keep the container from emptying itself. This is not the case with the other two common safety relief devices, the frangible disc and the fusible plug. The frangible disc (frangible means capable of bursting) will rupture at a predetermined pressure and will allow the container to empty itself completely. The fusible plug is designed to melt at a predetermined temperature, and, when it does so, the container will also empty itself.

The only reason for the use of safety relief devices is to prevent the container's violent rupture, usually with such explosive force that shrapnel is created, causing injury to any nearby personnel; of course, property damage is possible too. The violent collapse of the container will also cause the contents to be widely scattered, spreading fire or whatever other hazard the contents may represent. There is also the possibility of the container's failing in such a manner that all or part of it would be converted to an unguided missile because of the "rocketing" action created by the forceful escape of the gas from one end or the other. It is an awesome sight to see a 3,000-pound section of a rail car or tank truck flying through the air, seeking a likely target to destroy and trailing flaming liquids or vapors (or poisons or corrosives, or worse) behind it. The much slower release of the contents via the safety relief device is far less dangerous than the instantaneous release of the entire contents, no matter what that may be, with one exception: Class A poisons. In this case, the contents of the container are so dangerous to man that containers holding Class A poisons have *no* safety relief devices of any kind. In just this one case, therefore, it is considered much better for no material to be released to the atmosphere unless the entire container fails. In addition, there is always the possibility that the cause of the rising pressure might be removed and thereby remove all danger of the material's escaping its container. (Poisons are discussed more fully in Chapter Fifteen.) Be advised that when you see a container of pressurized gas, which has no safety relief device, it contains an extremely hazardous material.

Pressure, Temperature, and Volume

What would cause the pressure within a container to rise? To answer this, you must be aware of the properties of gases and how they respond to various influences. These influences are pressure, temperature, and volume. As these conditions change, the gases, particularly pressurized gases, are affected. Let us look at these interrelationships.

When the temperature of a gas goes up, its pressure goes up. All these interrelationships will be considered to occur within the container, for our examples. This rise is consistent with the kinetic molecular theory, since as energy is added to a group of molecules, each molecule begins to move more rapidly than it did before. Because the molecules are moving faster, they are colliding with each other and with the sides of the container more often. By doing so, they are causing the pressure gauge to show a higher pressure, since all that a pressure gauge does is to count the number of collisions against the sides of the container and on its own end. Conversely, as a gas is cooled (that is, as heat is removed from the gas), its pressure will drop. The kinetic molecular theory states that as energy is withdrawn from molecules, their movement slows down. With fewer collisions against the end of the pressure gauge, the gauge will show a lower pressure.

Look at the situation from another aspect. If the temperature of a gas is known, what will happen if its pressure is changed? As the pressure of a gas is increased (that is, as more gas is added to the container), there is an increase in the number of collisions. You will recall that one of the ways heat is generated is by mechanical means, or friction. As the number of intermolecular collisions increases, heat is generated by this friction; therefore, the temperature rises. If gas is removed from the container (that is, its pressure is lowered), less heat is generated, so the gas cools. Consider it also from the standpoint of decreasing the volume of the gas, which is what happens as you add gas to a container. If you consider the container as a cylinder with an adjustable volume and then compress the gas by pushing the ends of the cylinder closer together, thereby decreasing the volume inside the container, the temperature of the gas will rise. If you expand the volume of the container by reversing your action, the temperature of the gas will fall; increasing the volume of a gas by this method is the same as allowing the gas to escape from the original cylinder to a larger container. By decreasing the volume, you also cause the pressure to rise, since the same amount of gas occupies a smaller space; as a result, the number of molecular collisions with each other and with the sides of the container will increase. By expanding the volume, you cause the pressure to decrease because the same number of gas molecules in a larger space will collide fewer times with the sides of the container or each other.

Therefore, you can cause the pressure of a gas to rise by increasing the temperature or decreasing the volume, and you can cause the pressure to drop by decreasing the temperature or increasing the volume. You can cause the temperature of a gas to rise by increasing the pressure or decreasing the volume; you can cause the temperature to decrease by decreasing the pressure or increasing the volume. You can increase the volume of a container (cause it to expand, like a balloon, or cause it to burst) by increasing the temperature of the gas or by increasing the pressure of the gas (by adding more gas); you can cause the volume of the container to decrease (the balloon shrinks) by decreasing the temperature of the gas or by decreasing the pressure. If you understand these relationships, you can actually calculate the resulting temperature of a gas when you alter the pressure and/or volume or the resulting pressure when you alter the temperature and volume. These calculations are beyond the scope of this textbook but they are *not* difficult. For our purposes, you must be aware of which values change in the same direction as the change in other values (which is directly proportional) and which values change in the opposite direction (which are inversely proportional). All these relationships are consistent with the kinetic molecular theory. If you are interested in learning these calculations, any chemistry or physics textbook will include discussions on Boyle's law, Charles's law, and the combined gas law.

A gas may be liquified in two ways, either by pressurizing it or simply by cooling it to a temperature below its boiling point. When a gas like butane, which has a relatively high boiling point of 31°F., is liquified, the cooling technique alone is used, since it costs relatively little to reach 31°F. Propane, however, which has a boiling point of −44°F., is liquified by using a combination of pressurizing and cooling.

Two properties of gases must be defined here. First is *critical temperature*, which is the temperature above which it is impossible to liquify a gas. Another way of stating this is to say that above the critical temperature, it is impossible for a gas to exist in the liquid phase. The other property is *critical pressure*, which is defined as the pressure required to liquify a gas at its critical temperature; that is, you may liquify any gas above its boiling point by the use of pressure alone, but the limiting factor of how high above its boiling point you can still use only pressure to liquify the gas is its critical temperature. At that specific temperature, the amount of pressure required to liquify the gas is called its critical pressure. Above the critical temperatute, there is *no* amount of pressure that can liquify that gas.

Hazards

Compressed gases, like other classes of hazardous materials, have certain hazards associated with them. No particular gas possesses all the hazards listed, but many of them have more than one. Your job is to realize that there is a list of hazards with which you must be familiar. These are the hazards associated with compressed gases:

1. They are under pressure.
2. They may be flammable.
3. They may be unstable.
4. They may be toxic.
5. They may be corrosive.
6. They may be oxidizing agents.
7. They may be subject to *Boiling Liquid, Expanding Vapor Explosion* (BLEVE) (liquified gases only).

Let us begin by looking at the group of compressed gases that are flammable. We will indicate whether they are usually found as pressurized gases or as liquified gases, or, in some cases, as both.

Flammable Gases

Natural Gas

The most common of the compressed gases is natural gas, which is almost always found as a pressurized gas. The boiling point of natural gas is so low that it is classified as a cryogenic gas, and we will discuss natural gas again in Chapter Ten, Cryogenic Gases. Natural gas is used as a fuel to heat our homes, commerical buildings, and industrial operations, cook our food, and heat our water (and, in a few cases, to fuel automobiles). It is usually found in a pipeline, under different amounts of pressure, depending on the pipeline's location. If the pipeline is in our homes, the pressure is usually only a few ounces per square inch gauge, while it is probably a few psig (pounds per square inch gauge) outside our homes, and considerably higher in transmission lines. Natural gas is made up principally of methane, and its properties are very similar to methane's because of that fact. Deposits of natural gas were formed by the anaerobic (no air present) decomposition of hydrocarbons and other organic materials. This decomposition is occurring continuously, wherever organic materials are decomposing in an environment that contains little or no oxygen, such as at the bottom of a lake or swamp or in animal intestines.

Methane has a molecular weight of 16 and therefore is lighter than air (which, as a mixture, has an *average* molecular weight of 29). You may calculate the "quick and dirty" vapor density of any gas by dividing its molecular weight by 29; if the answer is less than 1.0, the gas will rise and disperse; if the answer is greater than 1.0, the gas will sink to the lowest spot it can, and it will "hang together" longer, the higher its vapor density. Since natural gas is a mixture, it contains some other gases such as ethane, but it is made up of so much methane (as high as 95 percent, depending on where on Earth it was found) that whatever is said for methane can also be said for natural gas. The ignition temperature of natural gas is anywhere from 1,000°F. to 1,200°F., again depending on the percentage of methane in the mixture. Its vapor density ranges from .55 to .65; its flammable range is from 4.0 percent to 14.0 percent. Natural gas has no color, taste, or odor; therefore, as a safety consideraiton, an odorant, similar to the chemical that makes skunks so unforgettable, is added to natural gas to make it detectable in air at a concentration as low as 1 percent, which is still considerably below its lower flammable limit. Natural gas is not toxic, incidentally, so it is very difficult to commit suicide by turning on an unlit natural gas oven.

Propane

After natural gas, the next most common flammable gas is propane, which, together with butane, make up the LP (liquified petroleum) gases. Both propane and butane are colorless, tasteless, odorless gases that must be odorized to warn of leaks. The LP gases are often described together, as if they were one gas, but it is important to discuss their properties separately. Propane is the lighter of the two gases but still has a vapor density of 1.6 (44/29), as opposed to 2.0 (58/29) for butane. Propane's boiling point is −44°F., its ignition temperature is 842°F., and its flammable range is from 2.1 percent to 9.5 percent. Butane's boiling point is 31°F., its ignition temperature is 550°F., and its flammable range is from 1.6 percent to 8.5 percent. When a leak occurs in a propane tank, the liquid, which has been stored under pressure at temperatures above its boiling point, immediately flashes to a gaseous state, with an expansion ratio of 270 cubic feet of gas for every one cubic foot of liquid. This rapid conversion from the liquid state to the gaseous state is accompanied by a visible vapor cloud, which results from the contact of the extremely cold vapors of propane with the air. This contact causes the invisible water vapor in the air to condense into visible water droplets. That cloud will tend to "hang together," barring winds, and will drift along the low spots in the ground, seeking out an ignition source. Do not make the mistake of believing that the visible portion of the vapor cloud marks the boundary of the propane vapors, because it does not. As soon as the cloud begins to move away from the leak, the water vapor starts to warm up and re-evaporate, thus shrinking the visible portion of the cloud, while the propane vapors are now on the outside of the visible portion of the cloud. Once the propane vapors reach an appropriate ignition source and ignite, the fire will flash back to the source of the leak and will form a torch, flaming out from the leak. The heat energy radiated back from this torch now introduces the next, and perhaps worst, hazard of liquified flammable gases, the BLEVE.

BLEVE is an acronym (a word or phrase made up of the first letter or letters of other words) that stands for *B*oiling *L*iquid, *E*xpanding *V*apor *E*xplosion. Liquified gases are gases that are stored as liquids at temperatures above their boiling points. When the pressure is released, as happens in a leak, the liquid, following the laws of physics, flashes to a vapor almost instantaneously. If the gas is a flammable gas, there is now a tremendous volume of fuel presented to the atmosphere; if there is a suitable ignition source, a tremendous explosion will occur as the gases expand. The BLEVE usually occurs because of a flame impinging on the vapor space above the liquified gas in the tank. As the metal of the tank weakens, the pressure inside the tank rises. The pressure may rise even if the safety relief device is operating; indeed, the radiated heat from the burning gases escaping from the spring-loaded valve may actually be the source of the heat, causing the weakening of the metal. Then a tear begins as the internal pressure of the expanding gases pushes against the weakening metal. This tear usually runs longitudinally along the tank for a short distance and produces a huge opening that allows, to all intents and purposes, the entire remaining liquid cargo of the tank to flash instantaneously to vapors, which immediately ignite, producing a huge fireball that may reach proportions of 500 feet or more in diameter. Of course, anything or anybody within that distance will be incinerated. National Fire Protection Association's movie, *BLEVE*, portrays the awesome destructive power of a *Boiling Liquid, Expanding Vapor Explosion*, including the blowing of a firefighter off a ladder within the range of the fireball. The rule used to be, "Never approach a tank containing a liquified flammable gas from the ends, but always from the sides." We have now incidents reported of firefighters being injured by shrapnel, as they approached from the side. The rule now should be, *"DO NOT APPROACH A CONTAINER OF LIQUIFIED FLAMMABLE GAS THAT HAS BEEN EXPOSED TO HEAT!"* If approaches *are* made, they should *never* be made within a distance that might be enveloped by a fireball; in some cases, that distance could extend along the ground for 2,000 feet or more.

The BLEVE is not limited to the LP gases but may happen with any flammable liquified gas; you might also have a pressure-relief explosion with a container of flammable liquids, if internal pressure overcame the design strength of the tank. In either case, if you are in the danger zone, your chances of surviving the ensuing explosion are virtually nil.

Acetylene

The next most common flammable gas is acetylene, which is extremely flammable and possesses the added hazard of being explosively unstable. Acetylene is a colorless, tasteless, odorless gas that must have an odorant added to it to warn of its presence in a leak. Acetylene, whose molecular formula is C_2H_2, is an alkyne, which means that its structural formula contains a triple bond between its two carbon atoms. This triple bond in a molecule as small as acetylene's is extremely unstable. In addition, the triple bond is loaded with energy just waiting to be released. Thus, when acetylene is burned in a controlled manner, the tremendous amount of heat released is put to work by man, usually as a metal-cutting torch or in welding. Care must be taken not to shock the molecule, however, or it will decompose with explosive force. Acetylene must never have any higher pressure than 15 psig applied to it, or the slightest shock or rise in temperature will then cause an explosion. For this reason, acetylene is never shipped for great distances but is generated near wherever it will be used.

Acetylene is generated from calcium carbide, CaC_2, by the simple act of adding water. Calcium carbide, a gray, solid compound, is shipped in unorthodox-looking metal containers on railroad cars. These containers are sift-proof and waterproof, and sometimes serve as the generators themselves, needing only the addition of water in one port and the removal of acetylene at another. The acetylene is then pumped into special cylinders, unlike any other compressed gas container. Such a cylinder contains a porous mineral that will hold a liquid without allowing it to "slosh" around. The acetylene that is being pumped into the cylinder then dissolves in the liquid, which is acetone. The solution of acetylene in acetone (which is itself a flammable liquid) is stable enough to permit the transportation of the cylinder over short distances by truck and to allow its use under normal circumstances. As is true of all compressed gases, care should be taken not to abuse the cylinder nor to cause any leaks.

Acetylene gets its tremendously high heat of combustion from the presence of the triple bond; its very wide flammable range and relatively low ignition temperature can be attributed to this same fact. When acetylene burns, it reaches a flame temperature in air of *4.217°F.*, which is far higher than most flammable gases. When combined with oxygen in a welding or cutting flame, the temperature reaches *5,710°F.* Acetylene's flammable range is from 2.5 percent to 83 percent. This widest flammable range of all the common flammable gases implies that virtually every time acetylene is released, you can count on it being within its flammable range, ready to explode! Its ignition temperature is only 581°F., which is well within the range of every common ignition source. Its vapor density is 0.9, meaning that it is lighter than air.

As in all gas fires, your number-one objective is to *stop the flow of fuel*. If you extinguish the flames, and the gas is still leaking, there is always the possibility that the gas will re-ignite; this time it will be accompanied by an explosion. The only time you should deliberately extinguish the acetylene flame is when you can *immediately* stop the flow of gas, because it was the presence of the flame that had prevented you from doing so.

Hydrogen

Another fairly common compressed, flammable gas is hydrogen. Hydrogen has the smallest atom of all the elements, and its molecule, H_2, is the smallest of all the molecules; therefore, it is hard to keep hydrogen within its container. It is also the lightest of all the gases, with a vapor density of only 0.1, which made it more useful than helium for lifting, as in ligher-than-air ships, but infinitely more dangerous because of its flammability. This fact was tragically proven in the *Hindenburg* explosion and fire. Hydrogen's flame temperature in air is also very high, 3,700°F., and in a cutting flame mixed with oxygen it rises to 4,820°F. Its flammable range too is very wide, from 4.0 percent to 75.0 percent, and its ignition temperature is 932°F.

There is just one single bond between the two hydrogen atoms, and yet hydrogen's heat of combustion is almost as high as acetylene. How can this be? The answer is in the way the energy is released during the bond-breaking. Whereas acetylene's energy is released as both heat *and* light, almost all of hydrogen's energy is liberated as heat. This means that hydrogen burns with an almost invisible flame, which is *not* visible under normal lighting conditions. Therefore, you must exercise extreme caution when approaching a hydrogen cylinder that shows no evidence of a leak or as a source of the fire, because there may be a long, invisible, intensely hot flame protruding from it, waiting to inflict some exceedingly painful injuries or even death upon any unwary firefighter who may approach it. Hold some combustible material in front of you as you approach a hydrogen cylinder, and don't be too surprised if it bursts into flame.

Ethylene Oxide

Another extremely hazardous flammable gas is ethylene oxide. It is technically a liquid because of its vapor pressure, but its boiling point is 51°F., so it is a gas at room temperatures. Ethylene oxide has an extremely wide flammable range, from 3.0 percent to 100 percent, meaning that it can burn inside the container, with no air present. Its ignition temperature is 804°F., its vapor density is 1.5, and the specific gravity of the liquid is 0.9. It is dangerously reactive, and it is subject to violent runaway polymerization, a form of explosion. Ethylene oxide is not only a fuel but also an oxidizing agent, because of its cyclical molecular structure.

There are many more flammable gases than the few we have discussed; they will be presented at the end of this chapter, along with many other gases. It will be your responsibility to look up their characteristics and hazards and list them on an appropriate Hazardous-Materials Data Sheet. Examples of these gases are carbon monoxide, hydrogen cyanide, and hydrogen sulfide, three deadly flammable gases that will be discussed further in Chapter Fifteen, Toxicity (poisons).

Non-Flammable Gases

Oxygen

The most common non-flammable gas of all the compressed gases is oxygen. The fact that it is so common is the only reason for mentioning it here, because oxygen's greatest hazard is not as a compressed gas, but as an oxidizing agent. It will be repeated here and in several more places that oxygen *does not burn*, but it supports combustion. Cylinders of oxygen are shipped for many reasons, one of which is to be used in conjunction with flammable gases in metal-cutting and welding operations, where you will find this fantastic oxidizing agent *literally chained* to a cylinder of acetylene, a tremendously flammable gas. You will also find cylinders shipped to other places where pure oxygen is used, such as hospitals, universities, and small manufacturing plants. Large users of oxygen have switched to liquid oxygen (LOX), which will be covered in Chapter Ten, Cryogenic Gases. The vapor density of oxygen is 1.1, just slightly heavier than air.

Other non-flammable gases are required by many other commercial and industrial installations, in addition to hospitals and universities (see figure 36). These gases include carbon dioxide,

Figure 36
Label and Placard for Non-Flammable Compressed Gases.

helium, nitrogen, ammonia, the halogen gases (fluorine and chlorine), and the refrigerant gases.

Ammonia

Ammonia is a particular problem, in that it is listed by the Department of Transportation (DOT) as a non-flammable gas, and the required placard is the green compressed gas placard. The problem is that ammonia *will* burn, but DOT says that for a gas to be classified as flammable, it must have a lower flammable limit no higher than 13.0 percent, and/or a flammable range of 10 percent. Ammonia's lower flammable limit is 16 percent and its flammable range is from 16.0 percent to 25.0 percent, a spread of 9 percent, rather than the required 10 percent. Another rationale is that no one will willingly stay in an atmosphere containing 16.0 percent ammonia, because it is a choking, pungent gas. This rationale may be true, but it breaks down when someone *must* enter an area where this concentration of ammonia exists. This individual may be protected from breathing the vapors but will not be protected from the explosion and fire that will result from the ignition of a "non-flammable" gas, someone such as a firefighter who is depending upon the accuracy of information on a placard.

An important property of which to be aware when handling ammonia that has been released to the air is that it is highly soluble in water. Therefore, a fine water fog presented in a sweeping motion will dissolve ammonia out of the air; then, however, care must be taken of the run-off, since that will now have become a solution of ammonium hydroxide—a solution with some caustic properties. The molecular weight of ammonia is 17, so it has a vapor density considerably less than 1.0 (0.6); therefore, it will rise and disperse rather quickly. A greater hazard arises from a spill of liquid ammonia, which will then boil away, producing large amounts of vapor. Ammonia's boiling point is −28°F. Even though its vapor density is less than 1.0, these vapors from liquid ammonia will be extremely cold and dense, so dispersion in the air is not as rapid as it would be if the vapors were warmer. Ammonia's ignition temperature is 1,204°F.

Refrigerants

The refrigerant gases are another example of non-flammable gases that present problems in a fire. The gases commonly used today are listed as being of low toxicity and as having no particular hazard, but when "exposed to high temperatures or a hot surface," the decomposition products are highly dangerous and extremely toxic. Of course, the only people exposed to these hazards are those who respond to a fire! The refrigerant gases all have vapor densities considerably higher than air, and so they are all ground- or floor-hugging.

Halogens

The halogen gases, fluorine and chlorine, are extremely hazardous. Fluorine is the most powerful oxidizing agent that exists (even more powerful an oxidizing agent than oxygen); it is a deadly poison and a powerful corrosive. We will meet it again as a cryogenic gas, since that is how it is most commonly shipped. Fluorine has a vapor density of 1.3 and is a pale yellow gas with a sharp, pungent odor. Chlorine, on the other hand, is shipped as a compressed gas in liquified form; its boiling point of $-29.3°F.$ is too high for it to be classified as a cryogenic gas. Chlorine is also a very powerful oxidizing agent, a toxic material, and a powerful corrosive; we will encounter it again in the chapters dealing with those particular hazards. Chlorine is a heavy gas, with a vapor density of 2.4, is greenish yellow in color, with a sharp, pungent odor and a sharp, bitter taste. Being a powerful oxidizing agent like fluorine, chlorine will react violently with most organic materials.

Acid Gases

The halogen acid gases, hydrogen fluoride, hydrogen chloride, hydrogen bromide, and hydrogen iodide, are all covalently bonded gases that are non-flammable and that ionize almost completely when they dissolve in water, which they do quite readily. When they dissolve in water, they form the strong acids, hydrofluoric, hydrochloric, hydrobromic, and hydriodic acids. The gases are used to create the acids and generally are not shipped as compressed gases. There are other gases that form acids when they dissolve in water, but since these are not very common, they will not be mentioned here.

Inert Gases

Helium is an inert gas, with a vapor density of only 0.2 and is much more valuable as a lifting gas for airships, because it will not burn. Like all the inert or "noble" gases, helium will not enter into any chemical reactions, including combustion. It is usually shipped as a cryogenic liquid, so we will encounter it again later. It is relatively non-hazardous as a gas, because of its vapor density.

Nitrogen is another essentially inert gas that is sometimes shipped as a compressed gas but more often as a cryogen. Its only hazard (aside from being under pressure) is that it *is* inert, its vapor density is very close to that of air, and so it could present a problem in excluding air and cause death by asphyxiation. Under some circumstances, nitrogen can be oxidized. This transformation can happen in very hot fires, because nitrogen is such a huge component of air. A series of toxic gases, known as the nitrogen oxides, can thus be formed. The nitrogen oxides will be discussed later in Chapter Fifteen.

Table 17   Other Compressed Gases

acetylene	formaldehyde	natural gas
ammonia	Freons®	neon
argon	germane	nitric oxide
arsine	halons	nitrogen
boron trifluoride	helium	nitrogen dioxide
1,3-butadiene	hexaethyl tetraphosphate	nitrogen peroxide
butane	hexafluoropropylene	nitrogen sesquioxide
1-butene	hexafluoropropylene oxide	nitrogen tetroxide
2-butene	hydrogen	nitrogen trifluoride
carbon dioxide	hydrogen bromide	nitrosyl chloride
carbon monoxide	hydrogen chloride	nitrous oxide
carbon oxysulfide	hydrogen cyanide	oxygen
chlorine	hydrogen fluoride	ozone
chlorine monoxide	hydrogen iodide	parathion and compressed
coal gas	hydrogen selenide	gas mixture
cyanogen	hydrogen sulfide	phosgene
cyanogen chloride	isobutane	phosphine
cyclobutane	krypton	propane
cyclopropane	liquified petroleum gas	propylene (propene)
deuterium (heavy hydrogen)	MAPP gas	propyne (allylene)
diborane	methane	sulfur dioxide
dichlorodifluoromethane	methylamine	sulfur hexafluoride
difluoroethane	methyl chloride	sulfuryl fluoride
difluoro-1-chloroethane	methyl chloride-methylene	tetraethyl pyrophosphate and
dimethylamine	chloride mixture	compressed gas mixture
dimethyl ether (methyl ether)	methyl mercaptan	tetrafluoroethylene
dinitrogen pentoxide	2-methylpropene	trifluorochloroethylene
ethane	monobromotrifluoromethane	trimethylamine
ethene	monochlorodifluoromethane	vinyl chloride
ethylene	monochloropentafluoroethane	vinyl fluoride
ethylene oxide	monochlorotetrafluoroethane	vinyl methyl ether
fluorine	monochlorotrifluoromethane	xenon

Another industrial gas that is sometimes shipped as a compressed gas is carbon dioxide, CO_2, which has a vapor density of 1.5, and is essentially an inert gas. The word "essentially" is used because, under certain conditions, carbon dioxide will actually support combustion, such as when it is used in attempts to extinguish a burning metal, magnesium, for example. The only hazard of carbon dioxide, in addition to its being under pressure, is its asphyxiation properties, which could cause death by excluding air.

Other Gases

Table 17 is a list of other gases that qualify for the classification of compressed gases. Some of them are found as pressurized gases, some as liquified gases, some as both, and some are cryogenic. No attempt will be made to separate them into those groups. It is your responsibility to look them up, in several references, to fill out Hazardous-Materials Data Sheets on them to add them to your store of knowledge. You should also, of course, do the same with those gases presented in the text, because there are many more properties of the various gases that you need to know in order to be able to handle them in an emergency, including extinguishing procedures.

Glossary

Acid Gas: A gas that forms an acid when dissolved in water.

BLEVE: Acronym for *B*oiling *L*iquid, *E*xpanding *V*apor *E*xplosion.

Compressed Gas: A gas that is under pressure, either still in the gaseous state, or liquified.

Critical Pressure: The pressure required to liquify a gas at its critical temperature.

Critical Temperature: The temperature above which it is impossible to liquify a gas.

Cryogenic Gas: A gas with a boiling point of $-150°F$. or lower.

Frangible Disc: A safety release device that will burst at a predetermined pressure.

Fusible Plug: A safety relief device that will melt at a predetermined temperature.

**Figure 37
Boiling Liquid, Expanding
Vapor Explosion.**

Gas: A state of matter defined as a fluid with a vapor pressure exceeding 40 psia at 100°F.

Liquified Gas: A gas that has been converted to a liquid by pressure and/or cooling.

Pressurized Gas: A gas that is still in the gaseous state, but under higher pressure than 14.7 psia.

Spring-Loaded Valve: A safety relief device that is set to operate by opening to relieve pressure, and then to reseat when the pressure drops below the spring's rated strength.

9

Flammable Solids

Introduction

The Department of Transportation (DOT) defines a flammable solid as any solid material, other than an explosive, which is liable to cause fires through friction, through retained heat from manufacturing or processing, or which can be ignited readily and when ignited burns so vigorously and persistently as to create a serious transportation hazard (see figure 38). This chapter will go far beyond this official definition and include *any* solid material that will burn, excluding explosives and plastics, both of which will be covered later, in separate chapters. We will include wood and wood products, paper and paper products, carbon and all solid carbon-containing products, elements, compounds, and mixtures. Some organic compounds, such as organic peroxides, which burn very vigorously, will also be covered in a separate chapter, because their hazards as oxidizing agents that burn are greater than their hazards as flammable solids. In other words, this chapter will cover Class A materials, Class D materials, and some flammable solids that are not classified.

Wood and Wood Products

The mixture we know as wood contains, as its principal ingredient, the polymer cellulose. A polymer is defined as a "giant" molecule made up of thousands of small molecules called monomers, which have the

**Figure 38
Placards for Flammable
Solids.**

capability of joining together to form a new compound called a polymer. Cellulose is an extremely long-chain compound that contains carbon, hydrogen, and oxygen. Together with various resins and other compounds, this mixture is known as wood, and it originates in plants that were alive at one time, mostly trees. The materials known as paper and paper products are almost always derived from wood, so they will be considered at the same time.

The flammability and combustibility (the terms will be used interchangeably) of wood and wood products and paper and paper products depend on their shape, moisture content, and whether they are free-standing, laminated, or otherwise bonded to some other material. Their shape is probably the determining factor in how fast the wood and wood products might be ignited and how fast the fire might spread. The reason ignition temperature is not listed as a primary dependent factor is because, although the ignition temperature of the same types of wood and paper is always the same, the time and energy input required to reach that ignition temperature vary according to the shape and moisture content of the material. The wood's proximity to other materials (that is, whether it is in intimate contact with another material that might conduct the heat energy away from the wood rapidly enough to prevent the material from reaching its ignition temperature) is important to know, especially if the wood or wood product under consideration is a finished product used in the construction of a structure.

Ignition temperatures of various common types of wood range from 378°F. for western red cedar to 507°F. for white pine. The way in which individual particles of wood reach their ignition temperatures is dramatically different. Therefore, the ignition temperature of the wood or wood product does not concern us so much as the ease of ignition of the material in question. The ignition temperatures of various woods can be determined easily enough, but what cannot be determined easily is how fast the material will reach that ignition temperature. Here the variables of shape, moisture content, and proximity to other materials come into play.

What is said for wood and wood products can also be said for paper and paper products. The shape of most paper products lends itself to ease of ignition, which is what concerns us. Needless to say, paper and paper products are so easy to ignite that they are very often the main actors in the initial stages of fires and in the rapid spread of the fire to other flammable solids.

The burning of wood and wood products, including paper and paper products, is not a simple, straightforward process. It is not always the wood itself that is burning but instead the flammable gases and vapors produced by the *pyrolysis* of the wood. As the cellulose molecule is broken down by heat, it produces much smaller molecules that come off the wood as gases and vapors that burn very easily. Not until after these flammable gases and vapors have been consumed in the flame (space burning) can the wood begin to glow (surface burning). Strictly speaking, woods *burn* (in contrast to liquids, which do not), by direct combination with the oxygen from the atmosphere. This combustion takes place, however, only after the wood has been heated sufficiently by the radiated heat from the flame, in which are burning the pyrolyzed gases and vapors produced by the original ignition source and the continuing radiated heat. If a particular sample of wood contains more moisture than another, it should be obvious that there must be some absorption of energy by the water as it evaporates (its latent heat of vaporization) from the sample. Therefore, although the ignition temperatures of both samples may be the same, the drier sample will ignite before the wetter sample, assuming their shapes are identical, simply because the wetter sample has been using energy from the ignition source to evaporate the water it contained, while the drier sample absorbed the energy directly and was raised to its ignition temperature faster than the wetter sample.

The concept of different shapes requiring different amounts of energy to reach the ignition temperatures of the same type of wood containing identical amounts of moisture really means nothing more than that the sample being heated must have a significant percentage of its mass raised to its ignition temperature before combustion will begin. This fact is illustrated in the difference between the ease with which a massive log of a particular type of wood will ignite, and the ease with which sawdust made from identical wood (perhaps even from the same log) with identical moisture content will ignite, assuming an identical ignition source. In the case of the log, considerable effort is needed to build up the amount of energy to be presented to the log (using paper and kindling) before the log will begin to produce vapors that burn, while sawdust from the log may be blown into the fireplace and ignite with explosive force. This case may be somewhat extreme, but it does illustrate the principle of shape, as it relates to ignition temperature. The ignition temperatures of both the log and the sawdust may be identical, but the ease with which the individual particles of wood reach their ignition temperatures is dramatically different. Therefore, it is not strictly the ignition temperature of the wood or wood product that is a concern but the ease of ignition of the material in question. The ignition temperatures of various types of wood can be determined easily enough, but what cannot be determined very easily is how fast each type will reach its ignition temperature. It is here that the variables of shape, mositure content, and proximity to other materials will have an effect.

Paper, on the other hand, has been made from wood and wood products, so that many of the other materials in wood have been removed. When paper is pyrolyzed, there is very little material left to burn by direct combination of oxygen (glowing); as a result, there is very little glowing combustion of paper, since it is almost totally consumed in the pyrolysis.

Elements That Burn

There are many elements that burn; the largest group are the metals, which are classified as Class D materials. Before we reach that group of elements, however, we will first look at the non-metals, of which the three that most concern us are carbon, phosphorus, and sulfur.

Carbon

Of these three non-metallic elements, carbon is by far the most important, since carbon and all solid carbon-containing products make up the largest group of flammable solids. Carbon itself is the basis of *organic chemistry*, the chemistry of compounds that have their origin in things that were once alive, and the duplication of those compounds by synthesis (that is, being made by man). Carbon itself is a non-metallic element with the atomic number 6 and an atomic weight of 12 atomic mass units. It sits in the center of the first long period of the

Periodic Table and can form four covalent bonds with other non-metallic elements. It is a solid and can assume many forms, called *allotropes*, different forms of the same substance. Carbon's allotropes include charcoal, coal, coke, carbon black, graphite, and diamond.

Carbon's biggest use to man in the elemental form is as a fuel, in the form of coal or charcoal. Coal exists either as bituminous or anthracite coal, with anthracite containing a larger percentage of carbon than bituminous. In both types of coal, the non-carbon portion is made up of other organic compounds that have not been compressed out by the heat and pressure created by the coal's covering with tons of earth and rock. Depending upon the amount of carbon and other compounds in the coal, it will have an ignition temperature somewhere between 600° and 1,400°F. Carbon's biggest hazards are present in the allotropes used as fuels, because they burn so very hot, and because tremendous quantities of carbon monoxide are generated because of incomplete burning of the carbon in the coal. Large quantities of carbon dioxide are also created, with these two oxides of carbon as carbon's only combustion products. There may be many other combustion products of coal, depending upon the presence of other elements and organic compounds in it. Coal may produce from 11,000 BTUs to 15,600 BTUs per pound, depending upon the amount of carbon in the coal; it therefore burns very hot. In fires within coal — *(A LOT OF SMOKE)* piles at mines and utility companies, the temperatures at the center of the burning pile can reach high enough to break down the water molecule, producing flammable hydrogen to be added to the burning coal, while oxygen too is generated from the water to support the combustion.

Another hazard of freshly cleaned coal (coal that has been washed to eliminate coal dust, which is dangerously explosive) and charcoal is their tendency to be subject to spontaneous ignition. The very nature of the carbon atom makes it ready to be oxidized at any moment; if conditions are right (as with other cases of spontaneous ignition), a very hot fire will occur when you least expect it.

Far and away the greatest hazard, however, is the carbon monoxide that is produced when carbon burns. Even though carbon monoxide is highly flammable, in fires involving large quantities of coal enough of it escapes combustion to be extremely dangerous. Self-contained breathing apparatus (SCBA) should always be worn when fighting carbon fires, and this dictum includes carbon fires outdoors. Carbon monoxide poisoning can be extremely misleading when it occurs out in the open.

Phosphorus

Phosphorus is an extremely hazardous material, especially the allotrope called white phosphorus, which is *pyrophoric* (reacting in air). White phosphorus is too dangerous to be shipped very far; when it is shipped for short distances, it is shipped in containers constructed so that the phosphorus is kept under water. When white phosphorus must be shipped long distances, it is converted to the allotrope called red phosphorus, which is not pyrophoric. When phosphorus (either red or white phosphorus) burns, it forms phosphorus pentoxide, which is a soft, white powder, highly corrosive to skin and tissue. Phosphorus pentoxide is water-reactive, evolving heat when it comes into contact with water. Because phosphorus begins to burn as soon as it contacts the air, it is useful in incendiary bombs and other pyrotechnic devices.

Sulfur

Sulfur, on the other hand, is rather harmless by itself. Its use by some of our grandparents as part of a spring tonic called sulfur-and-molasses attests to its non-toxicity in the elemental form, but once it burns, the story is different. Sulfur dioxide, the principal combustion product of sulfur, is a choking gas that can be toxic in large quantities. Sulfur is an important element, used mainly to manufacture many other important chemicals, the most important of which is sulfuric acid. Another fact about sulfur is that no matter how safe *it* is, all of the organic compounds formed from it are foul-smelling and toxic in some degree or another. Any organic compounds containing sulfur will liberate sulfur dioxide, among other gases, when they burn.

Metals

The metals are the elements to the left and below the line in the Periodic Table. The atoms of metals all have one, two, or three electrons in their valence rings and, in an ionic reaction, give up those electrons to become positively charged cations. Most metals will combine with oxygen to form oxides and therefore will burn. When metals burn, they do so with tremendous heat. Whether or not they can be ignited easily depends on their shape. Since metals conduct heat very efficiently, any heat energy applied to a piece of metal will be conducted away from the source and spread evenly over the entire mass of the metal. Before metal will ignite, almost the entire piece must be raised to its ignition temperature. Just as in the case of wood and wood products, it is much easier to ignite a small or thin piece of metal than a large chunk, and the explosive hazards of metal dusts are even worse than those of sawdust. Both substances are explosive in the "right" mixture of dust and air, but metal dusts will detonate, which means the explosive power of metal dusts is several times greater than that of wood dusts.

Aluminum

Aluminum is one of the most common metals used by man today and is the most abundant metal in the earth's crust. It is light in weight and strong and is non-toxic. It melts at 1,220°F. and ignites at about the same temperature, or some 20°F. lower. Aluminum burns with a brilliant white light; its dust is a very powerful explosive in its own right and is often added to high explosives to increase their power.

Magnesium

Magnesium is another very common metal. Because it has so many uses, it is often found in machine shops being cut, turned, and having all sorts of operations performed on it to convert it into useful shapes and products. This constant working of magnesium presents several opportunities for the metal to become overheated and to ignite at about 1,200°F. Once large pieces of magnesium are ignited, they become very difficult to extinguish. Magnesium burns with so hot a flame, and its demand for oxygen becomes so great, that it will seek oxygen from anywhere and everywhere, including water and carbon dioxide. The burning magnesium literally rips the water and carbon dioxide molecules apart to get at the oxygen; therefore, if you use either of these as a fire-extinguishing agent on burning magnesium, you stand an excellent chance of actually feeding the fire. Powdered magnesium is an even greater hazard. It will ignite with explosive force, and there have been occasions when it has been reported that magnesium powder, dusts, chips, and shavings mixed together and dampened by water have actually detonated when shocked.

Other Metals

There are many metals that pose similar hazards, especially when in powder form. Some, like lithium, sodium, and potassium, will be discussed in Chapter Eighteen, in the section on water-reactive materials. Others, like zinc, calcium, barium, strontium, iron, steel (an alloy of iron), chromium, nickel, manganese, tin, lead, cadmium, antimony, mercury (the only liquid metal), and bismuth present different hazards, particularly when they burn. The last five mentioned are toxic metals, so care must be taken in handling emergencies involving them. When metals burn, they do so by *direct combination* with the oxygen in the air and therefore are surface-burning materials; that is, there is no pyrolytic action by metals, and therefore they do not produce flammable gases when heated. The burning of metals involves no flaming combustion.

Many metals are attacked by strong acids, and, when they are, they liberate the extremely flammable gas, hydrogen. The heat of the reaction is, in many such cases, high enough to ignite the hydrogen, so care must be taken to prevent strong acids such as sulfuric, hydrochloric, and nitric acids from coming in contact with most metals.

If most metals burn with an extremely hot flame and can cause hydrogen or steam explosions when water is used on them as they burn, and carbon dioxide sometimes worsens the situation much as nitrogen might do, what can be used to control and/or extinguish a metal fire? The noble gases will work, but they are rarely on hand. The problem has led to the development of the dry powders, the Class D extinguishers. These dry powders are usually nothing more than sodium chloride mixed with a small amount of polyethylene, which melts quickly and carries the salt with it as it flows around the burning metal, forming a solid cake around the metal and thus excluding air from it. Even if you are successful in totally covering the burning metal, you must not disturb it for some time, because the metal was extremely hot, and it is now well insulated. Any action which disturbs the caked salt around the piece of metal will cause it to begin burning again.

Some metals, when reduced to ultra-fine powders, become pyrophoric. These include sodium, potassium, magnesium, and calcium. Even if the powders that you encounter are not of these metals, remember that all metal powders possess explosive power if dispersed in air and allowed to come into contact with an ignition source.

Other Flammable Solids

Many other solid materials will burn, and they generally come under the label of "other organic compounds." They are mostly carbon-containing products, and they may be pure compounds or mixtures. Paraffin wax and asphalt are good examples of mixtures. Paraffin wax is a mixture of alkanes and has an ignition temperature of 473°F., while asphalt is a much heavier mixture of alkanes and other materials, including aromatic hydrocarbons and sulfur. Both are solids that burn and that will flow as they burn.

Cellulose Nitrate

Cellulose nitrate (nitrocellulose) is a compound that is highly hazardous. It is fairly common, and it ignites readily, then burns explosively. It may be in the form of pyroxylin, a type of plastic used to make engineers' drawing aids (triangles, straightedges, and the like), eyeglass frames, and some other products that must be strong and transparent. In addition to its very rapid rate of burning, cellulose nitrate will liberate nitrogen oxides when it burns—as will many nitrogen-containing materials that include a particular molecular combination of nitrogen and oxygen. These reddish brown gases are extremely toxic and will frequently produce delayed results, often 24 to 48 hours after exposure. Symptoms of poisoning by the nitrogen oxides are exactly the same as those of a heart attack, so that many overweight, out-of-shape firefighters and officers who have died the day after working a large fire, from apparent heart attacks, may well have died from the damage done their lungs by the nitrogen acids formed from the nitrogen oxides breathed at the fire. It is not necessary for nitrogen oxides to originate in a burning material that contained nitrogen and oxygen, although that is a common source. Remember that nitrogen makes up some 78 percent of our air, and at a very hot fire, the relatively inert nitrogen can be oxidized. Nevertheless, cellulose nitrate, and any other organic nitrate, will liberate those deadly gases, along with the carbon monoxide formed by the carbon in the cellulose part of the molecule. Of course you remember that in ionic chemistry the nitrate ion in a compound makes that compound an oxidizing agent. While cellulose nitrate is a covalent compound, the presence of the nitrate *functional group* is enough to make cellulose nitrate an oxidizing agent that burns because of the organic portion of the molecule.

Solids with Flash Points

Naphthalene is a very unusual solid which does not burn but does give off volatile vapors that do burn. This means that naphthalene has a flash point, in spite of the rule that only liquids have flash points. Indeed, at 174°F., naphthalene gives off vapors sufficient to form a mixture that will ignite with the air (the definition of flash point). The flammable limits of those vapors are from 0.9 percent to 5.9 percent, and the ignition temperature is 979°F.

Another solid that fits into the unusual class of those having a flash point is paraformaldehyde, a white solid that possesses the irritating and pungent odor of formaldehyde. Whenever paraformaldehyde is heated, it releases formaldehyde, which has a flammable range of 7.0 percent to 73.0 percent, and the process occurs at 160°F., the flash point of paraformaldehyde. The vapors have an ignition temperature of 572°F., and those vapors are soluble in water.

Table 18 / Other Combustible Solids

all metals	methyl parathion
antimony pentasulfide	nitrophenol
arsenic trisulfide	nitrotoluene
barium azide	oxalic acid
beryllium	phenylmercuric acetate
calcium carbide	phosphorus pentasulfide
calcium phosphide	phosphorus sesquisulfide
calcium resinate	phosphorus trisulfide
o-chloronitrobenzene	phthalic anhydride
m-chloronitrobenzene	picric acid
cobalt resinate	potassium sulfide
cyanoacetic acid	sodium aluminum hydride
decaborane	sodium amide
3,4-dichloroaniline	sodium hydride
2,4-dinitroaniline	sodium hydrosulfide
dinitrobenzene	sodium hydrosulfite
dinitrotoluene	sodium methylate
hydroxylamine	sodium phosphide
lead thiocyanate	sodium picramate
lithium acetylide-ethylene diamine complex	sodium potassium alloy
lithium aluminum hydride	sodium sulfide
lithium amide	stannic phosphide
lithium borohydride	trinitrobenzene
lithium ferrosilicon	trinitrobenzoic acid
lithium hydride	trinitrotoluene, wet
lithium nitride	urea nitrate
lithium silicon	zirconium hydride
magnesium aluminum phosphide	zirconium picramate
N-methyl-N'-nitro-N-nitrosoguanidine	

Ammonium Nitrate

Ammonium nitrate is such a hazardous material that it will be discussed in greater detail in Chapter Seventeen, Explosives. Ammonium nitrate is also an oxidizer. Its great hazard necessitates its mention here. Ammonium nitrate *will* burn, and since it is also an oxidizing agent, it will provide its own oxygen to support that combustion. It is manufactured principally as a fertilizer, which means that many people are unaware of its additional hazards.

Other Solids

There are many other solids that burn; some of them will be covered in other chapters that highlight a greater hazard than their combustible quality. The others are in table 18, and you should include these in your growing list of Hazardous-Materials Data Sheets.

Glossary

Allotrope: One of several possible forms of a substance.

Ignition Temperature: The minimum temperature to which a substance must be raised before it will ignite.

Polymer: A "giant" molecule made up of thousands of tiny molecules linked together in a long chain.

Pyrolysis: The breakdown of a molecule by heat.

10

Cryogenic Gases

Introduction

You will recall from the chapter on pressurized gases that there were two methods of liquifying gases; one is to apply pressure at temperatures above the boiling point of the gas, and the second is simply to cool the gas to a temperature below its boiling point; a third method uses a combination of the first two. There are also two major reasons that make the liquification of gases economically desirable.

The first is conservation of space. The expansion ratio of liquid to gas in a cryogenic gas is even larger than in a non-cryogenic liquified gas. An example would be the 270-to-1 expansion ratio of liquid to gas for liquid propane, as opposed to the 700-to-1 expansion ratio of liquid to gas for liquid helium. The implication of this tremendous liquid-to-gas ratio is that the user of the gas can store so much more of the product in the same space; the container is different, however, as we will see.

The second reason for liquification of gases is the ability to transport so much more of the product in the liquid phase than in the compressed gas phase. The cost of transportation of these gases has dropped significantly because so much material can be moved in a single shipment (refer to table 19).

The only difference between a normal liquified gas and a cryogenic gas is the boiling point. The definition of the term cryogenic is the study of the behavior of materials from −150°F. down to absolute

Table 19 / Important Properties of Cryogenics

Cryogenic	Boiling Point	Expansion Ratio	% by Volume of Gas in Air
liquid argon	−302°F.	840 to 1	0.93
liquid fluorine	−306°F.	980 to 1	—
liquid helium	−452°F.	700 to 1	0.0005
liquid hydrogen	−423°F.	848 to 1	0.00005
liquid krypton	−243°F.	695 to 1	0.0001
liquid natural gas	−289°F.	635 to 1	0.0002*
liquid neon	−411°F.	1,445 to 1	0.0018
liquid nitrogen	−320°F.	694 to 1	78.00
liquid oxygen	−297°F.	857 to 1	20.95
liquid xenon	−163 °F.	560 to 1	0.000008

*as methane

zero (−459.67°F.). Absolute zero is the temperature at which all molecular motion, and, consequently, all life ceases, according to the kinetic molecular theory. Absolute zero has never been reached (except perhaps in outer space), but it *has* been approached by the temperatures of some cryogenic gases. Strange and interesting things happen at these low temperatures, as we will see when we discuss liquid helium.

Production of Cryogenic Gases

Cryogenic gases are not manufactured, since the gases already exist in our atmosphere and in mixtures with other naturally occurring gases. The company that wants to sell cryogenic gases to its customers retrieves these gases from the atmosphere or other mixtures and cools them to the temperatures needed for liquifying them. The method by which this is accomplished is quite simple.

Let us assume that the goal is to produce liquified oxygen and nitrogen. The most obvious source of these gases is the atmosphere, and that indeed will be our raw material—air pulled in from the outside. First, we must draw the outside air through a system of driers and filters to remove the moisture and particulates that are usually present. We then lead the air into a container that will be able to withstand the pressures at which we will be working. We continue to add air to this container, compressing it until we reach 2,000 psi. The kinetic molecular theory states that as the pressure of a gas goes up, its temperature will also go up; therefore we now have hot air at 2,000 psi in our container, and our next job is to cool it down. Pipes have been built into our container for this purpose. Now we begin to circulate cold water (33°F.) through these pipes, which will cool the compressed air down to the same temperature. As the temperature of the gas falls, its pressure also drops, and more air is pumped in. This of course, causes the temperature of the gas to rise; therefore, the cooling water circulation is continued. This procedure goes on until we have our container filled with air at 2,000 psi and at 33°F. The water is then bled out of the cooling system, and liquid ammonia or some other coolant is now circulated, reducing the temperature of the air to −28.3°F. (the boiling point of ammonia). More air is introduced as the pressure drops, until we have 2,000 psi and −28.3°F. as the pressure and the temperature of the air in our container.

This procedure is an excellent example of using the laws of nature to our economic advantage, for now we are going to allow the cold, highly pressurized gas to escape into a larger container, which means we are going to allow its pressure to drop rapidly. The combined gas law (available in any chemistry textbook) will tell you exactly what the resulting temperature and pressure of the gas will be, given the original volume, temperature, and pressure. The size of the second container is designed to produce a specific drop in pressure so that the resultant temperature of the gas is *below* –320°F., the boiling point of nitrogen. When the air is allowed to expand into this second container, its temperature drops to ⟨ − 320°F., and every gas that is present in the mixture we call air with a boiling point *higher* than − 320°F. will *condense* into a liquid. This liquid mixture will contain nitrogen, oxygen, carbon dioxide, carbon monoxide, and any other gas produced by our society and released into the air (such as hydrocarbons and sulfur dioxide, generated by the burning of gasoline and other fuels) whose boiling point is − 320°F. or higher.

Now we have a liquid mixture where we formerly had a gaseous mixture; our job is to separate it into its component gases. Again, we use a law of nature, that each of these compounds has a different boiling point. We allow the temperature to rise very slowly, and as each individual gas boils off at its own boiling point, it is collected and sent to a different container, where the same process is used to re-liquify it; since the gas is already very cold, the process of condensing the gas again will not take much energy. We can quite easily collect the liquid nitrogen and the liquid oxygen in this manner and allow the other gases to escape back into the atmosphere, unless, of course, we have a customer for them.

We have completed our job in providing liquid nitrogen and liquid oxygen for our customers, but suppose that they wanted to buy liquid hydrogen and liquid helium from us. How can we get down to the boiling points of these gases (−423.0°F. for hydrogen and −452.1°F. for helium)? Simple! We only need sources of these gases, and we can produce them in the liquified form just as we did the others. The hydrogen needed is generated in other chemical manufacturing operations, such as petroleum distillation and chlorine manufacture. Helium is found in minute quantities in air and in larger quantities in natural gas. With the hydrogen and helium thus collected and sent to the compressing unit, the procedures described earlier are repeated, with the exception that previously produced liquid nitrogen is used to cool the pressurized gases down to −320°F. before the gases are allowed to expand to the volume of containers of predetermined size, where expansion drives the temperature of the mixture down to −453°F., and both gases are condensed. They are separated and re-condensed as before, and we now have quite pure liquid hydrogen and liquid helium. Production of all cryogenic gases is similar, and some of the liquified gases with higher boiling points are produced as byproducts of the cryogenics.

Uses of Cryogenics

Why do we make the effort to produce these extremely cold liquids, and take the chance that something will go wrong with them? The answer is the same as that at the beginning of this chapter: to save money in transporting and storing these gases in liquified form rather than in a compressed, gaseous form. Moreover, our customers may want to use the liquified gases in their liquid form, rather than allow the gases to vaporize and then use them, as some other customers do. We will explore the use of each gas separately, and you will see that some customers actually use these gases in the liquid form.

Liquid Oxygen (LOX)

Liquid oxygen is used primarily as the oxidizer for liquid fuels in the propellant systems of rockets and missiles. Its only other use as a liquid is to be transported and stored in the liquid state so it can then be allowed to vaporize for use by the ultimate consumer. For this reason hospitals are large users of liquid oxygen. In the chemical and petroleum industries, oxygen is used in the production of synthetic gas from coal, natural gas, or liquid fuels to produce gasoline, methanol, and ammonia, and in other processes to make aldehydes and alcohols, among other chemicals.

Liquid Nitrogen

Liquid nitrogen has a few uses in its liquid form, mainly as a refrigerant in the trucking industry, for food preservation, and to preserve biological specimens. It is also used in a method of attaching railroad-car wheels to their axles so that they will seldom or never come off, even in a derailment. The wheel is heated so that the inside diameter of the hole into which the axle is inserted is just slightly larger than the outside diameter of the axle (the original inside diameter is *smaller* than needed to accommodate the axle), the axle is inserted, and the whole end of the axle with the wheel in place is dipped into liquid nitrogen; the subsequent shrinking of the wheel forms a bond stronger than the metal itself.

Another use for liquid nitrogen is in cryogenic grinding of materials that would tend to melt if the grinding operation took place at normal temperatures. A prime example is the cryogenic grinding of plastics, since the heat generated by the grinding process would melt the plastic and prohibit it from being reduced in size to a powder. In its gaseous state, nitrogen is used as an inert gas in electrical systems, the chemical industry, and the food-packing industry, among others.

Nitrogen itself is not a toxic gas, for it makes up the largest part of our breathing air (78 percent by volume). Wherever liquid nitrogen is being used, however, there is a chance that enough of it will be released into the air we breathe to lower the oxygen content to a level below that necessary to sustain life. In this context, nitrogen can be considered a simple asphyxiant and as such would be considered a hazardous material.

The hazards of asphyxiation may be present in such areas as cryogenic grinding operations where there is no effort to contain the release of nitrogen into the surrounding space and in refrigerated trucks or other vehicles that use liquid nitrogen as the refrigerant. In this latter case, the liquid nitrogen is allowed to escape slowly into the cargo area. Cooling is accomplished by absorbing whatever heat is present in the cargo area to allow evaporation of the liquid nitrogen, plus the introduction into this area of the extremely cold (−320°F.) gas. Anyone entering the cargo area must be aware that the atmosphere present is very nearly pure nitrogen.

Other Cryogenic Gases

Liquid natural gas (LNG) is used mostly as a fuel in its vapor state. It can also be used as a source of hydrogen, but, in any event, it is liquified only for transportation and storage purposes.

Liquid fluorine is the most powerful oxidizer known, and its use is limited to those operations where its oxidizing power is needed, mostly as an oxidant for rocket propellant systems. As a gas, it is useful anywhere its oxidizing power will be useful and also as a source of raw material to produce the fluoride ion; it is utilized in several chemical processes.

Liquid argon, liquid krypton, liquid xenon, and liquid neon are used primarily in gaseous form wherever an inert gas is needed, such as in filling incandescent lamp bulbs and in filling special bulbs and display tubes to obtain special effects. Argon is also used as an inert gas shield in arc-welding processes to prevent oxidation of the metals being welded.

Liquid hydrogen is used in large quantities as a primary rocket fuel, in conjunction with liquid oxygen and liquid fluorine, for the propelling of rockets, and as a propellant for nuclear-powered rockets and space vehicles. Gaseous hydrogen is widely used in the chemical and food industries, along with the metallurgy and semiconductor industries.

Liquid helium is used to produce gaseous helium for use as an inert shielding gas in welding and as an inert gas atmosphere in the production of reactive metals, such as titanium and zirconium. It is also used as the lifting gas in dirigibles and balloons.

In liquid form, helium is used to run experiments in superconductivity and investigations of low-temperature physics. This work has produced some rather spectacular results. After removal from a bath of liquid helium, coils of copper wire with electrical currents induced in them have proved to have the current *still flowing* after the current inducer has been removed. These superconductors can produce magnetic fields that will float a railroad train *above* the track and result in unbelievable speeds, with the magnetic fields as the force that propels the train. Speeds attainable by the train are limited only by safety considerations or by the curvature of the earth, whichever is overcome first!.

In the low-temperature experiments, steel ball bearings of the same composition and specific gravity have been dropped into liquid helium; some of the ball bearings floated, while others sank and still others fluctuated between the two extremes! When liquid helium is allowed to reach its boiling point, no bubbles of vapor are present as there are in other boiling liquids. It seems that within six or seven degrees of absolute zero, strange things can happen.

Transportation and Storage

Cryogenic gases are transported and stored in what amounts to giant thermos bottles. When the container is small, it is called a Dewar's flask (figure 39). These do not resemble other containers for pressurized liquids or gases because the vapor pressure above a cryogenic liquid is very low, often less than a few pounds per square inch! The containers are really one container within another, using an air space between the container walls as insulation. Cryogenic liquids are basically self-insulating. You will recall that heat energy is required whenever evaporation occurs (the latent heat of vaporization). This means that for the cryogenic gas to convert from a liquid to a gas, heat must be absorbed by the liquid so that some of it may evaporate. The heat must be absorbed from somewhere, and, since the container is insulated, and the cryogenic liquid is so cold, there is very little heat energy to supply for the evaporation process. With so little energy available, evaporation occurs very slowly. Even when a container of a cryogenic liquid is opened to the atmosphere, evaporation occurs so slowly that it is difficult to realize that the liquid is boiling.

**Figure 39
Dewar's Flask.**

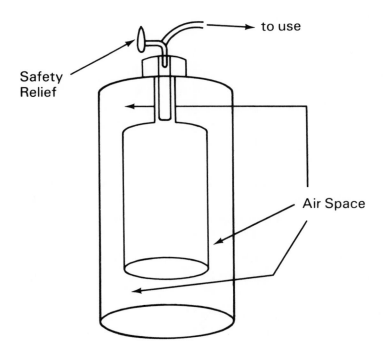

to use

Safety
Relief

Air Space

Safety relief devices are used in all cryogenic liquid containers, with one in the internal container to vent to the atmosphere what little gas has evaporated, and another in the outer container to vent any gas that may have accumulated between the walls of the inner and outer containers. If this second safety relief device vents, it is usually because there is a leak in the inner container. An alarm is normally connected to the outer safety relief device to alert workers to the leak, which could be extremely hazardous. The hazards, of course, depend on the nature of the gas. In any event, any gas leaking will be extremely cold.

Because of this extremely cold gas venting through the safety relief device, a unique problem could develop for first responders answering an emergency call where cryogenic gases are used. Any time that firefighters respond to a fire where cryogenic containers are present, care must be taken to keep water from contacting any safety relief device. Because of the extremely low temperature of the escaping gas, any water that contacts the safety relief device, *or any area immediately surrounding it,* will immediately freeze and build an ice deposit on the device that might prevent it from operating. The incident commander must be aware of this fact and make a decision as to whether or not he wants to chance the use of water near the venting device.

This introduces another difference in the way emergencies with cryogenic gases are handled, as compared with "normal" liquified gases. With impinging flame or large amounts of radiated heat affecting a container of non-cryogenic gas, one of the objectives is to cool the container to prevent a catastrophic failure, particularly the BLEVE associated with liquified flammable gases such as the LP gases (and many others not yet discussed, such as monomers). Water will be effective in cooling a container of non-cryogenic liquified gas but not a container of cryogenic liquid. You must remember that the outer shell of this container is covering air space only, so cooling water will have little effect on trying to regulate the temperature of the cryogenic liquid, which is in the inner container. More important, you must realize that even if the water is at 33°F., it will still be a minimum of 183°F. *above* the boiling point of the cryogenic; therefore, any water contacting a cryogenic liquid will freeze immediately and also cause the cryogenic to produce vapors more rapidly by raising its temperature, since the water is so much *warmer* than the cryogenic. The danger of a BLEVE is present but is unlikely because of the outer container insulating the inner container; that is, the outer container protects the inner container from the impinging heat. Of course, anything that causes a massive release of the cryogenic will present a tremendous hazard, depending on the particular hazard of the gas itself.

Another danger of which to beware when dealing with cryogenics is the tremendous amount of the product present in the container. The huge expansion ratios mean that even a small container, one cubic foot in volume, for example, will contain anywhere from 560 cubic feet to 1,445 cubic feet of gas (depending, of course, upon which cryogenic liquid is in the container). Any release of the liquid will produce enormous quantities of gas.

The economies allowed by the liquification of cryogenic gases mean that more and more such products will be transported; these huge tank wagons carrying the liquid pose a serious hazard if they are involved in traffic accidents. Your first response should be to identify the material present and then to determine the specific hazards posed by the release of this material. The type of breach of the container *must* be known, because this information will tell you whether the material is escaping as a gas or as a liquid. Again, aside from the extreme coldness and the very great expansion ratio of the liquid, the specific hazard of the gas itself must be known to determine what hazards are present.

Hazards of Cryogenics

The hazards of cryogenic liquids fall into three main areas: the extreme coldness of the liquids and resulting vapors, the tremendous vapor-to-liquid expansion ratio, and the hazards of the particular chemical itself. The first two are self-explanatory, but the third needs a further explanation. The hazards of a cryogenic liquid must include all the hazards possessed by the gas itself, before it is liquified, since in any accidental release, you must be concerned with what the gas itself will do during or after its evaporation. Those hazards include the flammability of the gas, its ability to support combustion, its toxicity, corrosiveness, or any other hazard possessed by the particular gas. These hazards can be used to group the cryogenics into three classes.

Flammable Cryogenics

This group includes liquid hydrogen and liquified natural gas. Needless to say, an accidental release of these materials will cause an extreme explosion hazard, even though the specific gravity of both gases is less than 1.0. Any spill will produce tremendous amounts of flammable vapor, and both gases are highly flammable; you should review the properties of these gases in relation to flammability. If you can protect the area of the spill and the area downwind from any ignition source, you may want to consider causing the cryogenic liquid to evaporate faster than it would naturally, by the addition of water, which to the cryogenic liquid will be relatively warm. Just be careful if you use this approach, recognizing that you are causing more vapors to be produced by your actions.

Hydrogen, of course, is the smallest of all the atoms in the Periodic Table, and, as a gas, it exists as the smallest of all molecules; its molecular formula is H_2. It is the lightest gas known (specific gravity, .07) and therefore will disperse very rapidly. This means that although hydrogen is *very* flammable due primarily to its extremely wide flammable range, the chance of accidental ignition is *relatively* less than that of liquified natural gas because hydrogen gas will rise from the spill rapidly and disperse quickly in the air. Its wide flammable range, however, does make it highly susceptible to ignition very close to the surface of the spill *and* high above the spill.

Liquid natural gas (LNG), on the other hand, will release from a spill gas that is heavier than hydrogen but still considerably lighter than air (specific gravity, approximately 0.55 to 0.6, depending on the percentage of methane in the mixture). This fact normally means that the gas will rise and disperse rapidly in air. At the extremely low temperature of the liquid ($-289°F.$), however, the gas will be somewhat denser and will not rise and disperse as rapidly as under "normal" conditions.

You may also want to consider the deliberate ignition of the gas to produce a controlled burn. Naturally, before selecting this method of handling the emergency, you would want to be sure of the path of any vapors already released from the spill and the length of time the burn will last, including the amount of radiated heat which will be released during the burn. This method should be considered only when there will be no threatened exposures and no danger of the controlled burn getting out of control. If the spill is liquid hydrogen, you must remember hydrogen's extremely high heat of combustion and high flame temperature. Also, you must consider the fact that there will be a considerable amount of radiated heat back to the surface of the spill. This heat will cause a more rapid evaporation of the liquid, as more energy is made available for use in the evaporation process. This process will in turn increase the intensity of the fire, and the "controlled" burn may no longer be controlled.

If the spill can be contained in a pit or diked area, or in a container of some kind, there is always the possibility that the liquid may be pumped back into the original container until another container of sufficient size and suitable physical properties can be supplied. Just remember that a container that can be used to hold "normal" flammable liquids might not withstand the extremely low temperatures of the cryogenic liquid. This situation is also true of any pump used to try to pump the liquid and of any hose and other tools used in the incident. It should not come as any surprise to you to see very pliable materials *and* very strong, hard tools (and fingers and any other human parts) literally shatter into thousands of pieces after being subjected to such low temperatures.

As in *any* hazardous-materials incident, the option of contacting the manufacturer of the material (along with the company that owns the tank wagon, if it is not owned by the supplier of the gas) should be considered for expert advice on how to handle the situation.

Cryogenics That Support Combustion

This group of cryogenic liquids includes liquid oxygen and liquid fluorine. You must realize that in a spill of either of these materials, you have a condition that has produced *more than 100 percent concentration* of an oxidizing agent; 100 percent would be pure oxygen and pure fluorine, but don't forget the expansion ratio of liquid to gas. The cryogenic liquid will also soak into anything present, forming a gel-like material that will not evaporate as rapidly as a normal liquid at its boiling point, thus forming an explosive mixture with any ordinary flammable material. There have been cases reported where liquid oxygen has spilled onto an asphalt street, and that section of asphalt *detonated* when an apparatus rolled over it. Be advised that both oxygen and fluorine are tremendous oxidizing agents in their own right (fluorine is an even stronger oxidizing agent than oxygen), but when they are liquified, the hazards are several orders of magnitude greater!

While the primary hazards of oxygen and fluorine are their capacities as powerful oxidizing agents, fluorine is associated with two even more dangerous hazards. Fluorine is a highly corrosive gas; any exposure to it of any material subject to corrosion will cause rapid destruction of that material. This includes metals, minerals—and humans. Any exposure of a person to fluorine will be extremely destructive to any exposed parts; care should be taken to protect oneself completely from any contact with fluorine.

In addition to its oxidizing and corrosive hazards, fluorine is a deadly poison. Positive-pressure SCBA must be used in conjunction with total-encapsulating chemical suits whenever fluorine is encountered.

Inert Cryogenics

This group of cryogenic liquids includes liquid helium, liquid argon, liquid krypton, liquid xenon, liquid neon, and, usually, liquid nitrogen. Liquid nitrogen is to all intents and purposes inert; you will remember that if it is released near a very hot fire, however, it *can* be oxidized, and it will vigorously support the combustion of some metals, like magnesium. Since these occasions are so rare, we can include liquid nitrogen in the class of inert cryogenics. The hazards of these gases, aside from their extreme cold, and the high liquid-to-gas expansion ratio, is that they will lower the concentration of oxygen in the air below life-sustaining levels and therefore are considered simple asphyxiants.

Glossary

Absolute Zero: The temperature at which all molecular motion ceases: $-459.67°F.$ or $-273.16°C.$

Asphyxiant: A gas that is essentially non-toxic, but can cause unconsciousness or death by lowering the concentration of oxygen in the air or by totally replacing the oxygen in breathing air.

BLEVE: Acronym for *Boiling Liquid, Expanding Vapor Explosion.*

Boiling Point: The temperature at which the vapor pressure of a liquid just equals atmospheric pressure.

Condensation: The process by which a gas or vapor passes into the liquid phase.

Cryogenics: The study of the behavior of matter at temperatures of $-150°F.$ and below.

Evaporation: The process by which molecules of a liquid escape through the surface of the liquid into the air space above.

Expansion Ratio: The determination of how many volumes of a gas or vapor are produced by the evaporation of one volume of liquid.

Latent Heat of Vaporization: The amount of heat a substance must absorb when it changes from a liquid to a vapor or gas.

11

Oxidizing Agents

Introduction

The class of hazardous materials that we call oxidizing agents (or, more simply, oxidizers) are really two classes of hazardous materials; one class contains the *inorganic* (that is, ionic in nature and not carbon-based) oxidizing agents, whose major hazard, as a class, is that they *are* oxidizers. The second class is made up of *organic* (that is, covalent in nature and carbon-based) oxidizing agents, many of which are really *unstable* hazardous materials. Table 20, the list of oxidizing agents, that appears at the end of this chapter contains both groups, and it will be your job to see that you know which is which. The organic oxidizing agents known as organic peroxides, however, will be covered later in Chapter Fourteen, covering these and another class of unstable materials called monomers.

An oxidizing agent is defined as a substance containing oxygen that gives it up readily, plus the halogens (fluorine, chlorine, bromine, and iodine). The halogens are included in the general definition of an oxidizing agent because they *will* support combustion.

The classic oxidizing agent is a substance that does not burn but will support combustion (see figure 40). This is true of only the inorganic oxidizing agents (with exceptions to be noted later), for the organic oxidizing agents will not only support combustion but are also flammable. Since the greater hazard of organic oxidizing agents is something other than merely being an oxidizing agent, they will not be

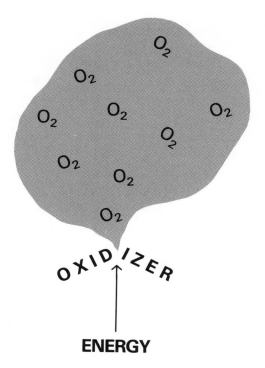

**Figure 40
The Oxidation Diagram.**

discussed here, but instead will be covered in the chapters on their most dangerous properties. The important part of the definition is that the oxidizing agent is a substance containing oxygen that gives it up readily. Many substances contain oxygen as part of the compound, but they do not qualify as oxidizing agents, since they hold on to their oxygen rather strongly. Examples of substances containing oxygen that do not give it up readily include water, carbon dioxide, and silicon dioxide (sand). In each of these examples, oxygen is the major portion of the compound by weight; yet they do not release it easily (witness the fact that all three are used as fire *extinguishers*).

Do not misunderstand the relative term *readily*, used in the definition. Water will indeed give up its oxygen in several instances, such as in the cases of being applied to lithium, sodium, or potassium, to burning magnesium, or to the center of a white-hot, burning coal pile. In each of these cases, the water molecules will be ripped apart to get at their oxygen, but the important thing to remember is that the water did not give up the oxygen *readily*. A tremendous amount of energy was required to remove the oxygen from the water, just as in the application of carbon dioxide or sand to burning magnesium and the subsequent removal of oxygen from these materials. The term *readily* means "at the slightest urging," such as the energy supplied by mechanical friction or by an increase in the temperature of the material, particularly when that material is in contact with an ordinary combustible material.

Those hazardous materials classified as oxidizing agents will decompose readily to yield oxygen when heated and therefore may react easily with other hazardous materials. Because of their ability to provide oxygen when heated, the hazard of oxidizing agents is greatly increased at higher temperatures; there may even be very violent reactions when they are intimately mixed with ordinary combustible materials and, in some cases, may even become explosive with ordinary combustible materials. The sensitivity of the mixture is dependent upon the ease with which the oxidizing agent gives up its oxygen, the flammability of the combustible material, and the amount of energy to which the mixture is subjected, so that each mixture will have to be individually judged as to its relative sensitivity. For this reason, care must be taken to prevent the accidental mixing of *any* oxidizing agent and *any* combustible material.

An additional danger of some inorganic oxidizing agents presents itself when you tend to consider harmless any chemical that you can handle with bare hands or even ingest in small quantities with no apparent ill-effects. What is important to remember is that these materials will liberate tremendous quantities of oxygen at the slightest urging; this result is precisely what you do *not* want to occur in the presence of a combustible or flammable material, particularly when there are many common ignition sources present (see figure 41).

Let us now look at the different inorganic materials that qualify as oxidizing agents under the definition (remember that although there are many organic oxidizing agents, the worst of those, the organic peroxides, will be covered later). We can break these inorganic materials into several groups, according to their anion (remember that we said some time ago we would be revisiting ionic compounds when we got

Figure 41
Oxidizer Placard and Label.

to the oxidizing agents). One of the clues that these compounds contain oxygen in their structure is that they are all inorganic compounds that end in -ite or -ate. Examples of such ions that classify as oxidizing agents are the nitrates and nitrites, the perchlorates, chlorates, chlorites and hypochlorites, the permanganates, the inorganic peroxides (those compounds having metal ions attached to the peroxide ion), the persulfates, the chromates and dichromates, the molybdates, and the perborates. In addition, there are certain inorganic compounds such as nitric acid that are powerful oxidizing agents. You should be aware that, in combination with specific metal ions, these complex ions (oxyanions) will form oxidizing agents of varying sensitivity. For example, potassium nitrate is a more sensitive oxidizing agent than calcium nitrate. It is not necessary to memorize a list of oxidizing agents to try to determine which materials are more dangerous than others, since you should treat any compounds containing these ions as equally dangerous until you have determined otherwise.

As a rule of thumb, you may consider the ionic compounds made up of the alkali metals (lithium, sodium, potassium, rubidium, and cesium) and the previously mentioned complex ions as more hazardous than those compounds containing the alkaline earth metal ions (beryllium, magnesium, calcium, strontium, and barium). Do not forget the ammonium ion, which has a + 1 charge, just like the metals of group I. Again, you should treat *all* inorganic compounds that end in -ite or -ate, or any compound that has per- as part of its name, as dangerous oxidizing agents until you can be sure that they are not. It is much better to err on the conservative side and perhaps to look a little foolish than to assume that some chemical or mixture is not an oxidizing agent, when it turns out in fact to be one. This means that although the phosphates, carbonates, borates, and sulfates are not oxidizing agents, you may mistake them for oxidizing agents because of the advice above. This mistake is much safer than assuming that the perborates and the persulfates are *not* oxidizing agents and then being injured in the ensuing fire or explosion.

Let us now consider the most common of the inorganic oxidizing agents, classified by the non-oxygen portion of the anion.

Nitrogen-Containing Oxidizing Agents

The most important of the inorganic oxidizing agents are the nitrogen-containing oxyanions, the nitrates and nitrites, and the chlorine-containing oxyanions, the perchlorates, chlorates, chlorites, and hypochlorites. The nitrates are probably the most important economically, because of the demand for their oxidizing power, for the nitrogen in the ion, which may also be needed as a fertilizer, and because the nitrates are found naturally in such great quantities. They are extremely useful as starting materials in the manufacture of many other chemicals (see figure 42).

Sodium nitrate, $NaNO_3$, is known as the most common of the nitrates and occurs naturally in mineral form (and is commonly called Chile saltpeter). It is *hygroscopic*, which means it will absorb moisture from anything containing water, particularly the atmosphere. It poses a special problem if it is packaged in any combustible material. Over time, sodium nitrate will absorb the moisture from this packaging material and increase its combustibility. For this reason, it is shipped in multi-walled paper bags (as many as five or seven layers of paper). It is also commonly used in salt baths, because of its relatively low melting temperature (586°F.). It is used in solid-fuel rocket engines as the oxidizing agent, in pyrotechnics, in manufacturing matches, fertilizers, and glass, and in countless other processes. Sodium nitrate is extremely common and shipped in very large quantities.

Sodium nitrite, $NaNO_2$, also has many uses as an oxidizing agent, including the manufacture of rubber and many chemicals and as a curative in food preservation. Although it contains one less atom of oxygen than the corresponding nitrate, it should nevertheless be treated as the dangerous oxidizing agent that it is.

Potassium nitrate, KNO_3, also known as saltpeter or niter, is used to manufacture many explosives, including black powder. It, and potassium nitrite (KNO_2), also have many uses in chemical manufacturing. You will find that, because of the amount of sodium and potassium in the earth's crust, the ions of these two metals are the most common in inorganic oxidizing agents, but nitrates and nitrites

Figure 42
Cautions for Nitrites and Nitrates.

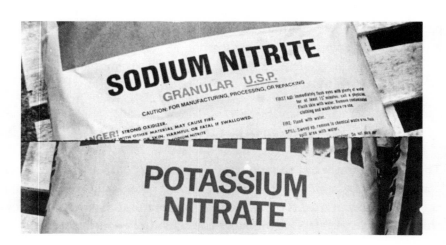

of many other metals also exist, and they are all oxidizing agents. The nitrogen-containing oxysalts are soluble in water.

There are many more combinations of metals and the nitrate and nitrite ions, and they all are oxidizing agents. Like all other oxidizing agents, they should never be allowed to come into accidental contact with combustible materials.

Another nitrogen-containing oxidizing agent is nitric acid, HNO_3. It is a very strong acid, has tremendous corrosive power, and is a colorless-to-yellow fuming liquid that is extremely hygroscopic. The fumes emitted by nitric acid are highly toxic by inhalation, and the liquid is corrosive to skin and mucous membranes. It is explosive with ordinary combustible materials, provided that an intimate mixture of oxidizing agent and fuel exists. It is used in the manufacture of nitrate fertilizers and explosives, in addition to its use in the synthesis of many organic chemicals. Nitric acid is shipped under *both* corrosive and oxidizing agent placards and labels.

Nitric acid is particularly dangerous because of the multitude of hazards that it presents. Since it is such a strong acid, and therefore a very powerful corrosive, sometimes its danger as an oxidizing agent is overlooked. Conversely, if great care is given to keeping the nitric acid from contacting any combustible material, its corrosiveness might be overlooked. In all situations, care must be taken to prevent the breathing of any vapors released by nitric acid. In the heat of a battle against an accidental release of a material such as nitric acid, one or more of its multiple hazards may be forgotten, and damage *will* be caused by the hazardous property against which the responder has *not* protected himself.

Chlorine-Containing Oxidizing Agents

The chlorine-containing oxidizing agents are valuable in many manufacturing processes for several reasons. These reasons include serving as a relatively safe source of chlorine, as bleaches, and as starting materials for other chemicals. The perchlorate ion, ClO_4^{-1}, contains four oxygen atoms and is therefore a very powerful source of oxygen, if it will yield it all. It does not do so on every occasion and therefore cannot be ranked as the most dangerous of the chlorine-containing complex ions, just as the others cannot. Therefore, under certain conditions, the hypochlorite ion, ClO^{-1}, may be the most dangerous of the four, even though the ion contains only one oxygen atom. You must treat all four (including the chlorate, ClO_3^{-1}, and the chlorite, ClO_2^{-1}) with just as much care and respect as you would any other oxidizing agent. Again, the most common perchlorates, chlorates, chlorites, and hypochlorites are sodium and potassium, simply because of the plentiful amount of these metal ions. All of the chlorine-containing complex ions form oxidizing agents with the metal cation. The ammonium ion is also very important in this group of oxidizing agents and is used in great quantities.

The hypochlorites are the bleaches used in the home and for commercial purposes, with the common liquid bleaches containing about 5 percent sodium hypochlorite, $NaClO$. Calcium hypochlorite, $Ca(ClO)_2$, is used to purify drinking water and is the additive used in swimming pools.

The acids made from these chlorine-containing oxyanions are also very strong oxidizing agents. They are perchloric acid ($HClO_4$), chloric acid ($HClO_3$), chlorous acid ($HClO_2$), and hypochlorous acid, ($HClO$). Perchloric acid is probably the strongest of all the inorganic acids and as an oxidizing agent is just as dangerous as nitric acid. The common chlorine-containing oxysalts are all soluble in water.

The Inorganic Peroxides

Inorganic peroxides are those compounds that have hydrogen or any metal attached *ionically* to both ends of the peroxide *ion*. The cation (hydrogen ion or metal ion) is the same on both sides of the peroxide ion. Organic peroxides, on the other hand, have organic radicals bound *covalently* to both sides of the peroxide *radical*. The organic radicals attached may be the same on both sides, or they may be different.

The most common of the inorganic peroxides is hydrogen peroxide, H_2O_2, a clear, colorless liquid which is soluble in water. Pure hydrogen peroxide is stable, but the slightest impurity will promote decomposition, often very violent. Concentrated solutions of hydrogen peroxide are highly corrosive and toxic. It is used as a bleach, deodorizer, in the manufacture of rocket fuel, as a source of the peroxide ion, and in the production of many other chemicals. Any heavy metal, such as iron and copper, will act as a contaminant, and cause the hydrogen peroxide to explode, so it has to be shipped in containers made of glass, aluminum, or polyethylene. Its structural formula is $H-O-O-H$. The hydrogen peroxide in your medicine cabinet—in an amber bottle to protect it from the ultraviolet rays of sunlight, which will also cause it to decompose—is about 3 percent in strength.

Sodium peroxide, Na_2O_2, is a yellowish solid and reacts in water to liberate oxygen and heat, which is the same reaction that potassium peroxide, K_2O_2, another yellow solid, undergoes. Both are powerful oxidizers and are used as oxidizing agents and bleaches in many chemical reactions. Sodium peroxide is used as a water purifier, as an oxygen generator in submarines and space vehicles, and in the manufacture of pharmaceuticals. Both sodium and potassium peroxide are dangerous fire and explosion risks in contact with water, alcohols, acids, powdered metals, and organic materials.

Lithium peroxide, Li_2O_2, is a white solid that is also a powerful oxidizing agent but is not water-reactive. All other metallic peroxides are powerful oxidizing agents and will be explosive with ordinary combustible materials.

Other Inorganic Oxidizing Agents

There are many other oxysalt ions that form oxidizing agents; they include the permanganate ion, MnO_4^{-1}, the persulfate ion, $S_2O_8^{-2}$, the chromate ion, CrO_4^{-2}, and the dichromate ion, $Cr_2O_7^{-2}$. Any ionic compounds with these last names will be oxidizing agents, with the alkali metal compounds, particularly sodium and potassium, again being the most common. Treat these oxidizing agents like any of the others, never allowing them to come into intimate contact with *any* combustible material. Storage areas in manufacturing plants should be inspected to insure that oxidizers and combustible materials are not stored near each other. Remember, these two represent the fuel and oxidizer legs of the fire triangle. All that is now needed to produce a fire is an ignition source, and sometimes the mixing of the two materials together *is* all that is needed. The heat generated by the reaction may be enough to cause ignition.

Ammonium Oxysalts

Ionic compounds that have oxygen-containing complex ions as the anions are all classified as inorganic oxidizing agents, and they do not burn. *Oxysalts of ammonium are an exception;* the common ones, like the nitrates and nitrites, will not only burn but will detonate! On April 16, 1947, a ship in Galveston harbor, containing 2,280 tons of ammonium nitrate, NH_4NO_3, *detonated* after burning for over 25 hours. This explosion caused other fires in warehouses and in other ships containing more ammonium nitrate, and a second explosion occurred in another ship. When all was done, a total of 468 people had been killed, including *the entire 27-man Texas City Fire Department*, all because the ammonium nitrate was treated as a fertilizer (which it is), not like a very effective oxidizing agent (which it is) and a material that might detonate at any time (which it does). Ammonium nitrate is an exception to the rule that inorganic oxidizing agents do not burn. So are ammonium nitrite, NH_4NO_2, ammonium acetate, $NH_4(C_2H_3O_2)$, ammonium chlorate, NH_4ClO_3, ammonium perchlorate, NH_4ClO_4, and ammonium permanganate, NH_4MnO_4. Many other ammonium compounds that contain oxygen may not be oxidizing agents, but they may be combustible. Be extremely careful with ammonium compounds.

The Halogens

Thus far, all the oxidizing agents we have discussed have been oxygen-containing salts, peroxides, or acids. There is another group of hazardous materials that are oxidizing agents but contain no oxygen to liberate. These materials are the halogens (fluorine, chlorine, bromine, and iodine). Fluorine and chlorine are gases, while bromine is a fuming liquid, and iodine sublimes (converts from a solid to a vapor, bypassing the liquid phase). It is the vapors from bromine and iodine, plus the fluorine and chlorine gases, that support combustion in exactly the same way that oxygen supports combustion. In other words, the halogens will participate in exactly the same chemical reaction as oxygen does when oxygen supports combustion. That means you do not need to have oxygen present to have a fire, as long as you have the halogens, a fuel, and the ignition energy. Fluorine is such a powerful oxidizer (that is, it will release its oxidizing power so easily) that the light from a light bulb might be sufficient to ignite certain fuels mixed with fluorine. Remember, therefore, when the halogens are present, they alone can support a fire. If oxygen is also present, the fire can be much worse—and if the concentration of the halogens is higher than what would be normal for oxygen, the fire could be explosive.

Oxygen

In our zeal to cover all the types of oxidizing agents, let us not forget the element for which the chemical reaction is named, oxygen.

Oxygen is by far the most common of all oxidizing agents, and most oxidation by this oxidizer (that is, the combination of oxygen with any material) occurs because of the oxygen in the atmosphere. Whether this oxidation is slow, as in the rusting of iron, or rapid, as in the combustion reaction, or instantaneous, as in an explosion, it is the most common type of oxidation. Other sources of pure oxygen are compressed oxygen and liquified oxygen. The rest of the sources of oxygen come, of course, from the materials we have been listing above, the inorganic oxidizing agents (excluding the halogens, of course).

Summary

The inorganic oxidizing agents generally are not flammable, with the exception of the ammonium compounds, which generally are flammable. The inorganic oxidizing agents are usually solids that are soluble in water. Hydrogen peroxide and the inorganic acids are liquid, and they are soluble in water. No oxidizing agents should ever to be allowed to mix with anything that will burn; in storage areas, oxidizing agents and any combustibles should be segregated. No oxidizing agents should ever be subjected to heat or organic contaminants.

Remember the clue in a chemical's name that tells you if oxygen is present in the compound. All such compounds end in the letters -ate and -ite. Of course, if the name includes "oxide" or "peroxide," oxygen is present. The "per-" prefix indicates more oxygen in the compound than in the normal state, so the combination of the prefix "per-" and the ending -ate, -ite, or -oxide indicates a "super" oxidizing agent and very probably one that is chemically unstable.

Following is Table 20, a list of additional oxidizing agents, most of which have not been listed in the text. You should determine whether they are organic or inorganic, and list their hazards on appropriate Hazardous-Materials Data Sheets. Many of them are organic oxidizing agents (organic peroxides) that will be covered later in Chapter Fourteen.

Table 20 / Additional Oxidizing Agents

acetyl acetone peroxide	ammonium chromate
acetyl benzoyl peroxide	ammonium dichromate
acetyl peroxide	ammonium perchlorate
aluminum nitrate	ammonium permanganate

Table 20 / Additional Oxidizing Agents (cont.)

amyl nitrate
barium chlorate
barium nitrate
barium perchlorate
barium permanganate
barium peroxide
benzoyl peroxide
beryllium nitrate
bromine pentafluoride
bromine trifluoride
butyl hydroperoxide
butyl isopropyl benzene hydroperoxide
butyl perbenzoate
butyl peracetate
butyl peroxyacetate
butyl peroxypivalate
calcium chlorate
calcium chlorite
calcium hypochlorite
caprylyl peroxide
chloric acid
chlorine dioxide hydrate
chlorine trifluoride
chlorobenzoyl peroxide
chlorosulfonic acid
chlorosulfuric acid
chromic acid
chromium anhydride
cobaltous nitrate
copper nitrate
cumene hydroperoxide
cupric nitrate
cyclohexanone peroxide
dibenzoyl peroxide
dibutyl peroxide
dichlorobenzoyl peroxide
dichloro-s-triazinetrione
dicumyl peroxide
diisopropylbenzene hydroperoxide
diisopropyl peroxydicarbonate
dilauroyl peroxide

dimethylhexane dihydroperoxide
ferric nitrate
guanidine nitrate
hydrogen peroxide
iodine peroxide
isopropyl percarbonate
lauroyl peroxide
lead nitrate
lead peroxide
lithium hypochlorite
lithium peroxide
magnesium nitrate
magnesium perchlorate
magnesium peroxide
mercuric nitrate
mercurous nitrate
methane hydroperoxide
methyl ethyl ketone peroxide
monochloro-s-triazinetrione acid
nickel nitrate
nitric acid
nitrogen peroxide
nitrogen trioxide
paramethane hydroperoxide
peracetic acid
perchloric acid
peroxyacetic acid
pinane hydroperoxide
potassium bromate
potassium chlorate
potassium dichloro-s-triazinetrione
potassium nitrate
potassium nitrite
potassium perchlorate
potassium permanganate
potassium peroxide
propyl nitrate
silver nitrate
sodium bromate
sodium chlorate
sodium chlorite

Table 20 / Additional Oxidizing Agents (cont.)

sodium dichloro-s-triazinetrione	succinic acid peroxide
sodium hypochlorite	tetranitromethane
sodium nitrate	thorium nitrate
sodium nitrite	trichloro-s-triazinetrione
sodium perchlorate	uranyl nitrate
sodium permanganate	zinc ammonium nitrate
sodium peroxide	zinc chlorate
strontium chlorate	zinc nitrate
strontium nitrate	zinc permanganate
strontium peroxide	zinc peroxide

Glossary

Halogens: The elements of group VIIA: fluorine, chlorine, bromine, iodine, and astatine.

Hygroscopic: Capable of absorbing moisture from its surroundings.

Oxidizing Agents: Any materials containing oxygen that will give it up readily, plus the halogens; also known as oxidizers.

Oxyanion: A complex anion containing oxygen and at least one other element.

Oxysalt: An ionic compound containing a metal and an oxyanion.

12

Plastics

Introduction

Plastics are a group of materials that belong to a much larger class of materials called polymers. A polymer is defined as a "giant" molecule made up of thousands of tiny molecules that have the unique property of being able to react with themselves to form larger molecules. The process by which this takes place is called polymerization, and the "tiny" (relatively speaking) molecules that have this unique property of combining with themselves are called monomers. Both words, monomer and polymer, come from Greek roots; *mono-* meaning one, *poly-* meaning many, and *-mer* meaning part. Therefore, monomer means one part, and polymer means many parts. Sometimes, the polymer can be made up of many tens of thousands of the monomer molecules, depending on the type of polymer the manufacturer (the polymerizer) wants to make (see figure 43). Before we get into the different types of plastics, however, we have to examine this much larger group of materials that are called polymers.

There are really two groups of polymers, the first being natural (existing in nature), and the other being synthetic, or man-made. If you combined all the manufacturing capabilities of all the companies in the world that make synthetic polymers (and there are quite a few companies making quite large amounts of synthetic polymers), this would seem a mere pittance compared to the amount of polymerization that Mother Nature contributes, on a regular basis. The natural polymers include cellulose (the major ingredient in wood and woody plants, and also cotton, which is nearly pure cellulose), leather, silk, wool, and skin. The cellulose from wood is also the major ingredient in paper and all paper products, including cardboard and cardboard products. You can plainly see the volume of polymers and polymeric products that nature manufactures, which man then uses to build shelters, make clothing, and the other comforts that are demanded by society.

Types of Synthetic Polymers

Synthetic polymers can also be divided into two groups: rubber and plastics. Although many will claim that rubber is really a thermosetting plastic (the term will be defined later), the consuming public probably accepts the fact that rubber and plastics are really two separate and distinct types of materials. You should be aware, though, that there are more and more plastics that have elastomeric properties (the ability to be stretched to twice the length and to snap back to approximately the original length) and more and more resemble rubber.

Polymerization

The chemical reaction we call polymerization is a very special type of reaction, which is undergone by only the few compounds called monomers. This special chemical reaction allows these molecules to react with themselves to form new molecules, made up of "repeating units" that resemble the original monomer molecule. Let us look at

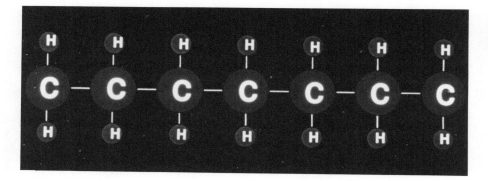

Figure 43
Diagram of a Part of a Polymer.

the most common of all the monomers, ethylene, to see how it reacts. The molecular formula for ethylene is C_2H_4, and its structural formula is

$$
\begin{array}{ccc}
\text{H} & & \text{H} \\
| & & | \\
\text{H}-\text{C} & = & \text{C}-\text{H}
\end{array}
$$

What makes ethylene and other monomers different from other compounds is what happens under special conditions of temperature, pressure, and the addition of special chemicals that begin and control the rate of reaction. When these monomers polymerize, the double-bond characteristic of the alkenes undergoes a remarkable change. Instead of both bonds breaking, as in the combustion reaction (which, of course, the monomer will still undergo under normal conditions), only one bond breaks and transfers both electrons of the broken bond to the outermost carbon atoms, giving them an unshared electron each. This happens to all of the monomer molecules in the reactor, and, as you know, a molecule with unshared electrons is in a very unstable condition which nature will not permit; these molecules can probably all be considered as free radicals. Since these molecules have all been prepared to react, the only other materials with which they can react are the other free radicals around them, which, of course, they do. In the polymerization reactor, this reaction is very carefully controlled, and the monomer molecules are not turned into free radicals all at once, but at a controlled rate. This control is necessary because there *are* bonds breaking, which causes the release of energy. If this process were allowed to proceed at an uncontrolled rate, the result will be a catastrophic explosion called "runaway polymerization"! Since the reaction *is* carefully controlled, the reaction proceeds at the rate called for by the chemical engineer, so that the polymer with the desired properties will be produced.

With the double bond now broken, and the two unpaired electrons now attached to the "other side" of the carbon atoms, the monomer now looks like this:

$$
\begin{array}{ccc}
\text{H} & & \text{H} \\
| & & | \\
-\text{C} & = & \text{C}- \\
| & & | \\
\text{H} & & \text{H}
\end{array}
$$

To keep the runaway polymerization explosion from occurring during transportation of the monomer, an inhibitor is mixed in with the monomer, which effectively prohibits the process called polymerization from beginning. If the inhibitor is not added to the monomer, or, after being added, is allowed to escape by evaporation during an emergency of some kind, protection against the instantaneous total polymerization of all the monomer in the tank truck or rail car is gone, and the situation becomes extremely dangerous. The monomer will begin to react with itself, liberating heat energy, which will cause the reaction to speed up, quickly getting out of control; the resulting explosion will resemble a BLEVE. Many monomers are easily liquifiable gases, so these materials are subject to BLEVE in any case. Nevertheless, the resulting explosion, no matter how it is classified, will kill anyone within the danger zone and cause tremendous property damage.

Classes of Plastics

There are two great divisions of plastics: thermoplastics and thermosets. The definition of a thermoplastic is a plastic that is formed by heat and pressure, and once formed, may be re-formed by the further use of heat and pressure. In other words, if a processor who changes plastic resins and compounds from powder, pellet, or cube form into a finished part by heating and applying pressure (in a machine called an extruder, injection molder, or similar equipment) is not happy with the part he has just made, he may grind it up, put it back through his equipment, and make a new part. He may do this once, or he may do it several times, depending on the particular thermoplastic and its properties.

A thermoset, on the other hand, may be formed by heat and pressure just once. Any further attempts to reheat a thermosetting plastic will result in thermal degradation (decomposition or even burning) of the product. Whether a particular plastic is thermoplastic or thermosetting depends upon the chemistry of its monomer and on the polymerization process. A plastic may be either a thermoplastic or a thermoset, however; it cannot be both—although there *are* groups of chemically similar plastics, such as the polyesters, which include both thermoplastic polyesters and thermosetting polysters. The thermoplastics (more than will be listed here) include polyethylene, polypropylene, polystyrene, polyvinyl chloride, polyurethane (there are also thermosetting polyurethanes), polyesters, acrylics, polyamides, cellulosics, and fluoropolymers. Thermosets include alkyds, phenolics, the epoxies, and the urea-formaldehyde plastics. Table 21 lists the names of different plastics.

Table 21   /   The Different Plastics by Name

acetal	polybutylene
acrylics (all types)	polybutylene terethalate
acrylonitrile-butadiene-styrene (ABS)	polycarbonate
acrylonitrile-chlorinated polyethylene-styrene (ACS)	polychlorotrifluoroethylene
allyl	polyesters (alkyd, aromatic, and unsaturated)
amino	polyetherether ketone
cellulose acetate	polyetherimide
cellulose acetate butyrate	polyethylene (all grades)
cellulose acetate propionate	polyethylene terethalate
cellulose nitrate	polyimide (both thermoplastic and thermoset)
cellulose triacetate	polymethylpentene
chlorinated polyethylene	polyphenylene ether (modified)
chlorinated polyvinyl chloride	polyphenylene oxide
epoxy	polyphenylene sulfide
ethyl cellulose	polypropylene
ethylene acid copolymer	polystyrene
ethylene-chlorotrifluoroethylene	polyurethane
ethylene ethyl acetate	polytetrafluoroethylene
ethylene tetrafluoroethylene copolymer	polyvinyl chloride
ethylene-vinyl acetate	polyvinyl fluoride
fluorinated ethylene-propylene copolymer	polyvinylidene chloride
furan	polyvinylidene fluoride
ionomer	silicone (all polymers)
nylon	styrene-acrylonitrile
perfluoroalkoxy resin	styrene-maleic anhydride
phenolic	sulfone polymers
polyacrylonitrile	thermoplastic elastomers—
polyamide-imide	and many alloys of the above

Thermoplastics

General

By far the largest group of plastics is the thermoplastics. This group contains all the common plastics, with whose names you will be familiar, plus all of the engineering plastics, so-called because they are called upon to perform some task other than packaging, or being decorative; they may be load-bearing, as are certain thermoplastics which serve as wheels, gears, slides, and other working and functional parts in many applications.

To understand the thermoplastics, you must know a little about their chemistry. It is not necessary to become a polymer chemist or a plastics engineer, but it is extremely helpful to know what these polymers are made of. Basically, the initial raw materials for the thermoplastics are monomers, those small molecules that can react *with themselves*, "hooking up" with each other in extremely long chains. For the most part, monomers are extremely reactive, and sometimes extremely unstable, hazardous materials that must be handled very carefully. Under the proper conditions of temperature and pressure in a reactor (a large sealed vat), and in the presence of an *initiator* (a substance to overcome the *inhibitor* added to the monomer in order to prevent premature polymerization and begin the polymerization process at the proper time) with, in some cases, a catalyst (a substance added to the reaction that speeds it up but which is not used up in the process), the polymer chemist or the chemical engineer knows that he can coax the molecules of monomer to "hook up" with each other so as to begin to form a long-chain molecule. He also knows that by varying the temperature, the pressure, and the length of time the reactants are subjected to such conditions, he can produce different variations on the same polymer. An example would be a plant that, with variations on the same equipment and processes, can produce linear low-density polyethylene (LLDPE), low-density polyethylene (LDPE), high-density polyethylene (HDPE), high molecular weight high-density polyethylene (HMWHDPE), and ultra-high molecular weight polyethylene (UHMWPE). All of these polymers started with the same monomer, ethylene (ethene), but ended as rather radically different plastics with widely varying properties. These different types of polyethylenes (PE) can be used for many different purposes, with each plastic selected having a different set of requirements.

Although each type of polyethylene in the preceding example had the same starting monomer, ethylene, polyethylene is the only thermoplastic you can make, beginning with ethylene alone. Therefore, each thermoplastic has its own monomer. For polypropylene, propylene (propene) is the monomer; for polystyrene, the monomer is styrene. To produce polyvinyl chloride, you must start with vinyl chloride monomer (VCM); to make ABS plastics, you need the three monomers, acrylonitrile, butadiene, and styrene, for which those plastics are named. Some other thermoplastics require more than one monomer, and they may not be monomers in the same sense that we originally defined them. For instance, some thermoplastic polyurethanes need an isocyanate and a polyol (a type of alcohol) to react to form the polymer; some thermoplastic polyesters may require polyethylene glycol and terephthallic acid to form PET (polyethylene terephthalate). Whatever the method and the monomers used, if the polymer has the ability to be reheated and re-formed without decomposing, it is a thermoplastic.

Specific Thermoplastics

1. ABS (Acrylonitrile-Butadiene-Styrene)

This is a thermoplastic terpolymer (a polymer made by the polymerization of three different monomers) that is produced in grades required by users with properties of the final thermoplastic designed-in by the proportions of each of the three monomers used. It is an extremely tough plastic that is used wherever rough treatment can be expected. ABS is very useful in structural parts such as camper bodies and tops, canoes, telephones, refrigerators and other appliances, sanitary ware, pipe, power tools, and interior aircraft parts. Whenever additional properties are needed and they cannot be achieved by altering any of the three monomers, other substances are added to the thermoplastic to impart these properties; when this situation occurs, the resulting thermoplastic material is called a compound. This statement is true of all thermoplastics, not ABS alone. ABS can acquire additional properties, such as flame suppression, color, and antioxidation properties, among many others. Unfortunately, ABS does not weather very well, so its use outdoors is limited, unless it is capped (a thin layer of a weatherable thermoplastic laminated over it).

ABS can be extruded (softened and forced through a die) or injection molded (softened and forced into a closed mold). If it is extruded into a flat sheet, it may then be formed (heated and drawn or forced over a shaped mold) into any desired shapes. It can be alloyed (other thermoplastics mixed with it) according to specific requirements, such as flame resistance or increased strength.

ABS is considered to be a tonnage (volume) thermoplastic, and yet it can be made so strong that certain grades are considered to be engineering thermoplastics.

2. Acetal

This thermoplastic was once considered an engineering thermoplastic only but is now receiving great acceptance as a decorative thermoplastic that can also be functional. It is a highly crystalline thermoplastic possessing a high degree of hardness, good chemical resistance, strength, and good electrical properties. It is a polymer of formaldehyde and, as such, is a good engineering thermoplastic. It can be extruded and injection and blow molded. Blow molding is a process in which a hollow tube of hot plastic is extruded into a mold and a blast of air causes it to conform to the shape of the mold. The process is used in making aerosol containers, pens, butane-filled cigarette lighters, and toys.

3. Acrylics

There are many types of acrylic thermoplastics, but the most popular, and therefore the most common, is the polymer based on the methyl methacrylate monomer: polymethyl methacrylate, or PMMA. This thermoplastic possesses excellent optical and weathering characteristics, is tough, resistant to heat and many chemicals, and exhibits low electrical conductivity. Because of their outstanding weathering and optical properties, acrylics are excellent candidates to replace glass in signs and glazing applications. They may be extruded or injection molded from what is called molding powder, but the vast majority are cast in a liquid form on to a moving belt or between glass plates to form sheets, which may then be used as such or may be vacuum- or compression-formed into other shapes. Products made from acrylics include signs, sanitary ware, spas and pools, residential and commercial glazing, automotive lighting lenses, and lighting fixtures.

4. Cellulosics

The thermoplastics known as the cellulosics are really a family of thermoplastics that are chemically different from each other in the basic make-up of the polymer. The starting point for these materials is the naturally occurring polymer, cellulose, mentioned in the introduction to this chapter. Generally, however, they are close enough to physical properties to be described together. They possess good optical clarity, good electrical resistance, strong surface characteristics to resist scuffing and marring, and medium water vapor transmission properties.

Cellulosics include cellulose nitrate, a plastic with good dimensional stability, but poor stability to heat and light, and very flammable if not properly stabilized. Its uses include T-squares, straightedges, and other engineering drawing tools, eyeglass frames, and decorative inlays on musical instruments.

Ethyl cellulose is compounded to possess high heat resistance and high impact properties. It is tough and somewhat flexible, and its toughness is outstanding at low temperatures. It is used to make parts for electrical applicances, fire extinguisher parts, and flashlight cases.

Cellulose acetate (CA) is the most common of the cellulosics, being used in electrical insulation, write-on pressure-sensitive tape, audio recording tape, lenses, microfilm, toys, and tool housings.

Cellulose acetate butyrate (CAB) is a good weathering polymer and is used as a laminate with metallic foil or is metallized itself.

Cellulose acetate propionate (CAP) has very good optical qualities and therefore is used to make such products as safety goggles, metallized flash cubes, packages and containers of many types, and steering wheels.

Cellulosics may be extruded or injection molded.

5. Fluoropolymers

Fluoropolymers, or fluoroplastics, are a group of plastics characterized by the substitution of fluorine for one or more of the hydrogens on an alkane, to form the monomer. We are discussing a group of plastics: there are several monomers that are produced, which, upon polymerization, each produce a unique polymer. As a group, the fluoroplastics offer great resistance to a wide variety of chemical environments and have good electrical resistance, high heat resistance, and extremely low coefficients of friction.

There are several fluoroplastics, which will all be mentioned by name, but their uses are determined according to their suitability for a particular job. The most familiar fluoroplastic is the one used to coat home cookware, polytetrafluoroethylene (PTFE, better known by its trademarked name, Teflon). The rest of the fluoroplastics family includes polychlorotrifluoroethylene, ethylene-chlorotrifluoroethylene copolymers, fluorinated ethylene-propylene copolymers, perfluoroalkoxy resins, ethylene-tetrafluoroethylene copolymers, polyvinyl fluoride, and polyvinylidene fluoride. (A copolymer, by the way, is a plastic made by the mixing together of two monomers, and the subsequent polymerization of the mixture; some copolymers may be produced by a "grafting" process, a totally different procedure.) These polymers may be difficult to pronounce, but they perform extremely important jobs in commerce, ranging from electrical insulation to earthquake protection for tall buildings. Fluoroplastics may be extruded or injection molded or may be powder coated, a process in which a layer of fine thermoplastic powder is made to cover the part to be coated, and the part is then heated so that the polymer flows evenly over the entire surface of the part.

6. Nitriles

The nitrile resins are really polyacrylonitrile (PAN), but it is easier to refer to them by the shorter name. These thermoplastics are known for their ourstanding gas transmission properties and are sometimes referred to as barrier resins; that is, bottles or containers made of nitrile thermoplastics are impervious to the passage of oxygen or carbon dioxide through the walls of the container, thereby serving as a barrier to either gas. In addition, these thermoplastics possess good resistance to hydrocarbon and chlorinated hydrocarbon solvents, good electrical resistance, and good mechanical properties. The acrylonitrile mono-mer's proper chemical name is vinyl cyanide. Polyacrylonitrile may be blow molded or injection molded.

7. Nylons

The family of thermoplastics known as the nylons belongs to a group of polymers known as polyamides. The chemistry of the nylons and the specific polymerization processes each one participates in is rather long and complicated; we will briefly describe only the general properties of the nylons, although you should realize that there are many different polymers possible, with different specific properties. Nylons, in general, have good mechanical and electrical properties, which originally put them in the class of engineering thermoplastics. Their outstanding resistance to wear enables them to continue to be used in many strictly engineering capacities, but the fact that they also are strong, stiff, and very tough, makes them useful in substituting for metals in the outer casings of power tools and several other applications. The nylons can be modified by compounding ingredients, by copolymerization, and even by *terpolymerization* (the polymerization of three monomers to make one thermoplastic). The largest growth area for the nylons today is the automobile, where they may be used in fifty or more different parts. Nylons are also widely used in the electrical industry, in consumer goods (clothing, carpeting, upholstery, household goods, packaging, and so on), and in a myriad of other products. Nylons may be extruded and injection molded.

8. Polycarbonate

Polycarbonate (PC) is a very important engineering thermoplastic whose strength enables it to be used in some structural applications. It has excellent high-impact strength, excellent dimensional stability, good electrical properties, and good heat stability. With its excellent optical properties, it is used as glazing material, automobile headlights and instrument panels, traffic light housings and lenses, protective covers for street lights, office water cooler bottles, and food containers. Polycarbonates may be extruded into film and sheet, and some small parts are injection molded.

9. Polyesters

The thermoplastic polyesters are the saturated (no multiple covalent bonds) polyesters: namely, polyethylene terephthalate (PET) and polybutylene terephthalate (PBT). PBT has good chemical, mechanical, and electrical properties and is usually injection molded. It is used to make electrical switches and connectors, fuse cases, TV tuners, automotive distributor caps and rotors, pump impellers, and other large automotive parts.

PET is the thermoplastic that has replaced glass in large soft drink bottles, mainly because of its light weight, clarity, and its action as a barrier to carbon dioxide. It is also very useful in producing film and fiber. Polyester clothing is made from PET fibers, as is the polyester carpeting, upholstery, and polyester cord used in automobile tires. It is also used as a coating over other materials, such as paper, and may be made into photographic and X-ray film. PET may be extruded, injection molded, blow molded, or extrusion coated (a process whereby the thermoplastic polyester is melted in an extruder and then applied in a thin coat over another material).

10. Polyethylene

The number-one thermoplastic in volume in the world today is polyethylene (PE), and the many variations that can be made from the ethylene monomer have been listed as an example at the beginning of this chapter. It is used in housewares, packaging, trash bags, milk bottles and other bottles and containers, film for use as vapor barriers in construction or as mulch in agriculture, electrical insulation, in pipe and conduit, and in literally hundreds of other applications. All grades of PE may be extruded, injection molded, blow molded, calendered (passed through a series of three or four mill rolls that press the thermoplastic into a sheet or film), or rotationally cast (a process whereby powdered resin is introduced to a mold that is revolving in all planes and heated, which causes the resin to fuse—partially melt—and flow into all parts of the mold, assuming the shape of the mold's interior). All the various grades of polyethylene added together make it, by far, the most widely produced plastic in the world.

11. Polypropylene

The monomer of polypropylene (PP) is propylene (propene), which is an alkene like ethylene (ethene), the monomer of polyethylene. If you recall, the double bond of the ethylene molecule breaks during polymerization, sending one unpaired electron to each of the carbons in the molecule. Propylene acts in the same way during its polymerization to polypropylene, but, since there are only two unpaired electrons available, and three carbon atoms in the molecule, the unpaired electrons always go to the end carbon atoms, giving each of them a "dangling bond," ready to form a full covalent bond with other unpaired electrons. The resulting "repeating unit" looks like this:

$$
\begin{array}{ccccc}
& H & & H & & H \\
& | & & | & & | \\
-\!\!\!\!& C & \!\!\!-\!\!\!& C & \!\!\!-\!\!\!& C & \!\!\!- \\
& | & & | & & | \\
& H & & H & & H
\end{array}
$$

The original name for the analogous series of unsaturated hydrocarbons containing one double bond that we know today as the alkenes, was the olefins. Consequently, polyethylene, polypropylene, and a rather small (but growing) volume thermoplastic called polybutylene (PB) are classed together as the polyolefins. They somewhat resemble each other as the finished polymer, and they burn alike, but that is the end of the close resemblance.

Polypropylene is tougher than polyethylene and exhibits good low-temperature impact resistance, good heat, chemical, and moisture resistance, and ease of processing. It may be extruded and injection molded and is used to make toys, housewares, luggage, furniture, packaging, bottles and containers, tubing, film and sheet, and many automotive parts.

12. Polystyrene

The third highest thermoplastic in volume behind polyethylene and polyvinyl chloride is polystyrene (PS). It has good optical clarity, dimensional stability, electrical and chemical resistance, and good colorability (all, of course, in the transparent and/or non-foamed grades). It is available as crystal grade, medium-impact polystyrene (MIPS), high-impact polystyrene (HIPS), and with major modifications produced by adding other monomers before polymerization. It is used in packaging, electrical parts, sheet and film, toys, appliance cabinets, refrigerator inner liners, audio and video cassettes, lids and containers, and countless other items. In its foamed grades (such as foamed hot-drink cups), it is a great insulating material, and may be used for that purpose in construction of residences and commercial buildings. Polystyrene may be extruded, injection molded, and blow molded.

13. Polyurethane

As with the other polymers previously mentioned (polyesters), there exist both thermoplastic and thermosetting types of polyurethanes; here we will discuss only the thermoplastic polyurethanes. They are usually very flexible, tough, and abrasion-resistant. They may be extruded or injection molded. The thermoplastic polyurethanes are used in clothing, shoes, diapers, athletic equipment and lettering, packaging, and automotive parts.

14. Polyvinyl Chloride

Although it is the number-two thermoplastic in the world as far as volume is concerned, polyvinyl chloride (PVC) is number one when it comes to the number of uses to which it is put. Its monomer is vinyl chloride (VCM), and once the resin is through the polymerization process, the PVC may appear in literally an endless number of compounds. PVC, as it comes out of the reactor, is useless as a thermoplastic. It is much too hard and brittle and may decompose the very first time anyone tries to extrude or mold it. A processing aid called a heat stabilizer must be added to get it through any process, and then it may have an off-white translucent or a straw-colored appearance. To correct this, pigments of one sort or another must be added to give it a pleasing color. Once you have added a stablizer, colorant, or *anything* to PVC resin, it is now a PVC *compound.* If only stabilizers or colorants have been added, it is a *rigid* PVC compound. To convert it into a flexible PVC compound, a *plasticizer* must be added to make it softer and more pliable. To lower the total cost of a compound, an inert mineral filler may be added; to help it get it through certain processes, a lubricant may be added. Different uses for PVC will dictate what degree of rigidity or flexibility is needed, and that will then determine the type and quantity of additive to use.

PVC may be processed by extrusion, injection molding, blow molding, calendering, rotational casting, and, after being extruded into sheet, may be thermoformed. It may also be transformed into a *plastisol,* a type of very viscous suspension of PVC particles in a plasticizer, which pours like a heavy paste. PVC has, as was said before, more uses than any other thermoplastic; these uses include shower curtains, phonograph records, flooring, wall covering, house siding, gutters and downspouts, windows, housewares, furniture, upholstery, automotive parts, electrical parts and insulation, and pipes and conduit.

Thermosets

Not much will be said about the thermosetting plastics because they do not represent a very large volume of materials, relative to the thermoplastics. Some, however, have importance to firemen in how they burn.

1. Probably the thermoset produced in largest volume is polyurethane, principally in its use as a foam. In the foamed grades, polyurethane is almost certainly one of the most valuable insulating materials available to us. It is found as an insulation in the walls of our homes, in industrial plants, and in refrigerators and freezers. It is also extensively used as a cushioning material and is present in upholstery in our homes and automobiles and in the mattresses on which we sleep.

2. Amino resins include the urea-formaldehyde and melamine formaldehyde resins. They are used for adhesives, laminates, coatings, and foams. Molding compounds are used to make sanitary ware, buttons, dinnerware, wiring devices, and appliance parts, while laminating resins are used to make countertops, tabletops, furniture, and wall paneling. The adhesive resins are used to bond together plywood, furniture, particle board, and flooring. Coating resins are used to greaseproof and shrinkproof textiles and to add fire retardancy to other plastics.

3. Epoxy resins have good electrical, thermal, and chemical resistance, low shrinkage, and good impact resistance. They are used mostly for coatings and finishes.

4. Phenolic resins are thermosetting plastics with good heat and chemical resistance, good dielectric properties, surface hardness, and good thermal and dimensional stability. Fillers are added to the resin to give specific properties. Phenolic resins are molded. They are used to make electrical receptacles and switches, automotive components, knobs, appliance parts, utensil handles, casters, rollers, and other load-bearing items.

5. Thermosetting polyesters are the liquid resins usually impregnated with glass fibers and called fiberglass. They are very tough and durable, being used for automobile and boat bodies, tanks, sanitary ware, construction panels, and recreational equipment.

Hazards of Plastics

General

Once the plastics are polymerized, they represent no known hazard in the use for which they are intended. As a matter of fact, many plastics are sanctioned by the government for use as food additives, and some have been cleared for use *inside* the human body, to replace a vital part that has worn out or become defective. It is *before* polymerization, when man has to deal with monomers, inhibitors, catalysts, and initiators that hazardous materials are involved. It is in this setting, either in the plant, or on the highway or railroad, that we can have problems with unstable hazardous materials. Like all organic materials, plastics burn, and, like all organic materials that burn, plastics liberate toxic combustion products. It is usually in this context that the firefighter becomes concerned about plastics, and well he should. Once he becomes aware of the hazards of burning materials, he can protect himself, and he can learn to extinguish fires involving plastics in a safe manner.

Combustion Products of Burning Plastics

To be able to understand the possible combusion products of burning plastics, you have to know of what they are made. That job can be simplified by grouping all the plastics we have discussed, both thermoplastics and thermosets, into similar types based on their molecular composition. We do not have to delve very deeply into the molecules, so long as we know what elements are present there. Knowing that, we can then calculate the *final* combustion products, along with some of the intermediate ones. Furthermore, knowing that plastics do not pyrolyze in exactly the same manner as wood, we can also tell something about intermediate combustion products. Intermediate combustion products are products formed in the initial stages of burning; they then, being flammable and combustible liquids and gases themselves, in turn will burn, and form the final combustion products, which are the products usually liberated in the largest quantities. In many cases, the intermediate combustion products mentioned, particularly the aldehydes, have *not* been found in burn tests done on certain plastics. Since the plastics are all very long-chain molecules, however, we will consider that these aldehydes *may* be present, if only in trace amounts.

Plastics Made Only of Hydrogen and Carbon

The plastics that contain only carbon and hydrogen in their molecules (polyethylene, polypropylene, polystyrene, and polybutylene) can have as final combustion products *only* water, carbon dioxide, carbon monoxide, and unburned carbon (which is what produces the blackness in black smoke). There may be some intermediate combustion products such as acrolein, formaldehyde, acetaldehyde, propionaldehyde, and butyraldehyde (hereinafter referred to as the first four aldehydes); some researchers may dispute this, since pyrolysis does not occur as readily with plastics as is does with wood, because of the melting of plastics and their possible flowing away from the heat source. Some claim that benzene is liberated as an intermediate combustion product when polystyrene burns, but it is clear that you can have no products containing nitrogen, chlorine, or fluorine from these products.

When polyethylene, polypropylene, or polybutylene burns, you will detect a faint odor of burning candle wax, since these polymers are indeed related to paraffin. These thermoplastics also burn with a dripping flame, so that you could observe a pool of molten polyolefin. On the other hand, polystyrene burns, liberating great masses of soot, along with a faint odor of natural gas, although there is no natural gas in polystyrene.

Plastics Containing Carbon, Hydrogen, and Oxygen

This group of plastics includes acetal, acrylics, cellulosics, polycarbonate, polyesters, phenolics, and epoxy. The only possible final combustion products are carbon, carbon dioxide, carbon monoxide, and water. There is always the possibility that acrolein and the first four aldehydes may be formed as intermediate combustion products, although the manufacturers of such plastics, particularly acetal, may argue otherwise. Tests done by one manufacturer showed no other combustion products than the final ones produced from burning acetal. With the long chains that are formed by polymerization, however, there may be certain existing combustion conditions that produce the intermediate combustion products in some of the other polymers, particularly the thermosets.

Plastics Containing Chlorine in the Molecule

There are at least four such thermoplastics, including polyvinyl dichloride (chlorinated polyvinyl chloride), polyvinylidene chloride (Saran™), chlorinated polyethylene (CPE), and polyvinyl chloride, that contain chlorine in the molecule as well as carbon and hydrogen. In this case, the possible final combustion products are carbon, carbon dioxide, carbon monoxide, water, and hydrogen chloride, an irritant gas that forms hydrochloric acid when it dissolves in water. The possible intermediate combustion products are the first four aldehydes plus acrolein. The biggest controversy over combustion products of thermoplastics in this group is with the thermoplastic of largest volume, polyvinyl chloride. It is a fact that hydrogen chloride is formed when polyvinyl chloride burns, but it is *not* true that phosgene and hydrogen cyanide are formed. Although phosgene contains chlorine, there have been countless tests performed on the combustion gases formed when PVC burns, and phosgene has never been found. As far as hydrogen cyanide is concerned, since there is no nitrogen present in the PVC molecule, hydrogen cyanide *cannot* be one of its combustion products, intermediate or final. PVC cannot be the source of ignition for a fire. PVC is very difficult to ignite and will not burn freely unless there is a supporting flame. Some highly plasticized PVCs will burn more readily than others, but, as the heat from another source contacts PVC, there is a breaking of covalent bonds, as in other organic compounds. When chlorine comes off the molecule as a free radical, it actually becomes a free radical scavenger, and, as such, will prevent further ignition and may even cause the fire to go out!

Plastics Containing Fluorine in the Molecule

The plastics in this group are the fluoroplastics, and there are several of them. Final combustion products (if, indeed, you can get the plastic to burn) will include carbon, carbon dioxide, carbon monoxide, water, and hydrogen fluoride; the intermediate products will be the first four aldehydes and acrolein.

Plastics Containing Nitrogen in the Molecule

This group of plastics includes acrylonitrile-butadiene-styrene (ABS), polyacrylonitriles (PAN), nylons, polyurethane, cellulose nitrate, and the amino plastics. When these materials burn, the final combustion products include carbon, carbon dioxide, carbon monoxide, water, and hydrogen cyanide; however, there is no hydrogen cyanide liberated when cellulose nitrate burns. Since hydrogen cyanide is flammable, the nitrogen oxides (NO_x) may also possibly be formed. Intermediate combustion products will include acrolein and the first four aldehydes. There is no plastic within this group that has chlorine in its molecule; therefore there are no plastics in this group that will liberate hydrogen chloride.

Extinguishment of Fires Involving Plastics

Plastics are Class A materials and can be extinguished by Class A extinguishing methods and agents, principally water. There may be an exception, when liquid plastics are involved, or when the burning thermoplastic melts (remember, thermosets do not melt, once they are formed into a shape of some kind) and begins to flow. At this point, you may wish to treat the molten plastic as a Class B (liquid) material, and use Class B extinguishing methods or agents. If not, water applied as a fog will quickly extinguish the burning molten plastic, and cool it so that it quickly resolidifies.

Locations Involving Burning Plastics

Plastics are found everywhere and are found in countless applications. In a fire involving residences, although there may be many articles in the home made of plastics, the fire may be treated as you treat any structural fire. If there are any plastics present that have chlorine or nitrogen in their molecules (and there will be), the chance of detecting hydrogen chloride in the fire gases is small, while the chance of detecting hydrogen cyanide is much higher, since there are more natural materials contributing to the liberation of hydrogen cyanide. In any event, the combustion gas that is known to be the killer in such fires is carbon monoxide; in such situations you should protect yourself by wearing your self-contained breathing apparatus (SCBA).

If you must respond to a fire where plastics are stored in great quantities, such as a warehouse, a plastics processing plant, or a plastics manufacturing plant, the vast majority of the material involved in the fire may be plastics, with very little natural material (as there would be in the residence fire). In such a case, there indeed may be dangerously high concentrations of hydrogen chloride in the fire gases, where burning polyvinyl chloride or other chlorine-containing thermoplastics are involved. In a case when large amounts of ABS, nylons, or other nitrogen-containing plastics are involved, there will be a significantly large amount of hydrogen cyanide generated, and a large amount of the nitrogen oxides present as the hydrogen cyanide burns. Again, as in any fire, wear your SCBA during the fire *and* during overhaul, to protect yourself.

Summary

It is difficult to answer specific questions about plastics without knowing which plastic is being asked about. Whenever the question is asked: "What burns hotter, a pound of plastic or a pound of wood?" the answer is, "That depends on what plastic and what wood, and are they in the same shape?" Another question is "What burns faster, a piece of plastic or a piece of wood?" Again, the answer is the same. "Which gives off more dangerous combustion products, wood or plastics?" The same answer: it depends on which plastic and which wood. There are some woods such as red oak that liberate significant amounts of hydrogen cyanide, and many plastics that liberate none, including *all* the plastics that contain no nitrogen. You must also take into consideration the compounding ingredients added to various plastics, and what their effect is on the resulting fire, if any. Some additives may add to the combustibility of a plastic (such as the addition of certain plasticizers to PVC), while others may aid the plastic in resisting fire (as in the addition of flame-resistant materials to polyurethanes).

Firefighters are mostly concerned about the amounts of plastics used in buildings today, and rightfully so. The very nature of residence fires has changed over the last forty years; it is noticeable in the way the fires appear when firefighters arrive. This should, however, be of no concern to the firefighter who uses the protective equipment that he is supposed to when he enters a burning building or in the overhaul procedure when the fire has been extinguished but may still be smoldering. He *must* wear his SCBA at all times, regardless of whether plastics are present or not. The stories that the old "smoke eaters" tell are greatly exaggerated; they may have indeed fought many fires inside buildings many years ago without SCBAs, but they do *not* tell you how many times they were carried from the building, overcome by smoke from natural materials. Nor do they tell you of how sick they became, if they were not overcome. Furthermore, have you noticed that the "smoke eaters" are all gone now? *Wear your masks!*

Glossary

Additive: Any material mixed with a plastic or resin to modify its properties.

Alloy: A blend of polymers and/or copolymers.

Atom: The smallest particle of an element that can still be identified as the element.

Blow Molding: A process in which hot plastic is forced against the inside of a mold by using a blast of air against it.

Calendering: The forming of a plastic into film or sheet by forcing the plastic through a series of rolls called a calender.

Casting: The forming of shapes by pouring a liquid plastic into a mold or on to a belt.

Catalyst: A substance used to control the speed of a chemical reaction but which is not consumed in that reaction.

Cellular Plastic: A plastic that has been processed to contain many empty cells throughout its mass: also referred to as foamed plastic.

Chain Length: The number of repeating monomer units in a polymer molecule.

Chemical Resistance: The ability of a material to resist chemical reaction by active chemicals.

Compound (Plastic): A mixture of resin and the necessary additives to give the resin the required properties of the finished part.

Compression Molding: A process in which a sheet of thermoplastic is placed in a molding cavity while heat is applied to the mold and pressure is applied to the sheet.

Copolymer: A compound formed when two chemical monomers are polymerized together.

Covalent Bond: The sharing of two electrons between the atoms of two non-metallic elements.

Cross-Linking: A process in which chemical links are set up between polymer chains. The process occurs mostly in thermoset resins, or, by cross-blending fillers and other additives with a thermoplastic resin, the conversion will be to a thermosetting resin.

Depolymerization: The process by which a polymer reverts to its monomer, or to a polymer of a lower molecular weight, usually occurring upon the polymer's exposure to high temperature.

Elastomer: A cured compound that in its final shape and at room temperature may be stretched to at least twice its length and will return to its original shape rapidly and with some force.

Extruder: A machine that forms continuous sheet, film, rods, or profiles by the action of a screw rotating (or, in an obsolete machine, a ram or reciprocating plunger) in a barrel, carrying forward and forcing the fused plastic mass through a die that imparts the shape to the plastic as it cools.

Extrusion: The process whereby a shape is imparted to a plastic material by forcing the molten mass through a die.

Extrusion Blow Molding: The process in which a parison, or hollow tube, is extruded and placed in a mold, and the parison is forced to assume the shape of the mold by a blast of air inside the parison.

Filler: A material added to a resin or plastic to alter its properties, but usually added to reduce its cost.

Film Blowing: The extrusion of a hollow tube followed by continuous inflation of the tube by internal air pressure (no mold is involved).

Film Casting: The pouring of a fluid plastic compound or resin on an endless carrier, followed by removal of the solidified film from the carrier.

Flux: To make fluid by melting or fusing.

Free Radical: An atom or group of atoms bound together chemically with at least one unpaired electron. A free radical is formed by the introduction of energy to a covalently bonded molecule, when that molecule is broken apart by the energy. It cannot exist free in nature and therefore must react quickly with other free radicals present.

Hardener: A substance that brings about the curing of a plastic.

Homopolymer: The polymer produced by polymerization of a single monomer.

Hydrocarbon Plastics: Plastics made from resins or monomers containing only hydrogen and carbon.

Impact Resistance: The relative ease with which plastic parts break under high-speed stress applications.

Inhibitor: A substance used to prevent a chemical reaction from starting; also called a stabilizer.

Initiator: A substance used to start a chemical reaction.

Injection Molding: The process in which plastic parts are formed by forcing fused resin or compound into a mold by means of a ram or reciprocating screw.

Inorganic Polymer: A polymer without carbon in its backbone.

Ionomer: A thermoplastic composed mainly of polyethylene, containing both covalent and ionic bonds.

Isomer: A compound with a molecular formula identical to another compound but with a different structural formula. That is, a compound may possess exactly the same elements, and exactly the same number of atoms of those elements as another compound, but those atoms are arranged in a different order from the first compound, having different molecular structures of these atoms, and therefore possessing different chemical and physical properties.

Laminate: A bonding together of two or more layers of materials, usually either all plastic or plastic and non- plastic materials.

Latex: A water dispersion of a polymeric material.

Lay-Up Molding: A process in which fluid resin is applied to a layer of reinforcing material, cured, and then formed with or without the use of pressure.

Linear Polymer: A polymer that has the appearance of a chain, with little or no side branching.

Macromolecule: The giant molecule formed by polymerization.

Mer: The repeating molecular unit in any polymer.

Mold: To shape a plastic by confining it in a closed cavity.

Molecular Weight: The total weight of all atoms in a molecule.

Molecule: The smallest particle of a compound that can still be identified as the compound; two or more atoms bound together chemically by covalent bonds and electrically neutral.

Monomer: A simple small molecule that has the special capability of reacting with itself to form a giant molecule called a polymer (*mono:* one; *mer:* part).

Olefin Plastics: The polyolefins (polyetheylene, polypropylene, and polybutylene).

Organic Peroxides: A group of highly dangerous hazardous materials; used as initiators for thermoplastics and curing agents for thermosets, they are highly reactive oxidizing agents that burn, and they can start their own decomposition process when contaminated, heated, or shocked.

Parison: In blow molding, the hollow tube that is extruded or injection molded, placed inside a mold, and forced to assume the shape of the mold by an internal blast of air.

Plastic: American Society for Testing and Materials definition: "A material that contains as an essential ingredient an organic substance of large molecular weight, is solid in its finished state, and, at some stage in its manufacture or in its processing into finished articles, can be shaped by flow." This definition excludes inorganic plastics and includes rubbers and elastomers.

Plasticizer: A material added to a resin or polymer to increase its flexibility.

Plastisol: A suspension of polyvinyl chloride in a liquid plasticizer.

Poly: A prefix meaning many; a polymer means many (*poly*) parts (*mer*).

Polymer: A "giant" molecule made up of thousands of tiny molecules linked together in a long chain.

Polymerization: The chemical reaction in which a special compound, called a monomer, combines with itself to form a long-chain molecule called a polymer.

Premix: A mixture of polyester resin and fillers.

Pre-Polymer: A polymer of intermediate molecular weight, somewhere between the monomer and the final polymer.

Prepreg: A mat of reinforcing fibers that have been impregnated with a thermosetting resin partially cured.

Pyrolysis: The breakdown of a molecule by heat; pyrolysis of polymers can produce shorter-chain (lower molecular weight) polymers of the original monomer.

Radical: An atom or group of atoms bound together chemically that has one or more unpaired electrons; it cannot exist in nature in that form, so it reacts very fast with another radical present, to form a new compound; also known as a "free" radical.

Regrind: Rejected plastic parts and any other parts of the finished molding removed from the part (or anything classified as scrap) that are ground up into small particles to be reprocessed with virgin material.

Reinforced Plastic: A plastic resin or compound to which fillers are added to greatly increase the strength or other physical properties of the finished part.

Resin (Natural): Solid or semi-solid viscous substances, that are secretions of certain plants and trees.

Resin (Synthetic): ASTM definition: "A solid, semisolid, or pseudosolid organic material which has an indefinite and often high molecular weight, exhibits a tendency to flow when subjected to stress, usually has a softening or melting range, and usually fractures conchoidally."

Rotational Casting: The forming of plastic articles by adding to a rotating mold a fluid plastic material, rotating the mold so the fluid is distributed on the mold walls, and heating until the plastic material has hardened.

Rotational Molding: The forming of plastic articles as in rotational casting, except that a dry powdered plastic is used.

Set: To change from an uncured state to a cured state.

Solid Casting: The process of pouring a liquid resin into a mold and curing the resin to a solid part.

Spray-Up: The application of resin and reinforcement into a mold by use of a spray gun.

Synthetic: Made by man, as opposed to being made naturally.

Thermoforming: The process of forming a plastic sheet into a finished part by clamping it on to a frame over a mold, causing it to soften by heating, and then applying pressure to make it conform to the mold.

Thermoplastics: Resins of compounds that may be formed over and over again upon applying heat and pressure.

Thermosets: Resins or compounds that may be formed and cured once; any subsequent heat will cause the plastic to degrade or burn.

Transfer Molding: The process of placing resins or plastic compounds in a heated vessel above the mold, where once fluxed, they can be forced into the mold by the pressure of a ram or plunger; usually used for thermosets.

Vacuum Forming: A process used in thermoforming, where the pressure applied is that of a vacuum.

13

Corrosives

Introduction

The corrosives, as a class of chemicals used by industry, probably represent the largest volume of materials used in the manufacturing portion of American business. The most valuable chemical, in terms of its usefulness to industry, not only in the United States, but also in the rest of the world, is a powerful corrosive, sulfuric acid, which is valuable precisely for its corrosiveness. There are several classes of corrosives, which we will look at one at a time. The different hazardous materials that we will be defining as corrosives are acids, bases, the halogens, and other chemicals which possess, as one of their hazards, a tendency to be corrosive. Before we examine these individual hazardous materials, let us define some of the terms we will need to discuss them.

A corrosive is defined as any material that will attack and destroy, by chemical action, any living tissue with which it comes into contact (see figure 44). There are many hazardous materials that fall into this definition, one broader than the DOT definition, which includes a statement about having a "severe corrosion rate on steel." There will be many corrosives we shall cover that will not attack steel, since our main concern is chemical action on humans. Even if there is no direct contact between a human and the corrosive material, however, the chemical action on the steel, or any other material, can have serious consequences for people, whether in the catastrophic collapse of a structure because of corrosion, or as the toxic byproducts of the chemical reaction between a corrosive and some other material.

An acid is defined as a compound that contains one or more "loosely held hydrogens," and is very active chemically. Admittedly, this is a very poor definition for such an important class of hazardous materials, but there really exists no good one anywhere that is not extremely technical, complicated, and difficult to understand. Some definitions include the statement that acids are compounds that contain the hydrogen ion, but that is not exactly correct. The ion that is formed when acids are dissolved in water is the *hydronium* ion, whose molecular formula is H_3O^{+1}, which, of course, looks like the hydrogen ion, H^{+1}, dissolved in water. That is why it is often referred to as the hydrogen ion. Even the definition of pH, which is a value that represents the *acidity* or *alkalinity* of a water solution, refers to the hydrogen ion, rather than the hydronium ion. Therefore, we will also refer to the hydrogen ion in our discussions, as long as you realize that it is not technically correct. We will, however, keep our rather poor definition of an acid, since it indicates that hydrogen is an extremely important part of the compound and is liberated as hydrogen gas when acids come into contact with many metals. Just keep in mind that there are other, more technically correct definitions of acids used by chemists. If you are interested in what these are, you should consult any college textbook on general chemistry.

Bases, more commonly known as *caustics*, are defined as compounds that contain the hydroxide ion OH^{-1} and are active chemically. In this case, the designation of the hydroxide ion is correct, in contrast to the confusion between the hydrogen and hydronium ions. The bases, since they are ionic compounds, and therefore inorganic compounds, are known as the chemical opposites of inorganic acids; the chemical reaction that demonstrates this is the *neutralization* reac-

Figure 44
Placard and Label for Corrosive Material.

tion, in which an acid and a base react to form a salt and water. Certain bases may be as powerfully corrosive as many acids, so do not make the mistake of limiting your concept of corrosives to acids alone.

The *strength* of an acid or base is defined as the degree of ionization of the acid or base in water. Remembering your ionic chemistry from the earlier chapters, you will recall that an ion is defined as an atom (a simple ion) or group of atoms bound together (a complex ion) that has collectively gained or lost an electron (or electrons) and therefore possesses an electrical charge equal to the number of electrons gained or lost. The positively charged ion is called a cation, while the negatively charged ion is an anion. The compounds that form acids are *covalently* bonded but *ionize* when dissolved in water. The bases are all ionic compounds to begin with and, when dissolved in water, the ions separate, as in the solution of any ionic compound. Therefore, when you have a typical *acid gas*, like hydrogen chloride, you will find upon examination that it, like all gases, is molecular in nature; that is, it is covalently bonded, so that there are no hydrogen ions or chloride ions in hydrogen chloride. As soon as the hydrogen chloride is dissolved in water, however, the solution instantly becomes ionic; once the gas is dissolved in water, there are hydrogen ions and chloride ions in tremendous quantities. As a result, we say that the *hydrochloric acid* formed by the solution of hydrogen chloride in water is a *strong* acid, since the ionization of the hydrochloric acid is so extensive (essentially 100 percent). Hydrochloric acid is an inorganic, or mineral, acid. On the other hand, an organic acid such as acetic acid, a covalent compound like hydrogen chloride except that it is a liquid, will ionize to a very slight degree in water (0.0001 percent), and it is therefore classified as a *weak* acid. The inorganic acids, all of which ionize to one degree or another, are generally much *stronger* acids than the organic acids. Most of the inorganic acids ionize nearly 100 percent, so that they are all *very* strong acids.

Another term generally used with acids is *concentration*, which is defined as the percentage of acid dissolved in water. The two terms, concentration and strength, are not related to each other but are often used interchangeably. You may have a low concentration of a strong acid, or a high concentration of a weak acid; the chemical activity (corrosiveness) of the two may be equal, or they may be significantly different. Generally, when you have equal concentrations of two acids, one of which is strong and the other weak, the strong acid will be more corrosive, since it will be more active chemically. But corrosiveness is the only hazard under examination here. Later we will see that there are many hazards of acids, and corrosiveness is the only one linked to strength or ionization.

Hazards of the Acids

The acids present one of the longest lists of hazards of any class of hazardous materials. Again, just as when lists of the hazards of other hazardous materials are presented, it is extremely rare, if not impossible, for a member of that hazard class to possess *all* the hazards attributed to the class. Since there exist both inorganic and organic acids, their hazards will naturally be different.

The hazards of the acids are the following. They may be:

1. corrosive
2. explosive
3. polymerizable
4. oxidizers
5. water-reactive
6. toxic
7. flammable
8. very reactive
9. unstable

As a general rule, the organic acids are flammable, while the inorganic acids will not burn. The inorganic or mineral acids generally ionize considerably more than do the organic acids, so the inorganic acids are stronger. Just do not be misled into believing that highly concentrated organic acids are not corrosive, since that is not true. What *is* true is that the inorganic acids are stronger acids than the organic acids *at the same concentration*. A few of the inorganic acids are oxidizing agents, while some of the organics are, also. The difference is that the organic acids are not used for their oxidizing power as the inorganic acids are; therefore, some of you might be surprised when the organics may provide oxygen to the reaction. As a rule, the inorganic acids are more chemically reactive than the organic acids, although some of the organic acids may actually polymerize. The organics are usually the toxic acids and the explosive acids. An inorganic acid may be involved in an explosion, but that is because the acid has supplied oxygen to the reaction, while some organic acids will explode. The water-reactive acids are usually inorganic. The inorganic acids are much more stable than the organic acids; one of the organic acids is actually an organic peroxide, and *all* organic peroxides are unstable.

Inorganic Acids

Sulfuric Acid, H_2SO_4

Far and away the most important acid and, indeed, the most important chemical in the world is sulfuric acid. More than 4,000,000 *tons* of sulfuric acid were produced in the last year for which there are records. No other chemical in the world is made in these quantities, and no other chemical is used in more processes than sulfuric acid. It is one of the strongest acids known and is used as a corrosive agent as well as a source of sulfates and for many other purposes. It is a highly water-reactive material and will liberate tremendous amounts of heat when it contacts water. A rule in the chemistry lab is that you may add concentrated sulfuric acid to water, but you must *never* add water to concentrated sulfuric acid. The reasoning is that with the first small amount of water that contacts the sulfuric acid, enough heat is generated to boil the mixture of liquids, and the vapor thus formed causes the mixture, highly concentrated sulfuric acid, to be forced up and out by the expanding vapor. Anyone nearby will be splashed with this powerfully corrosive material. On the other hand, if sulfuric acid, even concentrated, is added to water, the resulting rise in temperature will not be so high nor so dramatically fast, since the amount of concentrated sulfuric acid that will contact the water is small, and water is a great absorber of heat. Remember that the specific heat of water is much higher than that of most other materials. Therefore, even if the boiling point of water is reached (which might happen quite rapidly), what will boil up is considerably weaker sulfuric acid, which, although it might still be corrosive, will not be nearly as concentrated as the sulfuric acid splashed out in the previous example. Does this mean that, in an emergency involving sulfuric acid, you will not be able to use water? The answer depends entirely on the circumstances of *that* particular incident, just as in any other hazardous-materials incident. If the amount of sulfuric acid released is small, and in an area where you need not be concerned about the run-off, you may use a large amount of water to dilute the acid to a harmless concentration—a method which may require a lot of water. Just be sure the water can be applied from a distance and that any splashing of the acid will not harm exposures.

If, on the other hand, the amount of sulfuric acid released is very large, you may not be able to dilute the sulfuric acid to a safe level, even with unlimited amounts of water at your disposal. It will take an enormous amount of water, perhaps 100 times the amount of the spill, to render the acid so dilute that it will be *relatively* harmless, and, in this case, you may have a problem with the run-off. If you cannot create a large enough pit to contain the resulting amount of dilute acid, you may not want to take this alternative of handling the emergency. It is a good practice *never* to allow run-off water that has been used to control a hazardous-materials incident to enter a sewer system. Remember that, many times, as in the case of corrosives, you are flushing a very active chemical into a sewer system that might possess other active chemicals; common sewage and all other organic materials that have previously entered the sewer are just waiting for an active chemical to react with. This approach suggested for sulfuric acid may be used with any of the strong acids, even though the others are not water-reactive like sulfuric acid (although there is at least one exception—chlorosulfonic acid).

Sulfuric acid has many uses, including processes involving fertilizers, pigments and dyes, petroleum refining, electroplating baths, iron and steel production, rayon manufacture, industrial explosives, laboratory reagents, non-ferrous metallurgy, etchants, and many other chemicals.

Sulfuric acid is considered toxic, mainly because its corrosiveness destroys so much tissue, and vapors released from it are strong irritants. It will do great damage to many metals (and, needless to say, to any living tissue); hydrogen will be released as a result of its reaction with metals. This release of highly flammable hydrogen when sulfuric, and most inorganic, acids, come into contact with most metals, creates a *very* dangerous situation. In many instances, the heat of the reaction may ignite the hydrogen; in other cases, an accumulation of hydrogen may occur, which, of course, presents a highly explosive situation.

In relatively small amounts, depending upon the use for which the sulfuric acid is intended, it is shipped in carboys (glass or plastic bottles encased in a wooden or cardboard crate). It is also shipped in bottles, drums, tank trucks, tank cars, and tank barges, using specially lined containers for shipments of the pure sulfuric acid. The implication here is frightening, since sulfuric acid may be encountered in transportation incidents in truckload quantities of many small glass bottles, each holding approximately a pint of the acid, up to rail cars loaded with thirty *thousand* gallons!

Sulfuric acid is colorless to dark brown, depending upon the purity of the acid. This is also a description of *oleum*, or *fuming* sulfuric acid, which is sulfuric acid that has sulfur trioxide dissolved in it. Sulfuric acid is made by dissolving sulfur trioxide in water, and *fuming* sulfuric acid is *concentrated* sulfuric acid with an *excess* of sulfur trioxide forced into solution. It is more hazardous than sulfuric acid, and the fuming that it undergoes produces highly toxic vapors; it is used in some of the same applications as sulfuric acid and is shipped in similar containers.

Nitric Acid, HNO₃

Nitric acid was the chemical of tenth highest volume in the United States in 1979 and is extremely important in the chemical industry and many other industries. It is extremely corrosive and an extremely powerful oxidizing agent. Nitric acid is a transparent to yellow liquid that will attack nearly all metals. It is used to manufacture nitrates for explosives, fertilizers, and any other purpose suitable for the nitrate ion; on occasion it will provide the nitration (the attachment of a nitro functional group onto a hydrocarbon backbone) for the production of such organic nitrates as dyes, drugs, and explosives. It is also used in metallurgy, photoengraving, etching steel, ore flotation, and in processing spent nuclear fuel. Nitric acid is shipped in carboys, bottles, tank trucks, and tank cars.

To repeat, nitric acid is very corrosive, highly toxic, and a powerful oxidizing agent, and it is almost 100 percent ionized. It should never be allowed to come into contact with any combustible material. Another name for it is aqua fortis. When nitric acid decomposes in sunlight, nitrogen dioxide is liberated, and the color of the acid may change to yellow.

There are two fuming nitric acids, white fuming nitric acid (WFNA) and red fuming nitric acid (RFNA). The vapors are the released oxides of nitrogen. Both fuming acids are more dangerous than regular nitric acid.

Nitric acid is an extremely powerful corrosive. It will do tremendous damage to most metals, and, when it does attack metals, there may be more than one reaction, in addition to the corroding (dissolving) of the metal. In case nitric acid contacts ferrous (iron-containing) metals, hydrogen gas will be liberated. Care must be taken to keep all ignition sources away from spills where inorganic acids may be contacting metals, because of the danger of ignition of the liberated hydrogen. In many cases, rather than a breakdown of the acid to release hydrogen, nitric acid may decompose, liberating the reddish-brownish group of toxic gases known as the nitrogen oxides (NO_x). Care must be taken not to expose anyone to these gases, for the effects of inhaling them are almost always delayed and sometimes fatal.

Although nitric acid is such a powerful corrosive, its major hazard is that it is an extremely efficient oxidizing agent; it is this property that makes it so valuable. Nitric acid is used principally as an oxidizing agent, rather than as a corrosive. It is regulated as both an oxidizer and as a corrosive.

The Halogen Acids

There are four halogen acids, corresponding to the four halogens: hydrofluoric acid, hydrochloric acid, hydrobromic acid, and hydriodic acid. They are very strong acids, nearly 100 percent ionized. The covalent gases (called acid gases) that are dissolved in water to form these acids are hydrogen fluoride, hydrogen chloride, hydrogen bromide, and hydrogen iodide. All four gases are extremely soluble in water; the danger to anyone caught in a highly concentrated atmosphere of these acid gases is that the gases will dissolve in any moisture on the person's body—the eyes, nose, mouth, underarms, crotch, or any other part of the body that is moist—forming the acid. All of these gases are covalently bonded but ionize upon contact with water.

Hydrochloric acid, HCl, is the most common of the four, being the chemical of twenty-fourth largest volume in the United States in 1979. It is a colorless to slightly yellow liquid with a pungent odor. Like all the other inorganic acids, hydrochloric acid is completely soluble in water and, again like all the other inorganic acids, is non-flammable. It is used as a chemical intermediate in many processes, including ore reduction, food processing, pickling and metal cleaning, as an alcohol denaturant, a lab reagent, and in many other manufacturing operations. It is shipped in carboys, glass bottles, rubber-lined steel drums, and tank cars. An old name for hydrochloric acid is muriatic acid. Hydrochloric acid is also the acid in your stomach that aids in digestion.

Hydrofluoric acid, HF, is a colorless, fuming liquid. It is extremely corrosive and will attack glass as well as metals. It is shipped in polyethylene bottles, polyethylene-lined steel drums, and tank cars. Hydrofluoric acid is used in the production of aluminum, fluorocarbons, pickling stainless steel, etching glass, acidizing oil wells, gasoline production, uranium processing, and in any chemical process requiring the fluoride ion, or the attaching of the fluoride functional group to a hydrocarbon.

Hydrobromic acid, HBr, is a colorless to yellow liquid that is sensitive to light. It is used as a solvent for ore minerals, in analytical chemistry, and in the manufacture of organic and inorganic bromides; it is shipped in carboys and glass bottles. Use of hydrobromic acid is considerably less than that of either hydrofluoric acid or hydrochloric acid.

Hydriodic acid is a colorless to yellow liquid and a very strong acid, as are the other halogen acids. It is used in the preparation of iodine salts, organic chemicals, disinfectants, analytical reagents, and pharmaceuticals. It is shipped in carboys and glass bottles.

The four halogen acids are all very strong, corrosive acids that are exceedingly important to commerce. They should all be treated with great care in handling and storage; emergencies involving releases of these acids will be very serious. The corrosive action of these materials on most metals will release hydrogen, as will the action of most other inorganic acids. Again, care must be taken to avoid ignition of this gas.

The Chlorine-Containing Oxyacids

All acids may be viewed as derivatives of salts, with the hydrogen replacing the metal ion. Indeed, the word acid originally came from a word that translated as "salt-former." The chlorine-containing oxyanions, the perchlorates, chlorates, chlorites, and hypochlorites will form the corresponding acids, perchloric acid ($HClO_4$), chloric acid ($HClO_3$), chlorous acid ($HClO_2$), and hypochlorous acid ($HClO$).

Perchloric acid is a fuming, colorless liquid, the strongest of the mineral acids, and the most corrosive. It is a stronger oxidizing agent than nitric acid and is extremely dangerous in contact with any combustible material. It is shipped in carboys and glass bottles and is used as a catalyst, in the manufacture of esters, for electro-polishing, and in the manufacture of explosives. If heated or shocked, perchloric acid will explode.

The other acids are all strong oxidizing agents, and are extremely corrosive. They are shipped in carboys and glass containers; each has a specialized use, which is its primary reason for being in the stream of commerce.

Phosphoric Acid, H_3PO_4

The chemical produced in ninth largest volume in the United States is phosphoric acid. It may be a colorless liquid or a transparent crystalline solid, depending on concentration and temperature. It is used in countless chemical processes, including the manufacture of fertilizers, soaps, detergents, pickling and rust-proofing metals, phar-

Table 22 / Other Inorganic Acids

arsenic acid	dithionic acid	orthoperiodic acid
arsenous acid	dithionous acid	orthosilicic acid
bromic acid	hypobromous acid	orthotelluric acid
bromous acid	hypoiodous acid	perchromic acid
carbonic acid	hyponitrous acid	periodic acid
cyanic acid	hypophosphorous acid	permanganic acid
chlorosulfonic acid	hyposulfurous acid	perrhenic acid
chromic acid	iodic acid	pertechnic acid
dichloroacetic acid	isocyanic acid	phosphorous acid
dichromic acid	manganic acid	rhenic acid
dichloropropionic acid	metaboric acid	selenic acid
difluorophosphoric acid	metaphosphoric acid	sulfoxylic acid
diphosphoric acid	metasilicic acid	sulfurous acid
diphosphorous acid	nitrous acid	technetic acid
disulfuric acid	nitroxylic acid	triphosphoric acid
disulfurous acid	orthoboric acid	

maceuticals, soft drinks, and for many other purposes. It is corrosive to ferrous metals and alloys and is shipped in carboys, drums, barrels, tank trucks, and tank cars.

Other Inorganic Acids

There are many other inorganic acids, and all are important in their own ways. They are all corrosive and present hazards in their individual ways. Table 22 is not an all-inclusive list, but it does attempt to include as many as possible.

Organic Acids

Carbolic Acid, C_6H_5OH

Carbolic acid is the common name for phenol, a white crystalline solid that absorbs moisture from the air and liquifies. It has a flash point of 172°F. and is soluble in water. It is highly toxic by ingestion, inhalation, and absorption and is a strong tissue irritant. It is used in the production of phenolic resins, epoxies, nylon, picric acid, pharmaceuticals, disinfectants, and many other chemicals. It is shipped in drums, tank trucks, and tank cars. For years doctors used carbolic acid as a disinfectant, washing their hands with it regularly. Eventually, its toxic qualities became known, and, although it still has use in commercial disinfectant solutions and medicines, doctors no longer use it on themselves.

Hydrocyanic Acid, HCN

Hydrocyanic acid is a solution of hydrogen cyanide in water. It is a deadly poison and has the smell of bitter almonds. It is shipped in bottles, steel cylinders, and tank cars. The tank cars, by agreement among the manufacturers of the acid, are painted white, with a red horizontal stripe and two vertical red stripes around the car. This "candy striper" is highly distinctive and can be seen from a distance. Hydrocyanic acid is highly toxic by ingestion, inhalation, and absorption. It is used in the manufacture of acrylonitrile, adiponitriles, cyanide salts, dyes, chelates, rodenticides, and pesticides. Hydrogen cyanide is a flammable gas; when its solution in water is not pure or not stabilized, hydrocyanic acid will polymerize violently.

Acrylic Acid, $CH_2CHCOOH$

Another acid that will polymerize violently is acrylic acid. Like all the organic acids, acrylic acid will burn, and its flash point is 130°F. It is a colorless liquid with an acrid odor and is soluble in water. It is toxic by inhalation and corrosive to the skin. Acrylic acid is a monomer and, as such, is unstable. It must be *inhibited* during shipment; otherwise, it will be subject to the runaway polymerization explosion typical of monomers. It is shipped in bottles, drums, and tank cars and is used as a monomer for polyacrylic acid, polymethacrylic acid, and other acrylic polymers.

Methacrylic Acid, $CH_2:C(CH_3)COOH$

Methacrylic acid is another acid that is a monomer and poses the same problems as acrylic acid. Methacrylic acid has a flash point of 170°F. and is a strong skin irritant. It is shipped in bottles, drums, and tank cars and is used as a monomer for the large-volume acrylic polymers. It, like all monomers, will polymerize violently if subjected to heat and loss of its stabilizing inhibitor.

Peracetic Acid, CH_3COOOH

Peracetic acid is a colorless liquid with a strong odor and is soluble in water. It has a flash point of 105°F., but, more important, the structural formula of CH_3COOOH shows that, in addition to being an organic acid, it is also an organic peroxide. This means it is a combustible liquid that carries its own oxygen supply with it (a combustible oxidizing agent!). DOT recognizes this and requires that peracetic acid be shipped as an organic peroxide. It is a strong irritant and a powerful oxidizing agent and is therefore extremely dangerous when in contact with combustible materials (which, of course, it is). It is shipped in carboys and aluminum drums. Peracetic acid is used to bleach (oxidize) textiles, paper, oils, waxes, and starches. It is a polymerization catalyst and is used to make bactericides and fungicides. It has too many other uses to mention in full.

Picric Acid, $C_6H_2(NO_2)_3OH$

Picric acid is a solid material in the form of yellow crystals that are soluble in water. It will melt at 219°F. and will *explode* at 572°F. It is a true explosive; picric acid is shipped in one- and five-pound bottles, 25-pound boxes, 100-pound kegs, and 300-pound barrels. It is toxic by skin absorption and will explode when heated and/or shocked. It is used to make other explosives, matches, picrates, electric batteries, and reagents.

Acetic Acid, CH_3COOH

Acetic acid is a clear, colorless liquid with a familiar pungent odor. It is soluble in water and, in a very dilute state, is the acid in vinegar. The pure acid (glacial acetic acid) has a flash point of 110°F. and is a strong irritant to skin. It is used in the manufacture of acetic anhydride (the word anhydride means "no water present"), cellulose acetate and vinyl acetate monomers, dyes, insecticides, pharmaceuticals, photographic chemicals, food additives, and for many other purposes. It is shipped in bottles, drums, and tank cars.

Other Organic Acids

There are many more organic acids, and as in the list of additional inorganic acids, Table 23 will not be all-inclusive. It will be your responsibility to recognize these materials and their hazards.

Bases

The bases, as stated before, are the chemical opposites of acids. This is true only in ionic chemistry, since the covalent bonding of the organic

Table 23 / Other Organic Acids

acetylsalicylic acid	iodoacetic acid
adipic acid	isobutyric acid
alkanesulfonic acid	isopentoic acid
benzoic acid	lauric acid
bromoacetic acid	monofluorophosphoric acid
butyric acid	oleic acid
capric acid	oxalic acid
caproic acid	nitrohydrochloric acid
caprylic acid	palmitic acid
chloroacetic acid	performic acid
chlorobutyric acid	persulfuric acid
chloroisocyanuric acid	phthalic acid
chloropropionic acid	phenylacetic acid
cyanoacetic acid	phenylstearic acid
etching acid, liquid, n.o.s.	propionic acid
fluoboric acid	salicylic acid
fluoroacetic acid	sebacic acid
fluorosulfonic acid (fluosulfonic acid)	stearic acid
formic acid	thioglycolic acid
hexafluorophosphoric acid	trichloracetic acid
hexanoic acid	valeric acid
hydrofluorosilicic acid	

acids presents a totally different set of reactions. The bases are the hydroxides, those ionic compounds that contain the hydroxide ion (OH^{-1}).

Sodium Hydroxide, NaOH

The most common of the hydroxides is sodium hydroxide, a very powerful corrosive. It is also known as caustic soda, lye, and white caustic. The term caustic is used exclusively with the hydroxides and may appear as part of the common names of other hydroxides. Sodium hydroxide is a white solid that is deliquescent (will absorb water and liquify) and soluble in water. It is extremely corrosive to skin and will generate a great amount of heat when dissolved in water. It is toxic, and is shipped in one-pound bottles, five-pound and ten-pound cans, drums, and barrels. It may be shipped dissolved in water in drums, tank trucks, tank cars, and barges. Sodium hydroxide is used in the manufacture of rayon, cellophane, detergents and soaps, oil refining, reclamation of rubber, as a laboratory reagent, and in many other processes and in the manufacture of other chemicals.

Potassium Hydroxide, KOH

The second most important hydroxide is potassium hydroxide, also known as caustic potash or lye. It is so very similar to sodium hydroxide that all its chemical properties are the same. It is shipped in the same way and is every bit as corrosive and toxic as sodium hydroxide. Many of its uses are also the same, and it is used in the manufacture of many potassium compounds. It, too, generates a lot of heat when in contact with water, and this property is utilized as a drain opener, when it is added, with a slight amount of water, to a grease-clogged trap.

Other Bases

There are many other hydroxides, but only a few of them are commercially important and therefore shipped around the country. Lithium hydroxide, LiOH, is used in photographic developers, as an absorber of carbon dioxide in space vehicles, and in the manufacture of lithium soaps and greases. Calcium hydroxide, $Ca(OH)_2$ is also known as caustic lime and slaked lime. It is used in many manufacturing processes, including mortars, plasters, and cements. Both lithium hydroxide and calcium hydroxide are very corrosive and powerful skin irritants.

Remembering your ionic chemistry, you may take the hydroxide ion and combine it with any of the metallic ions in group I, II, and III, and you will form the corresponding base. Most of these bases are not commercially valuable, so they will not be shipped around the country in any appreciable quantities. Therefore, the chances are that you will not encounter them in a transportation incident, and we will not discuss them further. Some of them, however, *may* be used in a manufacturing site or may be stored in a warehouse in your jurisdiction, and, therefore, upon inspection of the premises, should be noted by you, along with the proper emergency handling procedures.

Other Corrosives

There are numerous other corrosive materials, many of which are included in other chapters under other hazards. For example, the halogens (fluorine, chlorine, bromine, and iodine) are highly corrosive materials, as well as very toxic and powerful oxidizing agents. The acid

Table 24 / Other Corrosives

acetyl bromide
acetyl iodide
acid butyl phosphate
acid, liquid, n.o.s.
acid, sludge
alkaline (corrosive) liquid, n.o.s.
alkaline battery fluid
alkaline battery fluid with empty storage
    battery
allyl trichlorosilane
aluminum bromide, anhydrous
aluminum chloride
2-(2-aminoethoxy) ethanol
N-aminoethylpiperazine
aminopropyldiethanolamine
N-aminopropylmorpholine
bis(aminopropyl) piperazine
ammonium bisulfite solution
ammonium fluoride
ammonium hydrogen fluoride
ammonium hydroxide
amyl acid phosphate
amyl trichlorosilane
anisole chloride
antifreeze compound, liquid
antifreeze preparation, liquid
antimony pentachloride
antimony pentafluoride
antimony tribromide
antimony trichloride
antimony trifluoride
benzene phosphorus dichloride
benzene phosphorus thiochloride
benzoyl bromide
benzoyl chloride
benzyl bromide
benzyl chloride
benzyl chloroformate
boron tribromide
boron trichloride
boron trifluoride-acetic acid complex
bromine pentafluoride

butyl trichlorosilane
calcium hydrogen sulfite solution
chloroacetyl chloride
chlorophenyltrichlorosilane
chromic anhydride
chromic fluoride
chromium oxychloride (chromyl chloride)
cupriethylene-diamine solution
cyclohexanyl trichlorosilane
dibenzoyl chloride
dichloroacetyl chloride
dichlorobutene
dichloroisopropyl ether
dichlorophenyltrichlorosilane
diisooctyl acid phosphate
dimethylchlorothiophosphate
dimethyl sulfate
diphenyl dichlorosilane
diphenyl methyl bromide
dodecyl trichlorosilane
ethyl chlorothioformate
ethylenediamine
ethyl phenyl dichlorosilane
ethyl phosrorodichloridate
ferrous chloride, solution
fumaryl chloride
hexachlorocyclopentadiene
hexadecyltrichlorosilane
hexanethylenediamine
hexamethyleneimine
hexyltrichlorosilane
hydrazine, aqueous solution
iodine monochloride
isobutyric anhydride
isopropyl acid phosphate
lead sulfate, solid
methyl dichloroacetate
methyl ethyl pyridine
methyl phosphonothioic chloride
methyl phosphorus dichloride
monoethanolamine
nonyltrichlorosilane

Table 24 / Other Corrosives (cont.)

octadecyltrichlorosilane	sodium monoxide
octyltrichlorosilane	sodium phenolate
phenyltrichlorosilane	stannic chloride
phosphoric anhydride	sulfur chloride (or monochloride)
phosphorus oxybromide	sulfur dichloride
phosphorus oxychloride	sulfur trioxide
phosphorus pentabromide	sulfuryl chloride
phosphorus pentachloride	1,2,3,6-tetrahydrobenzaldehyde
phosphorus tribromide	tetramethylammonium hydroxide
phosphorus trichloride	thionyl chloride
potassium bifluoride	thiophosphoryl chloride
potassium fluoride	tin tetrachloride, solid
potassium hydrogen fluoride solution	titanium sulfate solution
sodium aluminate	titanium tetrachloride
sodium bifluoride	trimethylacetyl chloride
sodium bisulfate	tris-(1-azirdinyl) phosphine oxide
sodium chlorite solution (not exceeding 42 percent sodium chlorite)	tungsten hexafluoride
	valeryl chloride
sodium fluoride	vanadium oxytrichloride
sodium hydrogen sulfate solution	vanadium tetrachloride
sodium hydrosulfide solution	zinc chloride solution
sodium methylate, alcohol mixture	zirconium tetrachloride

gases named earlier (hydrogen fluoride, hydrogen chloride, hydrogen bromide, and hydrogen iodide) must be classified separately as corrosives since they convert to corrosive acids upon contact with moisture, especially in and on a human being. In addition, many of the oxides and peroxides are corrosive. Other corrosives are listed in table 24. It will be your responsibility to become familiar with their properties and the proper emergency handling procedures when you discover them in your inspections. This knowledge will also prepare you for handling transportation incidents.

Emergencies Involving Corrosives

Unfortunately, there are no set guidelines for handling all corrosives when they have been released from their containers. Perhaps the best way to consider them would be to break them up into groups, according to some of their characteristics.

Inorganic Acids

The primary hazard here is corrosiveness, but you must also be aware of the possibility of the presence of oxidizing agents—and beware of all acid fumes.

There are several ways to handle these materials. We have already mentioned dilution. You may be limited because of inadequate water supply, inability to contain the run-off, or reaction to water. The conditions must dictate whether dilution is a viable alternative. Another handling procedure is neutralization. Here you want to render the acid harmless by adding another chemical. The strongest neutralizers for strong acids are the strong bases: sodium hydroxide, potassium hydroxide, and calcium hydroxide. Remember, however, that these are also hazardous materials and are so dangerous that their use as a neutralizer for inorganic acids is restricted to the laboratory, under controlled conditions. You may also use soda ash, which is sodium carbonate or baking soda, which is sodium bicarbonate; calcium carbonate, or ground limestone, is another candidate. The fizzing caused by these materials is the release of carbon dioxide by the acid's reaction with these materials. You will be limited here by the size of the release and the amount of neutralization material available, but use of sodium carbonate, calcium carbonate, or sodium bicarbonate is the recommended procedure. Every apparatus that has a chance of responding to a hazardous-materials incident should carry a few hundred pounds of one or more of these materials.

Absorption is another technique, as long as you remember that, by this method, you are not neutralizing the acid and its corrosiveness, but instead you are just transferring it to another medium, which might be easier to handle. You must still dispose of the solid material holding the acid. Here you will be limited by the size of the spill, the amount of absorption material available, and whether or not the acid is an oxidizing agent. If you add an oxidizing agent to an organic material, you will be exposed to a possible violent explosion and/or fire. There are several commercial sorbents available, but in some incidents you may be able to use such common materials as soil, sand, or clay, depending, of course, on the material to be absorbed.

Bases

The primary hazard here is corrosiveness. The same techniques employed for inorganic acids may be used with bases, but the neutralization agents for the bases will be the inorganic acids. Again, be careful of the neutralization agent, since the most efficient ones may be too dangerous to handle. You must not use soda ash on a base, but you may use sodium bicarbonate.

Organic Acids

The primary danger here will not be corrosiveness, but toxicity, oxidizing agents, explosives, and flammable and combustible materials. You will not attempt to neutralize these materials, but you will try to prevent the hazards of these materials from getting out of control. Absorption may be your best alternative, if fire has not occurred, but you are much better advised to handle each of these materials on its own merits. You will have to consult handbooks or the manufacturer for recommended emergency procedures.

Other Corrosives

The hazard with *most* of the "other" corrosives is the release of toxic fumes when water is applied. Some neutralization agents work on these materials, but you may have to adsorb them with soil, dry sand, or an approved adsorbing material. You should be prepared in advance for these corrosive materials by consulting handbooks and the manufacturers; when you are not sure what will work, practice with small amounts of the hazardous materials to see what works best. Unfortunately, in many cases, this is the only method that you may have at hand to learn how to handle these and other materials.

Glossary

Absorption: A process in which one material actually penetrates the inner structure of another; contrast with *adsorption*.

Acid: A chemical compound containing one or more hydrogen ions that will liberate hydrogen gas on contact with certain metals and is very active chemically.

Acid Gas: Any gas that forms an acid when dissolved in water.

Acidity: The degree to which the pH of a substance is below 7.

Adsorption: A process in which one substance is attracted to and held on the surface of another; contrast with *absorption*.

Alkalinity: The degree to which the pH of a substance is above 7.

Anion: A negatively charged ion.

Base: A chemical compound that contains the hydroxide ion, is the chemical opposite of an inorganic acid, and is very active chemically.

Carboy: A glass or plastic bottle encased in a wooden crate.

Cation: A positively charged ion.

Caustic: Any strongly alkaline substance that has a corrosive effect on tissue; usually refers to bases.

Chemical Action: The reaction of a chemical with another substance.

Concentration: The percentage of an acid or base dissolved in water.

Corrosive: Any material that will attack and destroy, by chemical action, any living tissue with which it comes into contact.

Covalent Bond: The sharing of two electrons between the atoms of two non-metallic elements.

Dilution: The act of adding water to a water-soluble material to lessen its concentration, thereby weakening it.

Functional Group: An atom or group of atoms, bound together chemically, that has an unpaired electron, which when it attaches itself to the hydrocarbon backbone imparts special properties to the new compound thus formed.

Halogens: The elements of group VII: fluorine (F), chlorine (Cl), bromine (Br), iodine (I), and astatine (At).

Hydrocarbon Backbone: The molecular fragment that remains after a hydrogen atom is removed from a hydrocarbon; the hydrocarbon portion of a hydrocarbon derivative.

Hydrogen Ion: H^{+1}; the hydrogen ion dissolved in water.

Hydronium Ion: H_3O^{+1}; the hydrogen ion dissolved in water.

Hydroxide Ion: OH^{-1}; the ion associated with bases.

Inorganic Acid: An acid that contains the anion of a salt; also known as mineral acid.

Ion: An atom, or group of atoms, bound together chemically, that have gained or lost one or more electrons and are electrically charged according to how many electrons were gained or lost.

Ionic Bond: The electrostatic attraction of oppositely charged ions to each other.

Ionization: The process by which an atom or group of atoms bound together chemically gain or lose one or more electrons and become an ion. In reference to an acid, it is the degree to which the acid gas or liquid acid forms ions when it dissolves in water.

Neutralization: The chemical reaction in which an acid or base reacts with another material, and the resulting pH is 7. The classic neutralization reaction is an inorganic acid plus a base:

$$HCl + NaOH \rightarrow NaCl + HOH \text{ (water)}.$$

Organic Acid: An acid containing the carboxyl (COOH) group; the exceptions are hydrocyanic acid and those benzene-ring compounds containing the hydroxyl (OH) functional group.

Oxyacid: An acid that includes an oxygen-containing anion.

Oxidizer: A substance containing oxygen that gives it up readily, or the halogens, which support combustion.

pH: Indication of the acidity or alkalinity of a substance. A pH from 1 to 7 is acidic, and from 7 to 14 is alkaline. A pH of 7 is neutral; technically, pH is the logarithm of the reciprocal of the concentration of hydrogen ions in solution.

Polymerizable: The ability of a small molecule, usually called a monomer, to react with itself to form a very large molecule called a polymer; when this occurs in an uncontrolled way, the monomer may react explosively.

Strength: The degree to which an acid ionizes in water; strong acids ionize nearly 100 percent.

Water-Reactive: The reaction of a substance when it comes into contact with water; this reaction could produce hazardous vapors or any other hazardous material or could release energy such as heat.

14

Unstable Materials: Organic Peroxides and Monomers

Introduction

There is no DOT classification corresponding to the term "unstable materials," but every hazardous material that will be discussed is listed by DOT somewhere else, under another hazard. What we will be dealing with in this chapter are hazardous materials so dangerous that their classification in another area will be insufficient to call the hazards to your attention. They really fall into two general types of products, organic peroxides and monomers. The definition adopted for unstable materials will be as follows: "those substances that decompose spontaneously, polymerize, or otherwise self-react.".

Most of the monomers are flammable gases that are easily liquifiable; the chance that a BLEVE may occur is very high, which will, quite naturally, cause you to try to guard against this type of devastating explosion. Meanwhile, the instant polymerization of the entire container could produce another type of violent explosion, called runaway polymerization. There are so many things that can go wrong with the monomers that it is felt they should be separated from the "normal" flammable liquids and gases and singled out as being *really* dangerous.

The same is true of the organic peroxides (see figure 45). While DOT has a placard and labeling system just for organic peroxides, that is simply not enough. The organic peroxides are so hazardous that it is not possible to convey all their hazards with a single placard or label. The following sections will attempt to cover *all* the hazards associated with these materials.

Organic Peroxides

General Description

The organic peroxides are a group of hazardous materials that are all man-made, having been created for one or more specific purposes. Generally, they are used as initiators or catalysts in a polymerization reaction. An initiator is a material that starts a reaction, while a catalyst is a material that controls the speed of a reaction but is not consumed in the reaction. Organic peroxides are covalently bonded, as their name suggests, and contain the peroxide functional group, $-O-O-$, in the molecule. Since there are two "dangling bonds" in each organic peroxide, there will be two hydrocarbon radicals in each molecule, one attached to each open bond. Therefore, when an organic peroxide has a name such as benzoyl peroxide, you will know that there are really two benzoyl radicals attached to the peroxide radical, and the proper name is dibenzoyl peroxide. All the organic peroxides have strange-sounding names; a few may not sound as if they are organic peroxides at all, but the vast majority will include peroxide or peroxy- as part of the name. Some materials are called acids, and some esters are actually organic peroxides.

If there ever was a group of hazardous materials into which man set out to pack as many hazards as he possibly could, it would be the organic peroxides. As the name states, the materials are organic, and

**Figure 45
Placard and Label
for Organic Peroxides.**

anything organic will burn. The presence of the peroxide radical is an intimate source of oxygen, which the organic peroxide will give up readily. This, you recall, is a definition of an oxidizing agent. Thus, two legs of the fire triangle are present, and all that is needed is the energy (heat) leg. Wanting to appear dissimilar to any other hazardous material, the organic peroxides even have the third leg present; this takes the form of an unstable molecule that can absorb energy from the environment to produce enough energy within the molecule to serve as its own ignition source!

The Self-Accelerating Decomposition Temperature (SADT)

The organic peroxides are generally liquids or white powders, depending on their individual chemistry. The peroxide functional group between two hydrocarbon radicals is extremely unstable; the slightest amount of energy may be enough to cause a violent molecular decomposition with the release of a tremendous amount of energy. What happens is this: the organic peroxide molecules are fairly stable at some low temperature, and they are usually kept at that temperature. But if ambient temperatures rise, and the organic peroxide is exposed to the temperature rise, the rather fragile peroxide linkage begins to disintegrate. This disintegration is actually covalent bond breakage, and energy is released from the broken bonds. This additional energy that is created and absorbed by the remaining organic peroxide molecules of course speeds up the reaction, which feeds upon itself until the organic peroxide seems almost to explode. It is not really an explosion but has been described by firefighters as a "firestorm." The absorption of energy which speeds up the disintegration of the molecule is known as its SADT, which stands for *Self-Accelerating Decomposition Temperature*. This SADT is a property of every organic peroxide; that is, every organic peroxide has its own *Self-Accelerating Decomposition Temperature*, which may be near 0°F. or may be higher than 50°F. Whatever it is, when that temperature is reached by some portion of the mass of organic peroxide, decomposition will begin, and it is irreversible. Once the reaction has begun, you have no chance of stopping it; your only hope is that the amount of organic peroxide involved is small, and that the exposures can be protected. It is important that you know where organic peroxides are used in your protection district, so that you can become familiar with the particular peroxide, its SADT, and its safe handling procedures.

Hazards of Organic Peroxides

The organic peroxides, as stated, have so many hazards that they do not appear to be useful to industry, considering all the risks involved. Industry does a pretty good job of handling these hazardous materials, however, and the occasional incident occurs mostly because someone was careless or lazy. The hazards of organic peroxides are as follows:

1. They are unstable.
2. They are flammable.
3. They are highly reactive.
4. They *may* be explosive (especially in a fire).
5. They are corrosive.
6. They may be toxic.
7. They are oxidizers.

The hazard of being unstable means that any sudden input of energy may cause the violent decomposition of the peroxide molecules. The organic peroxides are so unstable that even slow inputs of energy may cause decomposition. The user of these materials must guard against *any* energy in *any* form reaching them. He is the most common form of energy that threatens the organic peroxide molecule, so protection must be provided from the common methods of generating and transferring heat.

Shock is another common form of sudden energy input, so containers must be protected from falling objects and rough treatment. Indeed, one method of disposal of organic peroxides is to have a sharpshooter fire a high-powered bullet into the container from a safe distance.

The flammability of organic peroxides as a hazard means that they should never be exposed to an ignition source. It may seem redundant to make this statement, since you know that the organic peroxides are so unstable when heated, but a situation *could* arise in which the materials are kept cool but may still be exposed to an ignition source of one sort or another. Flammable materials should never be stored next to oxidizing agents, and vice versa.

The fact that organic peroxides are also oxidizing agents makes the above statement seem ridiculous, which should lead you to some clue as to the storage problems of organic peroxides!

The reactivity of organic peroxides means that they will enter into chemical reactions very easily. This makes them extremely useful to industry, but, as a first responder, you must recognize that chemical reactions *outside* the environment in which they were intended to be carried out can be deadly!

The explosiveness of organic peroxides is related to the speed with which energy is introduced to the material. If it is as rapid as the above-mentioned bullet, or as intense as an approaching flame front, most organic peroxides will indeed explode.

Not all organic peroxides are corrosive, but several are, particularly those that include *acid* as part of the name. Since it would be a monumental job to classify peroxides as to their relative corrosiveness, it is best to treat them all as corrosive.

The same may be said for the toxicity hazard of the organic peroxides. Some may be extremely toxic, while others may be relatively harmless from this aspect. Concentration of the peroxide is an important consideration in its toxicity.

Storage and Handling

Because of the SADT (which, of course, relates to instability) of the organic peroxides, they must be protected from high temperatures, and in most cases this means they must be refrigerated. They are usually shipped in ten-pound boxes; it would be a rare occasion if one were found in a larger container. Manufacturing technology is advancing so rapidly today, however, that some of the newer organic peroxides may be safely packaged in larger weights. The key for the inspector would be to observe if a quantity larger than ten pounds is in the manufacturuer's package, not in another container to which the user has transferred it. A user of organic peroxides should never have more material inside the building where it will be utilized than the amount for immediate use in the batch that is about to be processed. In other words, organic peroxides are not to be stored in the same building where they will be used. Regulations call for a "peroxide storage building" separated from all other buildings ("isolated"), with no windows, no inside lights, no inside electrical switches, sprinklered, and refrigerated to keep it cool. If there are organic peroxides stored that must be protected below 32°F., a "dry pipe" system will be required.

Since the peroxide storage building is some distance from the process that uses it, people will try to short-cut the system and bring enough organic peroxides back to the reactor for several batches, perhaps enough for the entire shift. They might also be sloppy enough to spill some of the peroxides on the way from storage. In such a case, the simple act of walking on the spilled organic peroxide will provide enough energy for it to decompose spontaneously. In either case, the results may be disastrous.

Transportation

As you can imagine, organic peroxides in transportation will have to be refrigerated or protected from heat in some way. This method of shipping offers another opportunity for accidents to occur; if an organic peroxide inside a trailer reaches its SADT, the entire shipment may not take very long to decompose spontaneously! Obviously, the fire occurring inside a truckload of organic peroxides will be quite different from an accident with a ten-pound package. Again, each different organic peroxide will have its own SADT and, therefore, its own level of danger.

It is so important to keep the cargo area cool, and often electrical or mechanical cooling systems are not dependable, so liquid nitrogen is used as the refrigerant. The liquid is allowed to boil away very slowly, thus using a law of Nature (the latent heat of vaporization) to do a job too important to leave to the ways man has devised.

This practice, of course, presents another hazard to anyone entering the cargo compartment: potential asphyxiation. The atmosphere will be almost pure nitrogen, and it will not be detectable. Anyone entering without self-contained breathing apparatus (SCBA) will be quickly overcome.

Emergencies Involving Organic Peroxides

Whenever you must respond to emergencies involving organic peroxides, it is probably too late to do anything about the situation, except hope to protect as many exposures as possible. If the problem is within the peroxide storage building, and the sprinklers (if, indeed, they are present) cannot handle it, nothing can. If the problem is in the polymerization reactor, you may not be able to handle that, either; being unable to handle it may be translated to mean evacuation and withdrawal. If the problem is outside the reactor and inside the process building, you *may* have a chance of doing something; that might range from saving the building to keeping yourself from being injured or killed.

The only way to handle an emergency involving organic peroxides safely is to prevent it from happening in the first place. Know where these materials are commonly used; they will almost always be found in a plant that manufactures—that is, polymerizes—plastics and other polymers. Close inspection of the operation should enable you to enforce safety regulations; constant updates on your part to become familiar with the particular organic peroxides in use at this plant will allow you to know what safety procedures must be enforced and how to handle an incident with each individual organic peroxide.

The Peroxides

Table 25, the following list of organic peroxides, represents most of the organic peroxides in use today. It is probably not 100 percent complete, and, since all organic peroxides are oxidizers, you will also find them listed in Chapter Eleven. Some of the names of these materials are almost unpronounceable, so don't feel you have to memorize them. The most common of them are benzoyl peroxide (and any peroxide with benzoyl as part of the name), methyl ethyl ketone peroxide, and cumyl hydroperoxide. Remember to look for the clues in

Table 25 / Organic Peroxides

acetyl benzoyl peroxide	t-butyl peroxycrotonate
acetyl cyclohexylsulfonyl peroxide	n-butyl peroxydicarbonate
acetyl peroxide	t-butyl peroxydiethylacetate
t-amyl peroxypivalate	t-butyl peroxyisobutyrate
t-amyl peroxy-2-ethylhexanoate	t-butyl peroxyisopropyl carbonate
t-amyl peroxyneodeconate	t-butyl peroxymaleate
benzoyl peroxide (dibenzoyl peroxide)	t-butyl peroxyphthalate
n-butyl-4,4-di-(t-butylperoxy) valerate	2,2-di-(t-butylperoxy) butane
t-butyl cumyl peroxide	2,2-di (cumylperoxy) propane
4-(t-butylperoxy)-4-methyl-2-pentanone	1,1-di-t-butylperoxy cyclohexanone
dibenzoyl peroxide	t-dibutyl peroxide
n-butyl-4-4-d-(t-butylperoxy) valerate	t-butyl peroxy-2-ethylhexanoate
di-t-butyl diperoxyazelate	t-butyl peroxyneodecanoate
di-t-butyl diperoxyphthalate	t-butyl peroxypivalate
t-butyl peroxy-2-ethylhexanoate	t-butyl peroxy-3,5,5-trimethylhexanoate
3-t-butyl peroxy-3-phenylphthalide	caprylyl peroxide
1,1-di-(t-butylperoxy)-3,3,5-trimethylcyclohexanoate	p-chlorobenzoyl peroxide
t-butyl isopropyl benzene hydroperoxide	a-cumyl hydroperoxide
OO-t-butyl O-isopropyl monoperoxycarbonate	cumyl peroxyneodecanoate
t-butyl peracetate	cumylperoxytrimethylsilane
t-butyl perbenzoate	cyclohexanone peroxide
t-butyl hydroperoxide	di-(2-butoxyethyl) peroxydicarbonate
t-butyl peroxyacetate	di-(4-t-butylcyclohexyl) peroxydicarbonate
t-butyl peroxybenzoate	di-n-butyl peroxydicarbonate

the name: peroxide (either separate or as part of another word in the name), peroxy-, and per- (especially in a name that ends in -ate or -ite).

Monomers

General Description

The monomers are a group of molecules, usually referred to as "tiny" in relation to the size of the final product formed, which have the unique capability of reacting with themselves to form a "giant" molecule called a polymer. The process in which this unique reaction occurs is called polymerization. The monomers are either easily liquifiable flammable gases, or low-flash-point flammable liquids. As liquified gases, they are subject to the rather spectacular BLEVE

Table 25 / Organic Peroxides (cont.)

di-sec-butyl peroxydicarbonate	di-(2-phenoxyethyl) peroxydicarbonate
di-t-butyl peroxide	di-n-propyl peroxydicarbonate
dicetyl peroxydicarbonate	1-hydroperoxy-1'-hydroxy-dicyclohexyl peroxide
di-(4-chlorobutyl) peroxydicarbonate	isononanyl peroxide
di-(2-chloroethyl) peroxydicarbonate	isopropyl percarbonate
di-(3-chloropropyl) peroxydicarbonate	lauroyl peroxide
dicumyl peroxide	p-methane hydroperoxide
dicyclohexyl peroxydicarbonate	methyl ethyl ketone peroxide
diethyl peroxydicarbonate	methylisobutylketone peroxide
di-(2-ethylhexyl) peroxydicarbonate	di-(1-naphthoyl) peroxide
1,1-dihydroperoxycyclohexane	p-nitroerbenzoic acid
di-(1-hydroperoxycyclohexyl) peroxide	para-methane hydroperoxide
di-(1-dihydroxycyclohexyl) peroxide	pelargonyl peroxide
diisopropylbenzene hydroperoxide	peracetic acid
diisopropyl peroxydicarbonate	peroxyacetic acid
di-2-methylbenzoyl peroxide	peroxymaleic acid OO-t-butyl ester
di-(3-methylbutyl) peroxydicarbonate	1-phenylethyl hydroperoxide
2,5-dimethyl-2,5-di-(t-butylperoxy) hexane	pinane hydroperoxide
2,5-dimethyl-2,5-di-(t-butylperoxy) hexyl-3	1,2,3,4-tetrahydro-1-naphthyl hydroperoxide
2,5-dimethyl-2,5-di-(2-ethylhexanoylperoxy) hexane	1,1,3,3-tetramethylbutyl hydroperoxide
3,5-dimethyl-3,5-dihydroxy-1,2-dioxolane	1,1,3,3-tetramethylbutyl peroxyphenoxyacetate
dimethylhexane dihydroperoxide	1-vinyl-3-cyclohexene-1-yl hydroperoxide
dimyristyl peroxydicarbonate	vinyltri-(t-butylperoxy) silane
2,4-dihydroxy-2-methyl-4-hydroperoxypentane	

(*Boiling Liquid, Expanding Vapor Explosion*). As monomers (whether liquified gases or liquids), they are subject to the equally spectacular explosion called runaway polymerization. Therefore, responding to an emergency involving monomers, you have at least two ways of being caught in an explosion.

The Polymerization Reaction

Most of the monomers are unsaturated hydrocarbons, with one double bond; this double bond is what makes monomers reactive and unstable. Under the proper conditions of heat and pressure, and with the proper initiator and/or catalyst, the double bond will "open up" and send an unpaired electron to both ends of the molecule, thus creating a unique free radical that can react at two positions on its molecule (and also creating the so-called "dangling bonds"). Since the molecule is surrounded by other molecules of monomers that are reacting in the same way, it is only natural that these free radicals react with each other, especially since Mother Nature will not allow them to remain as free radicals for any appreciable amount of time. Since the only other free radicals the molecules can react with are similar to themselves, the reaction goes forward, with the chemical engineer controlling the process so that he can select the length of the "chain" formed and, therefore, the properties of the resulting polymer. Once the reaction has been completed to his satisfaction, the polymer is removed from the reactor, dried, and sent through an extruder to be pelletized. This process is repeated for most thermoplastics, while some thermoplastics and some thermosets are polymerized in a slightly different manner.

Hazards of the Monomers

All monomers are not alike but most of those to be described here that polymerize to thermoplastics have similar hazards. These are as follows:

1. They are unstable.
2. They may be toxic.
3. They may be corrosive.
4. They may be under pressure.
5. They are highly reactive.
6. They are flammable.
7. They may be subject to BLEVE.

The instability of the monomers is quite different from the instability of the organic peroxides, although both are caused by the type of covalent bonding in their corresponding molecules. Whereas the organic peroxide contains the very weak peroxide $(-O-O-)$ configuration, the monomer has a *relatively* more stable carbon-to-carbon double bond $(-C=C-)$ structure. Notice the word *relatively*. Although monomers are not as unstable as the organic peroxides, they *will* react explosively if too much heat energy is allowed to reach them. This type of explosion, the result of a large amount of material undergoing a chemical reaction uncontrollably and instantaneously, is called runaway polymerization.

The reactivity of the monomers is due to the presence of the double covalent bond, and it is at this site in the molecule that controlled reactions are brought about.

Those monomers subject to BLEVE are the flammable gases that are easily liquified. Other monomers are flammable liquids.

Transportation, Storage, and Handling

Being extremely unstable materials, monomers must be stabilized before shipment; that is, something must be added that will prevent them from unexpected and violent polymerization during transportation and, afterward, in storage. If an accident occurs en route and for some reason the stabilizer, called an inhibitor, is allowed to escape from the monomer, a runaway polymerization explosion may be imminent. The monomer user must be careful that the monomer remain inhibited during its time in his inventory, before it is sent to the reactor. Once in the reactor, if everyone has done his job properly, the inhibitor remains in the monomer, and the desired polymerization process cannot begin until the inhibitor is overcome. That is the reason for the use of organic peroxides to overcome the inhibitor and initiate the polymerization process—hence the name initiator for organic peroxides. All this sequence must be worked out by the chemical engineer so that the proper temperatures and pressures are maintained, and generated heat can be removed. Reactor explosions are not common, but they can and do happen.

Emergencies Involving Monomers

The monomers should be treated like the hazardous materials that they are; every effort should be made to prevent the escape of the inhibitor from the monomer. Even if all proper care is taken, and a rail car or tank truck becomes involved in an accident, any impinging flame or other source of heat may drive out the inhibitor, and the runaway polymerization explosion may occur. Even if the inhibitor remains, the container of monomer is subject to a BLEVE. The main objective then would be to keep the tank as cool as possible, with flooding amounts of water applied from as far away as possible. The danger zone of a tank truck or rail car might be as wide as a 2,500 foot radius, and *that* estimate may be conservative. Do not attack close up, and, if the incident should take place far away from any exposures, it might be best to make no attack at all.

The Monomers

Acetaldehyde is a colorless, fuming liquid with a pungent fruity odor. It is water-soluble and is a skin and respiratory irritant. Its flash point is −40°F., and it boils at 60°F.

Acrylic acid is the monomer of some acrylic plastics. It is a colorless liquid with an acrid odor. It is toxic and corrosive, has a flash point of 130°F., and boils at 304°F.

Acrolein is a colorless or yellowish liquid, whose vapors produce a choking sensation. It is a powerful irritant, has a flash point below 0°F., and boils at 95°F.

Acrylonitrile is the monomer of polyacrylonitrile, is toxic by ingestion and inhalation, and is a carcinogen. It is a colorless liquid that is only partially soluble in water. It has a specific gravity of 0.80 and a flash point of 32°F., and it boils at 139°F. The chemical name for acrylonitrile is vinyl cyanide; when it burns it will liberate hydrogen cyanide gas. It is also one of the monomers of ABS (acrylonitrile-butadiene-styrene) plastic.

Adipic acid is a solid monomer that has a flash point of 385°F. It is one monomer in a system to produce nylons.

1,3-butadiene is a colorless gas with a mild aromatic odor. It is easily liquified and has a boiling point of 24°F. It is one of the monomers for ABS plastic and is flammable.

Butylene is the monomer of polybutylene, and it is one of the LP gases. It is used as a monomer rather than as a fuel.

Chloroprene is a colorless liquid that is toxic and flammable. It is the monomer of polychloroprene, better known as Neoprene.

Epichlorohydrin is a colorless liquid that smells like chloroform. It is the monomer of some epoxies and phenoxy resins. It is only slightly soluble in water, has a specific gravity of 1.18, a flash point of 93°F., and boils at 207°F. When it burns, it liberates hydrogen chloride.

Ethyl acrylate is another acrylic monomer. It is a colorless liquid with a penetrating, acrid odor. It has a flash point of 60°F., and boils at 211°F. It is slightly soluble in water and has a specific gravity of 0.92.

Ethylene is the monomer of polyethylene and is an easily liquifiable gas. Since it is the monomer of the thermoplastic used in largest volume, you would expect it to be the monomer that is shipped around the country in the largest quantities, and so it is. In 1979, it was the chemical of sixth largest volume in the United States. It is a colorless gas with a sweet odor and taste and an ignition temperature of 1,009°F. It is also used as a refrigerant, a fruit ripener, and in the production of many other chemicals.

Ethylene oxide is an extremely hazardous flammable gas that is used as one of the monomers for epoxy plastics. It is particularly dangerous because of its extremely wide flammable range, from 3 percent to 100 percent. The upper limit means that ethylene oxide will burn with no oxygen present. Any leaking from its container and burning present the additional hazard that the flame might back up *into the tank or cylinder* and explode.

Formaldehyde is the monomer of acetal plastics, is a colorless, toxic gas with a strong, pungent odor that is very soluble in water, and is highly flammable. It is toxic by inhalation and a strong skin irritant. Its ignition temperature is 806°F.

Methacrylic acid is the monomer of some of the acrylic plastics. It is a colorless liquid with an acrid, disagreeable odor and is corrosive and combustible. It has a flash point of 170°F., is soluble in water, and is a strong skin irritant.

Methyl acrylate is the monomer of another of the acrylics, a colorless liquid with an acrid odor, which is sometimes a polymer itself.

Methyl methacrylate is the monomer of polymethyl methacrylate, the most common of the acrylic plastics. Polymethyl methacrylate is a liquid resin. It forms a crystal-clear sheet when cast that is good enough to be used as glazing material. It has a flash point of 50°F. and is slightly soluble in water, with a specific gravity of 0.94, and is readily polymerizable by heat or light. It is a colorless liquid with an acrid odor.

Phenol is a toxic, combustible solid that liquifies easily. It is a monomer for phenol-formaldehyde and phenol-furfural resins.

Propylene is the monomer of polypropylene and is a colorless, easily liquifiable gas. It has an ignition temperature of 927°F. and is a simple asphyxiant.

Propylene oxide is a colorless, flammable liquid that is used as one of the monomers for epoxy plastics.

Styrene is a colorless, oily liquid with an aromatic odor and is the monomer of polystyrene. It is insoluble in water (being a hydrocarbon) and has a specific gravity of 0.9. It has a flash point of 88°F. and an ignition temperature of 914°F. It is toxic by ingestion and inhalation

Tetrafluoroethylene is a flammable, colorless gas that is the monomer for polytetrafluoroethylene (PTFE) resins. The most common is Teflon®.

Vinyl acetate is the monomer of polyvinyl acetate, is a colorless liquid with a flash point of 30°F., and an ignition temperature of 800°F. It is insoluble in water and has a specific gravity of 0.9345. It is toxic by inhalation and ingestion.

Vinyl chloride monomer is the monomer of polyvinyl chloride. It is an easily liquifiable gas that is subject to a BLEVE. It is insoluble in water and has a specific gravity of 0.9121. It is toxic by inhalation, is a suspected carcinogen, and liberates hydrogen chloride when it burns.

Vinylidene chloride is the monomer of polyvinylidene chloride and is a colorless liquid that liberates hydrogen chloride when it burns. Vinylidene chloride is better known as Saran®. It is a colorless liquid with a flash point of 14°F.

Summary

The monomers are a class of unstable materials that are used to form polymers through the chemical reaction called polymerization. When controlled (that is, when in the reactor vessel under the proper conditions of temperature and pressure and with the right additives), the polymerization reaction is a rather tame one. Let the inhibitor escape from the monomer, however, and then subject the monomer to heat, shock, or pressure, and the entire mass will instantly convert to the polymer, in an explosion called runaway polymerization. If the monomer is a liquified gas, the explosion resulting from impinging flames or radiated heat might be *either* runaway polymerization or a BLEVE. Care must always be taken when monomers are involved.

Glossary

Catalyst: A substance used to control the speed of a chemical reaction; but which is not consumed in that reaction.

Inhibitor: A substance used to prevent a chemical reaction from starting; also called a stabilizer.

Initiator: A substance used to start a chemical reaction.

Monomer: A simple, small molecule that has the special capability of reacting with itself to form a "giant" molecule called a polymer.

Polymerization: The chemical reaction in which monomers combine with themselves to form giant molecules called polymers.

SADT: Acronym for *Self-Accelerating Decomposition Temperature;* a property of most organic peroxides; the temperature at which decomposition of the molecule becomes self-feeding and irreversible.

15

Toxicity

Introduction

General Definitions

The term toxicity may be defined in many ways. One definition for toxicity is the ability of a chemical molecule or compound to produce injury once it reaches a susceptible site in or near the body. Another definition is the quality or condition of being toxic, with the word toxic defined as harmful, destructive, deadly, or poisonous. A third definition of toxicity is anything that causes harm to living tissue by chemical activity. We will use the first definition in this chapter.

The study of the nature, properties, and effects that chemicals have on living systems is the definition of pharmacology; the branch of this science that deals with toxicity is toxicology. Toxicology may also be defined as the science of poisons, their detection, effects, and antidotes. A poison is defined as any substance that causes injury, illness, or death to living tissue by chemical activity.

There are many types of poisons. Those that affect the circulatory system are called hemotoxins. Poisons that attack a specific organ are called cytotoxins, and those that attack the nervous system are known as neurotoxins. We will be further classifying toxins as asphyxiants, systemic poisons, and upper and lower respiratory poisons.

There is a difference between toxicity and hazard. Many materials are toxic to humans in certain quantities. At levels below that which will cause *immediate* harm, however, there appears to be little or no hazard, or so many people believe. The toxicity of a material may be one thing, but the hazard it presents to humans is certainly another. There are many deadly poisons in cigarette smoke, but many people are not apparently harmed by inhaling the smoke into their lungs; of course, there are many people who *are*. There is no question as to the toxicity of many of the combustion products of tobacco, but those who inhale those combustion products are not poisoned, at least not apparently. The point is, the quantity of the material brought into the body is all-important in determining whether there will be any damage. The deadly poisons inhaled by smokers are present in very small quantities and produce no apparent damage at the time of the inhalation.

There is a definite difference between hazard and risk. The hazards of many toxic materials are well known, but the risk of exposure may be negligible due to the amount of the material present at the time of exposure, the rate of absorption, and the rate at which the body eliminates the material or the material into which it has been converted by the body. Exposure time is also very important, because some toxic hazardous materials must be present for a considerable time before any ill effects are produced. The exposure may be acute, which is defined as a single dose or exposure; chronic, which is defined as repeated does or exposures; or the exposure may be subacute, which is defined as just two or three exposures (obviously somewhere between acute and chronic). The action of the poison must also be considered. It may be local, which is defined as action at the site of contact, or it may be systemic, which is defined as transmission through the skin, lungs, or mucous membranes to the site of activity by way of the bloodstream.

Entry Routes

The manner in which a poison enters the body is also important in determining the potential damage it will cause. Normally, toxic materials enter the body in one of four ways: inhalation, ingestion, absorption through the skin, and entry through a wound. The method of entry may determine how rapidly the toxic nature of the poison will produce its effects. The classic case may be that of the convicted killer sentenced to death in the gas chamber, which used hydrogen cyanide as the method of execution. If he is given one last request, and that request is for a self-contained breathing apparatus with an unlimited supply of uncontaminated air, the granting of this request will extend his life for only a few minutes. Hydrogen cyanide, a deadly poison when breathed in relatively low concentrations, is also absorbed through the skin, and death will occur even if the person exposed to the hydrogen cyanide breathes none of it.

Methods of Measurement

There are many ways to determine the level of exposure that will cause damage to living tissue. The following methods include most of them.

1. Lethal Dosage (LD_{50})

This measurement of the toxicity of a material is defined as the lethal dose, expressed in milligrams of toxin per kilogram of body weight (mg/kg) of the laboratory animal exposed, that *kills half the animals* during an observation period after the exposure. In other words, a number of white mice, or other selected animals, are fed or injected with a dosage of the material under examination. The amount administered is not the same for each animal, because of differences in body size. Each animal, however, *is* fed or injected with an amount that is proportionately equal, based on each animal's weight. The animals are then observed for a period of time, usually two weeks. If more than half the animals die in the observation period (which, of course, may be some other length of time), the experiment is repeated with a smaller dosage; conversely, if less than half die, the dosage is increased. The objective is to establish, as closely as possible, what amount of the material will kill exactly half the animals in the observation period. This amount is then reported as the LD_{50}

2. Lethal Concentration (LC_{50})

This measurement of toxicity is used to determine the toxicity of gases, vapors, fumes, and dusts in air. It is defined as the lethal concentration of gases, vapors, fumes, or dusts in air to which the animals have been exposed for a specified time, and which then kills half the animals during the observation period. The animals are usually secured in a container of some sort. The gases, vapors, fumes, and dusts are then introduced into the atmosphere, in a specified ratio of parts per million by volume of air, and breathed by the animals for a specified amount of time. The exposed animals are then observed for a specified observation period, during which a count of fatalities is kept. The LC_{50} is the lethal concentration that kills half the animals during the observation period.

3. Threshold Limit Value (TLV)

The threshold limit value is defined as the upper limit of a toxic material to which an average person in average health may be exposed repeatedly on a day-to-day basis with no adverse effects. These limits are expressed in milligrams per cubic meter (mg/m^3) for gases or vapors in air and in micrograms per cubic meter $(\mu g/m^3)$ for fumes and mists in air. These standards are set and revised annually by the American Conference of Governmental Industrial Hygienists (ACGIH) for concentrations of airborne substances in workroom air. The standards are time-weighted averages based on conditions which, it is believed, workers may be repeatedly exposed to during their normal work week with no ill effects. The values are *not* intended to serve as definitive lines between safe and dangerous concentrations, but instead they are intended to serve as guides in control of health hazards.

4. Maximum Allowable Concentration (MAK)

In contrast to the threshold limit value, the maximum allowable concentration is an easier measurement to understand. Whereas the TLV is set as a time-weighted average (TLV may be sometimes referred to as TWA) of concentration over a work day and work week, the MAK sets the absolute maximum exposure at any one given time. It is usually expressed in parts per million (ppm), just as TLV is.

5. Other Measurements

There are many other measurements of toxicity, but the first three mentioned are the major measurements listed by most references. Others include short-term exposure limits (STEL), effective concentration (EC_{50}), incapacitation concentration (IC_{50}), and immediately dangerous to life and health (IDLH). There will indeed be more measurements, as researchers try to define the effects for which they are looking.

Classification of Toxic Materials

Irritants

The class of hazardous materials known as irritants is defined as those corrosive materials which attack the mucous membrane surfaces of the body. It must be pointed out that these materials are classified as irritants by definition, but if they are encountered in very high concentrations, they will produce death, as noted in table 28 at the end of this chapter, rather than just irritation.

1. The water-soluble irritants are those that are *very* soluble in water and therefore will dissolve in the first moisture they meet, usually in the eyes, or in the mouth, nose, and upper respiratory tract; therefore, it is in these areas of the body that they will be most dangerous. These irritants include the halogen acid gases (hydrogen chloride [HCl], hydrogen fluoride [HF], hydrogen bromide [HBr], and hydrogen iodide [HI]), sulfur dioxide (SO_2), and ammonia (NH_3). The halogen acid gases convert to the halogen acids (hydrochloric acid, hydrofluoric acid, hydrobromic acid, and hydriodic acid, all with the same chemical formulas as the halogen acid gases), as you will recall from Chapter Thirteen, Corrosives. These covalent gases ionize 100 percent when they dissolve in water and therefore become very strong acids. Although they *are* strong acids, the concentrations they form when they dissolve in the moisture of your body are not very high; however, they will be high enough to cause discomfort and pain and thereby earn their name, irritants.

Sulfur dioxide will form sulfurous acid in the same way that the halogen acids are formed; hence its irritating properties are similar, if not as severe.

Ammonia will form a caustic (basic) solution when dissolved in water. It converts to ammonium hydroxide, NH_4OH. Ammonia is a very pungent gas, that, like the others, will force you to seek fresh air if you encounter it.

Although the above-named materials are irritants so far as toxicologists are concerned, DOT considers them all as non-flammable gases, and that is how they are placarded when they are transported. The major "fooler," however, is ammonia, which will burn within a flammable range of from 16 percent to 25 percent in air.

2. The moderately water-soluble irritants do not dissolve in water as rapidly or as easily as the materials in 1., above, do, so their effects will be felt farther along the respiratory tract. These will act in the upper respiratory tract and in the lungs. They are the halogens (fluorine, chlorine, the fumes of bromine, and the vapors of iodine), ozone (O_3), phosphorus trichloride (PCl_3), and phosphorus pentachloride (PCl_5). These materials are all corrosive and will cause irritation to sensitive areas in small amounts but will cause severe damage and even death in high concentrations.

3. The slightly water-soluble irritants will bypass all the moist areas that the first two groups of irritants attack and will do their

damage in the lungs, attacking the alveoli, and destroying them by chemical action. In small quantities, these irritants will not do great damage, but in high concentrations they will be fatal, usually by delayed effects. In the case of the nitrogen oxides (NO_x), the delay is anywhere from four to 48 hours. The oxides of nitrogen are nitrous oxide (N_2O), nitric oxide (NO), nitrogen dioxide (NO_2), nitrogen trioxide (N_2O_3), dinitrogen tetroxide (N_2O_4), and dinitrogen pentoxide (N_2O_5). Of these, nitrous oxide is a non-irritating gas, sometimes used as an analgesic (laughing gas). The other material sometimes classified as an irritant in this class is phosgene ($COCl_2$). Be aware that DOT says phosgene is a *Class A poison.*

Another irritant that is slightly soluble in water is trichloroethylene, C_2HCl_3. Its vapor is extremely dense (vapor density, 4.53), which makes it extremely dangerous in close quarters, where it will displace air and cause death either by asphyxiation or by its own action.

Asphyxiants

Asphyxiants are gases that interfere with the oxidation processes in the body.

1. Simple Asphyxiants

Simple asphyxiants are those materials that are not toxic in their own right, but which will kill by diluting or replacing the oxygen in the air needed for breathing. The most common of the simple asphyxiants is carbon dioxide, an inert gas produced in great quantities as a product of oxidation (combustion) of carbon-based materials. Although carbon dioxide is considered a very mild toxic material, when it does cause death it usually does so by acting as a simple asphyxiant. Carbon dioxide (CO_2) is heavier than air and therefore is most dangerous near the floor. Beware of fire rooms after the dumping of carbon dioxide fire-extinguishing systems, especially in basements and other below-grade rooms. You may effectively extinguish the fire *and* anyone else in the room, especially if they cannot escape.

Another simple asphyxiant is nitrogen, another relatively inert gas. The nitrogen molecule (N_2) is of almost the same molecular weight as air, and therefore it will not disperse quickly. Care must be taken wherever pure nitrogen is being used, as it will quickly dilute the oxygen content of air. This may be in tractor-trailer bodies where liquid nitrogen is used as a refrigerant (the liquid nitrogen is simply allowed to boil away slowly, thus cooling the trailer's entire inside) or in any cryogenic grinding operation in industry where liquid nitrogen is the cryogenic material.

Hydrogen, dangerous as a simple asphyxiant only in air-tight rooms or containers because it is so light that it rises in air and disperses rapidly, is another simple asphyxiant, but it has an added hazard. Hydrogen is one of the most flammable and hottest-burning gases known, so you must take care not only to prevent it from diluting oxygen, but you must also protect yourself from the explosion and fire hazard.

The noble gases (helium, neon, argon, krypton, and xenon) are all simple asphyxiants, merely because of their inertness. Helium is very light and will rise and disperse rapidly like hydrogen, and neon, whose molecular weight is 20.18, will also rise and disperse; again, the assumption is that you are not in a closed space. The other noble gases, argon, krypton, and xenon, are all heavier than air and will seek out low spots. Be careful anywhere these inert gases are used, such as in arc-welding operations. They may not enter any chemical reactions, but their inertness can be fatal to you if you are caught in an atmosphere dominated by any one of them.

The saturated hydrocarbons are another group of simple asphyxiants. These gases, methane (and, of course, natural gas), ethane, propane, and butane, are all non-toxic, but they *are* flammable. Methane and natural gas are lighter than air, but ethane, propane, and butane are heavier than air. The vapors of the liquid alkanes will, similarly, be non-toxic.

2. Blood Asphyxiants

Blood asphyxiants are those materials that combine with the red blood cells and render them incapable of combining with oxygen and thereby carrying the oxygen to the cells of the body. The normal process involved after breathing is the formation of a compound in the red blood cell called oxyhemoglobin. This weak compound serves as the vehicle for oxygen to be carried to the cells, where the oxygen is "dumped," and carbon dioxide is picked up and brought back to the lungs for disposal. Anything that interferes with this process by preventing the formation of the oxyhemoglobin is a blood asphyxiant.

By far the most common of the blood asphyxiants is carbon monoxide, the deadly gas formed in all fires involving carbon-based combustibles. Carbon monoxide is a colorless, odorless, tasteless gas. The hemoglobin that ordinarily picks up the oxygen for transmission to the body cells has an affinity for carbon monoxide that is up to 200 times stronger than the affinity it has for oxygen. The implication of this fact is that if you are in an atmosphere that contains a ratio of 199 parts of oxygen and 1 part of carbon monoxide as a portion of the normal atmosphere, your blood will preferentially absorb the carbon monoxide over the oxygen, and you may be poisoned.

Carbon monoxide reacts with the hemoglobin in your blood to form carboxyhemoglobin (COHb). This compound is stronger than the oxyhemoglobin formed when oxygen is present, and the red blood cell will not "dump" the carbon monoxide as it would the oxygen when it came to a cell waiting to exchange carbon dioxide for oxygen. Therefore, the red blood cell bypasses the waiting body cell, leaving it with its carbon dioxide, and depriving it of its needed oxygen. The red blood cell makes its return trip to the lungs where it is supposed to dump its load of carbon dioxide and pick up fresh oxygen. Since the cell is still carrying its load of stable carboxyhemoglobin, it cannot pick up oxygen, so it passes through the lungs and begins another trip through the circulatory system, only to disappoint more waiting body cells. As this process continues, more and more red blood cells become loaded with carbon monoxide and unable to pick up oxygen to exchange for carbon dioxide with oxygen-starved body cells, and the body cells begin to die.

It is imperative to get oxygen to these cells before irreparable damage can be done and, of course, to prevent death. The carboxyhemoglobin may be able to hold on to the red blood cell for four to five hours (this might be called the "half-life" of carbon monoxide in the blood), so merely giving the victim pure oxygen may not save him. It may be necessary to use a hyperbaric chamber to speed up his acceptance of the tissue-saving oxygen.

Two other blood asphyxiants are aniline ($C_6H_5NH_2$) and nitrobenzene ($C_6H_5NO_2$). Both are toxic by inhalation, ingestion, and absorption. Aniline is used in rubber manufacture, the production of dyes, photographic chemicals, isocyanates, explosives, pharmaceuticals, herbicides, and fungicides. Nitrobenzene is used in the manufacture of aniline and many other chemicals.

Tissue Asphyxiants

Tissue asphyxiants are those materials that are carried by the red blood cells to the cells of the body, given up to those cells in exchange for the carbon dioxide held by those cells, and render the body cells incapable of ever again accepting oxygen from the red blood cells. Unlike carbon monoxide, which attaches itself to the red blood cell so tightly that it will not let go and renders the *red blood cell* incapable of picking up oxygen, the tissue asphyxiant will allow itself to be "dumped" to the receiving body cell just as oxygen does. Upon acceptance of this material from the red blood cell, however, the body cell itself is poisoned so that it can no longer accept oxygen.

By far the most common tissue asphyxiant is hydrogen cyanide, HCN. When hydrogen cyanide dissolves in water, it becomes hydrocyanic acid, or prussic acid. Both the acid and the gas have the smell of bitter almonds, and the gas is colorless. It is produced when certain materials, such as wool, silk, polyurethane, nylon, or ABS (acrylonitrile-butadiene-styrene) plastics, are burned. Hydrogen cyanide is used in the manufacture of many plastics, cyanide salts, dyes, and pesticides. It is flammable, and under certain conditions it will polymerize violently.

Respiratory Paralyzers

Respiratory paralyzers are materials that upon entering the body will short-circuit the respiratory nervous system.

The most common respiratory paralyzer is hydrogen sulfide (H_2S), a flammable, colorless gas with an overpowering smell of rotten eggs. It is a familiar smell to anyone who has experienced it once. It is produced in the process of manufacturing coke from coal, is liberated at many oil wells and refineries, and is also liberated by malfunctioning automobile catalytic converters. Hydrogen sulfide is very specific in its attack on the body in that it will paralyze the olfactory nerve (the nerve that transmits the sense of smell to the brain) first. You may at first smell the characteristic hydrogen sulfide odor, and then it will disappear. This absence does not mean that the hydrogen sulfide has gone; in reality, the sense of smell has been eliminated, and the toxic material is still present. The next effect is to paralyze totally the nerves that control the breathing process, and the process then ceases. Hydrogen sulfide is used to purify certain acids, such as sulfuric and hydrochloric acids, to precipitate the sulfides of metals, and as an analytical reagent.

Another respiratory paralyzer is carbon disulfide (CS_2), also called carbon bisulfide. It is a clear, colorless liquid with the highly disagreeable odor of rotten cabbage. It has a flash point of $-22°F.$ and the remarkably low ignition temperature of 212°F. With a tolerance of only 10 ppm, it is extremely toxic. Carbon disulfide is used in the manufacture of rayon and carbon tetrachloride, but its biggest use is as a solvent and degreaser. It is a very efficient solvent and relatively inexpensive, so there are great quantities of it shipped around the country.

Other respiratory paralyzers include those materials that have an anesthetic or narcotic effect on the body. They include acetylene (C_2H_2), a highly flammable and unstable gas; ethylene (C_2H_4), another flammable gas that has many uses, as well as being a very popular monomer; diethyl ether, $(C_2H_5)_2O$, a popular anesthetic that is very flammable; acetone, $(CH_3)_2CO$, a flammable solvent; and ethyl alcohol, C_2H_5OH, a flammable liquid that is the alcohol in alcoholic drinks. If you surpass the safe amounts of these materials to which your body can be exposed, the nerves that control the breathing process will be paralyzed, and you will simply stop breathing. As you can see from the hazardous materials that we have mentioned here, you will be unconscious from one cause or another when you stop breathing forever.

Systemic Poisons

A material is a systemic poison if it interferes with any vital bodily processes. We will classify the following types of systemic poisons by the vital organ whose function is disrupted.

Disruption of the Function of the Liver and Kidneys

The most common poison that operates in this manner is arsenic. Arsenic is a non-metallic element whose atomic number is 33, with an atomic weight of 74.92. It is in group V of the Periodic Table, below nitrogen and phosphorus. It is a silvery gray solid that has several allotropes. Arsenic is used in manufacturing many metal products, in semiconductors, and as an alloying agent.

The heavy metals such as lead, cadmium, and mercury are also systemic poisons that will attack the liver and kidneys. These metals are very common and therefore are extremely popular in their use. Lead compounds were used as paint pigments and as gasoline additives for many years. Cadmium pigments are still highly popular. Mercury compounds are not as common, but much mercury is discarded in many metal reduction operations, and it has become a major water pollutant.

A third group of systemic poisons includes all the halogenated hydrocarbons. This is an extremely large group of hazardous materials, and many manufacturers will take exception, claiming that their materials are not hazardous. There is a great body of research which shows conclusively, however, that after enough exposure to halogenated hydrocarbons, you will develop problems with your liver and kidneys.

Attack on Bone Marrow

The most common hazardous material that attacks bone marrow is benzene, C_6H_6. Benzene is one of the most popular solvents used in the rubber industry, and it has many, many other uses. For a long time it was suspected of attacking bone marrow because of the high incidence of anemia among workers exposed to benzene. Later, it became a suspected carcinogen when many exposed workers developed leukemia. Today, benzene is classified as a carcinogen, and strict exposure limits are set.

Among other materials that are suspected to be attackers of bone marrow are toluene, $C_6H_5CH_3$, xylene, $C_6H_4(CH_3)_2$, and naphthalene, $C_{10}H_8$. Toluene and xylene are very popular solvents that have replaced benzene. Naphthalene is a white crystalline solid that sublimes; it has the characteristic odor of moth balls and is used in making insecticides, fungicides, smokeless powder, lubricants, antiseptics, and many other chemicals.

Attack on Muscles

By far the most common systemic poison that attacks the muscles is strychnine, $C_{21}H_{22}N_2O_2$, a white powder with a bitter taste. When it attacks the muscles, it causes extremely powerful spasms that use all of the power of each muscle. In severe poisoning, the victim will experience such powerful muscle contractions that he will break most of the bones in his own body. Strychnine is the favorite material of dog poisoners. Its only use is as a poison, and it is used primarily to control rat populations.

Interference with Vital Nerve Impulses

The largest group of poisons that interfere with vital nerve impulses are the organic phosphates, materials commonly used in pesticides. It is not possible to list all the nerve impulses with which they interfere, but in high enough dosages the resulting death will be quite unpleasant.

Carbon disulfide, a material we covered as a respiratory paralyzer, will also affect nerves other than those of the respiratory system.

Methyl alcohol, CH_3OH, also known as methanol, is an extremely common material that interferes with vital nerve impulses, being very specific in its attack on the optic nerve. It is not rare for a hospital to treat someone for blindness caused by the drinking of denatured alcohol, of which only a small portion is methanol. Methyl alcohol is also known as wood alcohol and may be added to "white lightning" by moonshiners, either accidentally or on purpose. It is a very common chemical and is available almost everywhere.

DOT Classification

Poison A is a classification that includes extremely dangerous poisons, those substances that are poisonous gases or liquids of such nature that a very small amount of the gas, or vapor of the liquid, mixed with air is dangerous to life (see figure 46).

Poison B is a classification that includes less dangerous poisons, substances, liquids, or solids (including pastes and semi-solids), other than Class A or irritating materials, which are known to be so toxic to man as to afford a hazard to health during transportation or which, in the absence of adequate data on human toxicity, are presumed to be toxic to man. (see figure 47).

Irritating materials are liquids or solid substances which, upon contact with fire or when exposed to air, give off dangerous or intensely irritating fumes, but not any poisonous material of Class A.

Tables 26 through 28 list those poisons as classified by DOT. Be advised that there are many other poisons that will kill you just as quickly as many of those listed here, but DOT has them classified elsewhere. Good examples of such poisons are carbon monoxide and chlorine. Table 29 gives the typical human reactions to different concentrations of toxic substances.

**Figure 46
Placard and Label for
Class A Poisons.**

**Figure 47
Placard and Label for
Class B Poisons.**

Table 26 / Class A Poisons

arsine
bromoacetone, liquid
chemical ammunition containing a Class A poison
chloropicrin and methyl chloride mixture
chloropicrin and non-flammable, non-liquified compressed gas mixture
cyanogen chloride containing less than 0.9 percent water
cyanogen gas
ethyldichloroarsine
gas identification set
germane
grenade without bursting charge, with Poison A gas charge
hexaethyltetraphosphate and compressed gas mixture
hydrocyanic acid (prussic) solution, 5 percent or more hydrocyanic acid
hydrocyanic acid, liquified
insecticide, liquified gas, containing Poison A or Poison B material
lewisite
methyldichloroarsine
mustard gas
nitric oxide
nitrogen dioxide, liquid
nitorgen peroxide, liquid
nitrogen tetroxide-nitric oxide mixtures, up to 33.2 percent nitric oxide
organic phosphate, organic phosphate compound, or organic phosphorus compound,
    mixed with compressed gas
parathion and compressed gas mixture
phenylcarbylamine chloride
phosgene (diphosgene)
phosphine
poisonous liquid or gas, n.o.s.
tetraethyl dithiopyrophosphate and compressed gas mixture
tetraethyl pyrophosphate and compressed gas mixture

Table 27 / Class B Poisons

acetone cyanhydrin
aldrin
aldrin mixture, liquid, with more than 60 percent aldrin
aldrin mixture, dry, with more than 65 percent aldrin
allyl alcohol
ammonium arsenate, solid
aniline oil drum, empty
aniline oil, liquid
arsenic acid, solid
arsenic acid, solution
arsenical compounds or mixtures, n.o.s., liquid
arsenical compounds or mixtures, n.o.s., solid
arsenical dip, liquid (sheep dip)
arsenical dust
arsenical flue dust
arsenical pesticide
arsenic bromide, solid
arsenic iodide, solid
arsenic chloride, liquid
arsenic pentoxide, solid
arsenic solid
arsenic sulfide (powder), solid
arsenic trichloride, liquid
arsenic trioxide, solid
arsenic, white, solid
arsenious acid, solid
arsenious and mercuric iodide solution, liquid
arsenious chloride, liquid
azinphos methyl
barium cyanide, solid
benzidine
benzoic derivative pesticide, liquid
benzoic derivative pesticide, solid
beryllium chloride
beryllium compound, n.o.s.
beryllium fluoride
bipyridilium pesticide, liquid
bipyridilium pesticide, solid
bordeaux arsenite, liquid
bordeaux arsenite, solid
brucine, solid (dimethoxy strychnine)

Table 27 / Class B Poisons (cont.)

calcium arsenate, solid
calcium arsenite, solid
calcium cyanide, solid, or calcium cyanide mixture, solid
carbamate pesticide (compounds and preparations), liquid
carbamate pesticide (compounds and preparations), solid
carbofuran
chemical ammunition, non-explosive, containing a Poison B material
4-chloro-o-toluidine hydrochloride
chloropicrin, absorbed
chloropicrin, liquid
chloropicrin, mixture (containing no compressed gas or Poison A)
cocculus (fishberry)
compound, tree- or weed-killing, liquid
copper acetoarsenite
copper arsenite
copper-based pesticide (compounds and preparations), liquid
copper-based pesticide (compounds and preparations), solid
copper cyanide
coumaphos
cyanide or cyanide mixture, dry
cyanide solution, n.o.s.
cyanogen bromide
dichlorvos
dinitrobenzene, solid, or dinitrobenzol, solid
dinitrobenzene solution
dinitrochlorobenzene
dinitrophenol solution
disinfectant, liquid
disinfectant, solid
disulfoton
dithiocarbamate pesticide (compounds and preparations), liquid
dithiocarbamate pesticide (compounds and preparations), solid
drugs, n.o.s., liquid
drugs, n.o.s., solid
endosulfan
endrin
ethion
ethylene chlorohydrin
ferric arsenate
ferric arsenite
ferrous arsenate (iron arsenate), solid

Table 27 / Class B Poisons (cont.)

flue dust, poisonous
grenade without bursting charge, with Poison B charge
hexaethyltetraphosphate, liquid
hexaethyltetraphosphate, dry (containing more than 2 percent hexaethyltetraphosphate)
hexaethyltetraphosphate, dry (containing not more than 25 percent hexaethyltetraphosphate)
hexaethyltetraphosphate, liquid (containing more than 25 percent hexaethyltetraphosphate)
hexaethyltetraphosphate, liquid (containing not more than 25 percent hexaethyltetraphosphate)
hydrocyanic acid solution, less than 5 percent hydrocyanic acid
insecticide, dry, n.o.s.
insecticide, liquid, n.o.s.
lead arsenate, solid
lead arsenite, solid
lead cyanide
london purple, solid
magnesium arsenate, solid
medicines, n.o.s., liquid
medicines, n.o.s., solid
mercuric acetate
mercuric ammonium chloride
mercuric benzoate, solid
mercuric bromide, solid
mercuric chloride, solid
mercuric cyanide, solid
mercuric iodide, solid
mercuric iodide, solution
mercuric nitrate
mercuric oleate, solid
mercuric oxide, solid
mercuric oxycyanide, solid
mercuric potassium cyanide, solid
mercuric potassium iodide, solid
mercuric salicylate, solid
mercuric subsulfate, solid
mercuric sulfatre, solid
mercuric sulfocyanate, solid, or mercuric thiocyanate, solid
mercurol or mercury nucleate, solid
mercurous acetate, solid
mercurous bromide, solid
mercurous gluconate, solid
mercurous iodide, solid
mercurous oxide, black, solid

Table 27 / Class B Poisons (cont.)

mercurous sulfate, solid

mercury-based pesticide (compounds and preparations), liquid

mercury-based pesticide (compounds and preparations), solid

mercury compound, n.o.s.

methyl bromide and more than 2 percent chloropicrin mixture, liquid

methyl bromide and non-flammable, non-liquified compressed gas mixture, liquid (including up to 2 percent chloropicrin)

methyl bromide-ethylene dibromide mixture, liquid

methyl bromide, liquid (including up to 2 percent chloropicrin)

methyl parathion, liquid

methyl parathion mixture, dry

methyl parathion mixture, liquid (containing 25 percent or less methyl parathion)

methyl parathion mixture, liquid (containing over 25 percent methyl parathion)

mevinphos

mexacarbate

motor fuel antiknock compound or antiknock compound

motor fuel antiknock compound or antiknock compound containing tetraethyl lead

nickel cyanide

nicotine hydrochloride

nicotine, liquid

nicotine salicylate

nicotine sulfate, liquid

nicotine sulfate, solid

nicotine tartrate

nitroaniline

nitrobenzene, liquid or nitrobenzol, liquid (oil of mirbane)

nitrochlorobenzene, meta or para, solid

nitrochlorobenzene, ortho, liquid

nitroxylol

organic phosphate mixture, organic phosphate compound mixture, or organic phosphorus compound mixture, liquid

organic phosphate mixture, organic phosphate compound mixture, or organic phosphorus compound mixture, solid or dry

organochlorine pesticide (compounds and preparations), liquid

organochlorine pesticide (compounds and preparations), solid

organophosphorus pesticide (compounds and preparations), liquid

organophosphorus pesticide (compounds and preparations), solid

organotin pesticide (compounds and preparations), liquid

organotin pesticide (compounds and preparations), solid

parathion, liquid

parathion mixture, dry

parathion mixture, liquid

Table 27   /   Class B Poisons (cont.)

perchloromethyl mercaptan
phenol
phenol, liquid or solution (liquid tar acid containing over 30 percent phenol)
phenoxy pesticide (compounds and preparations), liquid
phenoxy pesticide (compounds and preparations), solid
phenylurea pesticide (compounds and preparations), liquid
phenylurea pesticide (compounds and preparations), solid
phthalimide derivative pesticide (compounds and preparations), liquid
phthalimide derivative pesticide (compounds and preparations), solid
poisonous liquid, n.o.s., or Poison B, liquid, n.o.s.
poisonous solid, corrosive, n.o.s.
poisonous solid, n.o.s., or Poison B, solid, n.o.s.
potassium arsenate, solid
potassium arsenite, solid
potassium cyanide, solid
potassium cyanide, solution
selenium oxide
silver cyanide
sodium arsenate
sodium arsenite (solution), liquid
sodium azide
sodium selenite
strontium arsenite
strychnine salt, solid
strychnine, solid
substituted nitrophenol pesticide (compounds and preparations), liquid
substituted nitrophenol pesticide (compounds and preparations), solid
tetraethyl dithiopyrophosphate, liquid
tetraethyl dithiopyrophosphate mixture, dry
tetraethyl dithiopyrophosphate mixture, liquid
tetraethyl pyrophosphate, liquid
tetraethyl pyrophosphate mixture, dry
tetraethyl pyrophosphate mixture, liquid
thallium salt, solid, n.o.s.
thallium sulfate, solid
thiophosgene
toluene diisocyanate
triazine pesticide (compounds and preparations), liquid
triazine pesticide (compounds and preparations), solid
zinc arsenate
zinc arsenite
zinc cyanide

Table 28 / Irritants

ammonia	hydrogen iodide
bromine vapors	iodine vapors
chemical ammunition containing an irritating material	irritating agent, n.o.s.
	nitric oxide
chlorine	nitrogen dioxide
chloroacetophenone	nitrogen tetroxide
dinitrogen pentoxide	nitrogen trioxide
diphenylaminechloroarsine	nitrous oxide
fluorine	ozone
gas identification set	phosphorus pentachloride
grenade, tear gas	phosphorus trichloride
hydrogen bromide	sulfur dioxide
hydrogen chloride	tear gas candle
hydrogen fluoride	xylyl bromide

Table 29 / Characteristic Human Response to Varying Concentrations of Substances

Compound and Concentration, ppm	Symptoms
Acetaldehyde	
0.07 to 0.21	Threshold of odor
25 to 50	Transient slight irritation of eyes after 15 minutes
100	TLV value
134	Slight irritation of respiratory tract (30 minutes)
200	MAK value, irritation of nose and throat
Acetic acid	
10	TLV value, not irritating
20 to 30	No danger to workers exposed for seven to 12 years
25	MAK value
60	Slight irritation of respiratory tract, stomach, and skin
800 to 1,200	Cannot be tolerated for more than three minutes
Acetone	
0.5 to 1,000	Threshold of odor; quick adaptation
1,000	TLV and MAK value
2,000	No symptoms on workers exposed over many years
5,000	First narcotic symptoms induced
9,300	Irritation of throat after five minutes

Table 29 / Characteristic Human Response to Varying Concentrations of Substances (cont.)

Compound and Concentration, ppm	Symptoms
Acetonitrile	
40	TLV and MAK, odor detectable, no symptoms
80	No symptoms after four hours
160	Feeling of slight bronchial tightness (five minutes)
Acrylonitrile	
20	TLV and MAK value
Acrolein	
0.1	TLV and MAK value
0.805	Lachrymation, irritation of mucous membranes
1.0	Immediately detectable, irritation
5.5	Intense irritation
10 and over	Lethal in a short time
24	Unbearable
Ammonia	
1 to 50	Detectable odor
25	TLV value
50	MAK value
57 to 72	Respiration not significantly changed
96	Slight irritation of eyes, nose, and upper restiratory tract
100	Working possible; adaptation
200	Irritation of mucous membranes
500 to 1,000	Strong irritation of upper respiratory tract
2,000	Fatal
Benzene	
25	TLV value
500	Slight irritation
1,500 to 4,000	Dangerous to life after several hours
8,000	Fatal after 30 to 60 minutes
20,000	Fatal after five minutes
1,3-Butadiene	
1,000	TLV and MAK value
Butane	
1,000	MAK value

Table 29 / Characteristic Human Response to Varying Concentrations of Substances (cont.)

Compound and Concentration, ppm	Symptoms
Butyl acetate	
150	TLV value
200	MAK, irritation of upper respiratory tract, eyes, nose, and throat
300	Strong irritation effects
900	Strong irritation, appearance of narcotic effects with sensation of vertigo
1,800	Deep narcosis, vertigo, and unconsciousness
Carbon monoxide	
1,250	Lethal to mice after four hours
1,500	25 percent COHb in mice after five minutes
2,100	25 percent COHb in rats after five minutes
4,670	LC_{50} for rats in 60 minutes
5,000	Minimum lethal dose for rats in 30 minutes
5,500	LC_{50} for rats in 30 minutes
6,100	LC_{50} for rats in 20 minutes
8,800	LC_{50} for rats in ten minutes
Ethyl acetate	
0.2	Limit of perception of odor
200	Strong odor perceived
350	Irritation of nose and eyes
400	TLV and MAK value
700	Narcotic effects without fainting
3,800	Distinct narcosis, fainting, and significant irritation
Formaldehyde	
0.05 to 1.0	Threshold of odor
0.08 to 1.6	Slight irritation of eyes and nose
0.25 to 1.6	Threshold of irritation of eyes
0.5	Threshold of irritation of throat
1.0	MAK value
2.0	TLV value
10	Conjunctivitis, rhinitis, and pharyngitis in few minutes
10 to 15	Dyspnea, cough, pneumonia, bronchitis
over 50	Necrosis of mucous membranes, spasm of larynx, edema of lungs
Hexane	
500	TLV and MAK value

Table 29 / Characteristic Human Response to Varying Concentrations of Substances (cont.)

Compound and Concentration, ppm	Symptoms
Hydrogen chloride	
1 to 5	Limit of detection by odor
5	TLV and MAK value
5 to 10	Mild irritation of mucous membranes
35	Irritation of throat on short exposure
50 to 100	Barely tolerable
1,000	Danger of lung edema after short exposure
Hydrogen cyanide	
0.2 to 5.1	Threshold of odor
10	TLV and MAK value
18 to 36	Slight symptoms, headache, after several hours
45 to 54	Tolerated for one-half to one hour without difficulty
100	Fatal after one hour
110 to 135	Fatal after one-half to one hour, dangerous to life
135	Fatal after 30 minutes
181	Fatal after ten minutes
280	Immediately fatal
Hydrogen Fluoride	
3	TLV and MAK value
3 to 5	Redness of skin, irritation of nose and eyes after one-week exposure
32	Irritation of eyes and nose
60	Itching of skin, irritation of respiratory tract from exposure of one minute
120	Conjuctival and respiratory irritation, just tolerable for one minute
50 to 100	Dangerous to life after a few minutes
Hydrogen sulfide	
10	TLV and MAK value
20 to 30	Conjunctivitis
50	Objection to light after four hours, lachrymation
50 to 500	Irritation of respiratory tract
150 to 200	Objection to light, irritation of mucous membranes, headache
200 to 400	Slight symptoms of poisoning after several hours
250 to 600	Pulmonary edema and bronchial pneumonia after prolonged exposure
500 to 1,000	Systemic poisoning, painful eye irritation, vomiting
1,000	Immediate acute poisoning
1,000 to 2,000	Lethal after 30 to 60 minutes
over 2,000	Acute lethal poisoning

Table 29 / Characteristic Human Response to Varying Concentrations of Substances (cont.)

Compound and Concentration, ppm	Symptoms
Nitrogen dioxide	
5	TLV, MAK value, threshold of perception by odor
10 to 20	Mild irritant to eyes, nose, upper respiratory tract
20 to 38	No adverse effects on workers over several years
50	Distinct irritation
80	Tightness of chest after three to five minutes
90	Pulmonary edema after 30 minutes
100 to 200	Very dangerous after 30 to 60 minutes
250	Death after a few minutes
Octane	
400	TLV value
500	MAK value
Propane	
1,000	MAK value
Styrene	
60	Threshold of odor, no irritation
100	TLV and MAK value, strong odor, tolerable
200 to 400	Intolerable odor
216	Unpleasant subjective symptoms
376	Definite signs of neurological impairment
600	Irritation of eyes
800	Immediate irritation of eyes and throat, somnolence, and weakness
over 800	Nausea, vomiting, and total weakness
Sulfur dioxide	
3 to 5	odor threshold
5	TLV and MAK value
8 to 12	Slight irritation of eyes and throat, closing air tracts
20	Coughing and eye irritation
30	Immediate strong irritation, very unpleasant
100 to 250	Dangerous to life
600 to 800	Death in a few minutes

Table 29 / Characteristic Human Response to Varying Concentrations of Substances (cont.)

Compound and Concentration, ppm	Symptoms
Toluene	
100	TLV value
200	MAK value
190 to 380	No complaints
500 to 1,000	Headache, nausea, momentary loss of memory, anorexia, irritation of eyes
1,000 to 1,500	Palpitation, extreme weakness, loss of coordination, impairment of reaction time
2,000 to 2,500	Dizziness, nausea, narcosis after three hours
10,000	Immediately fatal
Toluene diisocyanate	
0.01	No irritation, no odor for 30 minutes
0.018 to 0.02	Odor threshold
0.02	TLV and MAK value
0.05	Slight irritation of the eyes
0.1	Tolerable irritation of eyes, nose, and throat
0.5	Heavy irritation of eyes, nose and throat
1.3	Heavy irritation, coughing,1 spasms of the bronchi, tracheitis lasting several hours after exposure

SOURCE: *Flammability Handbook for Plastics, Third edition.* Carlos J. Hilado (Westport, Connecticut: Technomic Publishing Company, 1982)

Glossary

Asphyxiant: A gas that is essentially non-toxic, but can cause unconsciousness or death by lowering the concentration of oxygen in the air or by totally replacing the oxygen in breathing air.

Blood Asphyxiant: A substance that interferes with the ability of the red blood cells to carry oxygen to the cells of the body and release it to those cells.

Cytotoxin: A poison that attacks a specific organ.

Hemotoxin: A poison that affects the circulatory system.

Irritant: A substance that is not classified as a poison but might produce dangerous or intensely irritating fumes.

LC$_{50}$: The lethal concentration in air of a gas, vapor, dust, or fume that kills half the test animals during the observation period.

LD$_{50}$: The lethal dose of a substance that kills half the test animals during the observation period.

Neurotoxin: A poison that attacks the nervous system.

Poison: Any substance that causes injury, illness, or death to living tissue by chemical means.

Respiratory Paralyzer: Any substance that will cause the respiratory system to shut down.

Simple Asphyxiant: Those materials that are not toxic, or that have a very low order of toxicity, but which will kill by diluting or displacing the oxygen needed for breathing.

Systemic Poison: Any substance that interferes with any vital bodily process.

Tissue Asphyxiant: Any substance that can be carried by the red blood cells to tissue cells and, upon release of such substance, will render the cells incapable of ever again accepting oxygen and thus killing them.

TLV: The threshhold limit value; the amount of a substance to which an average person in average health may be exposed in a 40-hour work week; the values may be averaged over time, and the TLV may also be referred to as the TWA, or time-weighted average.

Toxic: Anything harmful, destructive, deadly, or poisonous to the body.

Toxicity: The ability of a chemical substance to produce injury once it reaches a susceptible site in or near the body.

16

Radioactivity

Introduction

With all the hazardous materials we have discussed thus far, and with all the hazardous-materials incidents to which fire department personnel have responded over the years, the one type of hazardous-materials incident that is still the most mysterious and most dangerous is the incident involving radioactive materials. Even though the problems with incidents involving radioactive materials are usually straightforward enough, there is still a lot of mystery attached to radioactivity and the methods by which we hope to bring an incident involving radioactive materials to a successful conclusion.

Perhaps the problem is one of understanding exactly what is going on inside the package that carries the sinister trefoil emblem (see figure 48), sometimes called the propeller, that signifies that a radioactive material is inside. Perhaps the stigma attached to radioactivity is what causes the mental block that first responders seem to exhibit when they discover that they are responding to an incident involving radioactivity. There are many answers to this problem, but the only answer that makes sense is that in almost every other type of hazardous material that the first responder may encounter, he can *see* the danger (the corrosive material eating away at the container, the flammable liquid burning, or the water-reactive material reacting with water, or he can smell the hazardous material and know he is in some immediate danger), but in the case of

Figure 48
Placard and Labels
for Radioactive Materials.

the radioactive material he cannot *see* the real danger, the invisible radioactivity reaching out to touch him and cause some unspeakable harm to him.

It will be the purpose of this chapter to try to break down some of the mysteries of radioactivity, explain what is going on within a radioactive material, and how you can protect yourself when handling a radioactive incident. This chapter will not be a complete course on radioactivity. All the training and equipment that is needed to prepare you adequately to handle an actual radioactive incident must be provided over a much longer period of time than it takes to cover one chapter in a textbook. Use this chapter as an introduction to the problem, and then proceed to acquire the training in the actual handling of an incident, so that you may become well-versed in using the detection and protective equipment necessary to protect yourself when a radioactive incident occurs.

What Is Radioactivity?

To answer this first question, we must go back to our concept of the atom, and the various particles from which it is made. You will recall that every atom has two subatomic parts, the nucleus and the electron. The nucleus is the subatomic particle at the center of every atom; it contains essentially all the weight of the atom and all of its positive charge. The electron is a subatomic particle that has essentially no weight and a negative electrical charge. Inside the nucleus we find two nuclear particles, the neutron and the proton. The neutron has an atomic weight of one and no electrical charge; the proton has an atomic weight of one and an electrical charge equal to that of the electron but opposite in sign; the electrical charge on the electron is referred to as − 1 and the electrical charge on the proton is referred to as + 1. The electrons, as we said earlier, are orbiting the nucleus in well-defined orbits or rings, much like planets revolving around a sun. We know that it is the electrostatically opposite charge of the protons that keeps the electrons in orbit, but how are the neutrons and protons kept together in the nucleus? This force, and what happens to the nuclear particles it holds together, is the answer to radioactivity.

Since like charges repel each other, and the only charged particles in the nucleus are protons, there must be some force that holds the nucleus together. Since the only other particles present in the nucleus are electrically neutral neutrons, nuclear scientists have surmised that the force must emanate from these neutrons. This force has been called the "strong" force, and it is strong enough to hold most nuclei together. There are, however, some nuclei of some elements (and of some isotopes of these elements) that the force cannot hold together; these nuclei begin to disintegrate, following a basic law of nature that unstable materials may not naturally exist for long and that they *must* do whatever they can to achieve stability. The only way these unstable nuclei can reach stability is to throw off from each nucleus something that was causing the instability. This throwing off of particles is called radioactivity. It is, in reality, nuclear decay, or, as it is more commonly known, radioactive decay. It is the way nature has devised for unstable atoms to reach stability.

Before we become further involved inside the nucleus, let us first consider radiation. Radiation can be classified into two types: ionizing and non-ionizing. Non-ionizing types of radiation are waves of energy such as radiant heat, radio waves, and light. The amount of energy in these waves is small when compared to ionizing radiation.

Ionizing radiation, on the other hand, involves particles traveling in wave-like motions. This process includes a "ray" that may also be a particle but is today believed to be energy similar to light, although of a shorter wavelength and higher energy. When these particles or rays collide with matter, the particles of that matter's atoms with which they collide are the electrons. In fission reactions, where the nucleus is to be "split," very careful aim must be taken, and the particle used to collide with the nucleus must be raised to a certain speed, all under very carefully controlled conditions with extremely specialized equipment in very specialized experiments. When the particles of our present discussion collide with an electron, the tendency is to remove that electron from the atom, thus leaving a particle that is no longer an atom, but an ion. This is where the term "ionizing radiation" originates. X-rays are different from radioactive rays in that X-rays are generated from the energy lost by the *electrons* of *stable* atoms, while ionizing gamma rays originate from the *nucleus* of *unstable* atoms.

To understand more clearly the damage done by ionizing radiation, you must recall that ions are electrically charged particles which are very active chemically. Recall also that all atoms and all molecules are electrically neutral. Whenever an ion is created, it must have another ion, of electrically opposite charge, with which to react, so that the natural law of stability may be enforced. If the ion created by the collision of a radioactive particle is within a molecule of living tissue inside the human body, it is trapped there and cannot seek another ion with which to react. That fact causes changes within the human body to accommodate this unnatural presence; the changes that occur, if they grow large enough and are plentiful enough, will cause radiation sickness and/or cancer, and eventually death.

There are certain fundamental properties of radiation that must be understood before we go any further. Each of the fundamental radiation particles has mass. (This statement will bring a howl of protest from those who claim the gamma ray is "pure" energy and has no mass, but, since it is similar to the photon, and I believe that someday the mass of the photon will be measured or calculated, I take the position that the gamma ray is a particle. Any of you who feel more comfortable with the concept that the gamma ray is massless may continue to hold that belief.) Two of the particles, the alpha particle and the beta particle, possess an electrical charge, and all the particles have energy. We will discuss each of these radiation particles and their activities separately.

Types of Ionizing Radiation

Alpha Particles

The alpha particle is the largest of the four types of radiation particles. It is composed of two neutrons and two protons and therefore has an atomic weight of four and an electrical charge of +2. It is similar to the nucleus of the helium atom; when it is emitted from the nucleus of one of the large, heavy atoms, it seeks out two electrons to make it electrically neutral. When it gains these electrons, *it indeed becomes an atom of helium.* Alpha particles are emitted by nuclei that are trying to reach stability by transmutation to another element. You will recall that the atomic number of an element is defined as the number of protons in the nucleus of all the atoms of that element. The number of protons in the nucleus of the atom of *any* element never changes. If the number of protons in the nucleus of an atom does change, the atom becomes the atom of another element. Therefore, any nucleus of any atom that emits alpha particles will lose two neutrons and two protons and will thus become an atom of the element whose atomic number is two less than the element whose nucleus just emitted the alpha particles. This type of radioactivity exists in the very heavy atoms (those that have high mass numbers or high atomic weight).

Beta Particles

The beta particle is a particle that has the same mass as an electron and will have in some cases the same charge (−1), and in other cases the opposite charge (+1). The determination of the electrical charge is dependent upon which type of nucleus is emitting the beta particles. Emission of beta particles is typical of nuclei that have a ratio of neutrons to protons that is either high or low. The ratio is brought closer to one by the release of a beta particle. If the nucleus has many more neutrons than protons, the force within the nucleus causes a neutron to convert to a proton. Since the atomic masses of the neutron and the proton are identical, while the proton has a +1 charge and the neutron is neutral, the only difference between the two is an electron, which the nucleus will then emit. If the ratio of neutrons to protons is low, a proton will convert to a neutron, by throwing off a particle with the same mass as the electron, but with a +1 charge. In this case the positively charged electron is called a *positron.*

It was the mystery of the negatively charged beta particle that led ultimately to the discovery of the "strong" force. Since the nucleus contained only positively charged protons, and electrically neutral neutrons, from where in the nucleus did the negatively charged beta particle originate?

If you are interested in following the work that led to the present understanding of radioactive particles, any recently published college textbook on physics will contain a fairly detailed explanation.

Gamma Radiation

The third method of radioactive decay is by the emission of gamma rays or, more simply, gamma radiation. In this case, the nucleus is unstable because it has "too much energy," and the nucleus can rid itself of this instability by emitting radiation in the form of a photon, which is usually defined as a massless packet of pure energy. Light waves are thought to be made up of photons, but the waves in which light moves are appreciably longer than gamma radiation and therefore possess less energy. It may be easier to define gamma radiation as electromagnetic radiation of a considerably shorter wavelength and a considerably higher energy than light, heat, radio waves, and X-rays, which are also forms of electromagnetic radiation. The fission reaction (the "splitting" of nuclei) is a prime source of gamma radiation. The damage done by gamma radiation is ionization, just as in the case of alpha and beta emission. In this case, the gamma radiation possesses so much energy that it will "grab" an electron and leave an ion behind.

Neutrons

The fourth form of radiation is neutron radiation. The neutron is the neutral particle in the nucleus and is released by the explosion of a nuclear device. It is an extremely rare form of radiation, but it will be present in the event of the discharge of atomic weapons. The "neutron bomb" is a weapon that will depend heavily upon neutron radiation. This type of radiation is usually neglected when types of radioactivity are discussed, because the preparation of first responders for handling a radioactive incident usually involves a transportation accident of some sort and does not take into account the response of firefighters after a nuclear incident, whether that incident is war or the accidental discharging of an atomic weapon.

Damage by Radioactivity

Alpha Radiation

The alpha particle is by far the largest of the particles of radiation. Because of its size, it has the potential to do the most damage by ionization, simply because the probability of collision of alpha particles with an electron is higher than with the other forms of radiation. However, also because of its size and the probability of collision with the electrons of other atoms, particularly those of air, the alpha particle will not travel very far: the distance is from three to four inches. Normally, an alpha particle will not penetrate the epidermis, the layer of dead skin surrounding our bodies, and it can be stopped by a thin sheet of paper. Normal turnout gear will stop alpha particles. Great care must be taken, however, not to ingest or inhale anything emitting alpha particles, nor to allow that material to enter an open wound. Once an alpha particle-emitting substance enters the body, it will seldom be more than three or four inches away from a vital organ. The damage done by alpha particles inside the body can be devastating. Because of our ability to stop alpha particles easily, and their inability to travel very far or very fast, alpha particles are usually considered to be the least damaging type of radiation. Do not be lulled into carelessness by this unusual opinion of alpha particles.

Beta Particles

The damage caused by the beta particle is the same as that caused by any of the other radioactive particles, ionization. The beta particle is quite small when compared to the alpha particle. You will recall that the beta particle is the same size as an electron, but that it sometimes has a positive electrical charge rather than the negative charge associated with an electron. Since the beta particle is such a small particle, it can travel much farther than the alpha particle before it collides with an electron and forms an ion pair. The beta particle can travel up to 100 feet in air and will penetrate the skin layer, but it can be stopped by a thin piece of metal or by an inch or two of wood. It was originally felt that full turnout gear would stop beta particles, but recent tests indicate that normal turnout gear may not be effective at all in stopping beta particles. To be on the conservative side, you must assume that your full turnout gear will *not* protect you from beta radiation. You may have to use special protective equipment.

Gamma Radiation

By far the most dangerous of the three common types is gamma radiation. Neutron radiation is not usually considered in discussions of "common" types of radiation. Gamma radiation is considered by many to be pure energy and therefore can travel great distances, perhaps miles, at the speed of light (186,000+ miles per second). Again, since many feel that gamma radiation is pure energy, the photon may travel this great distance without an electron collision; the term "collision" may or may not be totally accurate; "capture" may be a more accurate descriptive term. When a collision does occur, however, ionization will be caused, just as in the other types of radiation. Because of its size (if any) and energy, gamma radiation is very deeply penetrating and will pass through any barrier that will routinely stop alpha and beta particles. This means that several feet of earth or concrete or several inches of lead will be needed to stop gamma radiation. The implication is that although your turnout gear protects you from alpha particles, and you may wear special protective clothing to stop beta particles, there is nothing you can wear to protect you from gamma radiation that will still allow you to move about.

Neutron Radiation

The neutron is a particle considerably larger than the beta particle but has only one-fourth the mass of the alpha particle. It is rare, being emitted by atomic weapons. It is very deeply penetrating and therefore highly dangerous, the type of radiation that will cause the most deaths among the survivors of the discharge of an atomic weapon. Lead shielding may be necessary for protection against neutron radiation.

Protection Against Radiation

There is only one "old adage" about protecting yourself from radiation; that is the phrase "Time, Shielding, and Distance." That adage is still true today. You must be concerned about exposing yourself to radioactive material, because of the damage the ionization process will cause; however, you must also be aware of becoming *contaminated*. There is a very big difference.

Exposure means that your body has been subjected to radiation emitted from the radioactive decay of an unstable nucleus. The damage done to your body by such exposure depends on the type of radiation to which you have been exposed, and the length of time the exposure occurred.

On the other hand, contamination means that the actual radioactive material has somehow attached itself to your body or clothing. Since radioactive materials are exactly like any other materials, except that they are emitting damaging radiation of some sort, they come in the same forms of matter as any other material; that is, radioactive materials may be gases, liquids, or solids. In the course of your normal activity at the scene of an incident, you may come into direct contact with the radioactive material and have it stick to your body or clothing. This situation means that you are now a source of radiation, wherever you go, as long as the material is on your person or clothing. No matter where you go, the radioactive material will continue disintegrating and producing harmful radiation. As long as this radioactive material remains in contact with you or your clothing, you are *contaminated*.

This does *not* mean that *you* are radioactive. You cannot become radioactive by contamination; you certainly will be exposed to radiation, and you will expose all others with whom you come in contact, but neither you or any others will become radioactive through either exposure or contamination. Radioactivity is a function of nuclear disintegration, not of exposure or contamination.

The implication of the dangers of contamination, however, should logically lead you to the conclusion that you must do something to protect against contamination, or to devise methods of *decontamination* at the site of any incident involving radioactivity. This task is done very simply.

This area will be called the "decontamination area." It will begin at some safe distance from the source of radioactivity and will extend the entire distance to the safe area, where all other incident-handling duties are being carried out. It will consist of three clearly marked zones, with strict supervision as to who may operate in which zone at any given time. Clear instructions must be given and followed as to how personnel and equipment may pass from one zone to another as they leave the source of radioactivity.

You must establish a "contamination zone," a "transition zone," and a "clean zone" in the decontamination area at every incident. Heavy disposable plastic sheets can be used, as well as any heavy sheets of any material; plastic sheets, however, are less expensive than tarps or other heavy cloth or fabric sheets. The sheets should be wide enough and long enough to allow some considerable distance between the clean end of the sheets and the end of the sheets in the contamination zone; you may have to use more than one sheet to get the distance required. The three zones should be roughly equal, and their boundaries clearly marked. All outer garments and equipment must be discarded in the contamination zone; this entire area should be well away from the radioactive material responsible for the radiation and the incident; no personnel should be in this zone except for those who have been exposed to the radioactivity. These personnel should then step into the transition zone to be checked for residual radioactivity whose presence would indicate some sort of contamination. If residual radioactivity is found, decontamination procedures should be carried out immediately; this information will be provided to you by expert instructors in specialized courses on radioactivity and radiation. No one involved in the incident should be allowed out of the contamination and/or transition zones until all traces of contamination have been eliminated, except when the contaminated individual must be rushed to a hospital. In this case, all personnel involved in this phase should alert the hospital of the emergency so that they may be prepared to handle it, as well as the hospital personnel themselves taking all necessary precautions against exposure and contamination.

Time

It should go without saying that the shorter the exposure time, the less the exposure. Since dosage rates (which will be explained later in this chapter) involve time, if you are exposed for half the time that another person is exposed, you will receive half the dose—assuming, of course, that you are dressed alike and were both exactly the same distance from the radiation source. It is absolutely mandatory that someone at the incident site be in charge of the management of time, and that all participants at the scene be informed of the exact time that they may have been exposed. All work be done quickly; if a task is unfinished, the exposed person must leave the area and be replaced by someone not yet exposed. Exposures to radiation are additive in their effect on the human body or on any other object.

Shielding

You will be able to protect yourself from beta particles by special protective clothing, and alpha particles will not penetrate normal turnout gear, but *you will not be able to protect yourself from gamma radiation with any sort of clothing.* An incident involving gamma radiation will cause serious radiation damage to anyone trying to handle the material, depending, of course, on the specific radioactive material and its level of energy. Try to keep *something* between you and a gamma radiator; the more dense the material and the more there is of it, the better. Of course, the best of all worlds is not to have to deal with any gamma ray-emitting substance.

Distance

The particles (or energy) emitted from a radioactive source will fall off (decline) as the distance away from the emitter increases. This is due to simple laws of nature and the symmetrical way the radiation occurs. As a matter of fact, a simple algebraic equation will determine *exactly* how the radiation declines as the distance is increased; this equation is expressed as the "inverse-square law." That is, the intensity of radiation falls off as the inverse square of the distance from the source. The algebraic expression is as follows:

$$I = I'/r^2$$

where I' = the original intensity of radiation
 I = the intensity of radiation at distance r
 r = the distance from the radiation source.

Stated simply, if you double the distance from the radiation source, the intensity is lowered by one-fourth. If you increase the distance ten times, the intensity is one one-hundredth of the original intensity; try it for yourself by solving the equation for I where $r = 1$ foot as the original distance from the radiation source and the second distance, r, = 10 feet. Let $I' = 1$ rad as the original intensity at one foot. A judicious use of time, distance, and shielding will dramatically lower the exposure rate of all personnel involved in the incident.

Units of Measurement

Exposure

The most commonly known unit of measurement of radiation exposure is the roentgen; however, the only type of exposure it measures is that of gamma radiation. Although the roentgen is a small amount of radiation, it is certainly significant. An exposure of 25 roentgens (abbreviated R) in a short period of time will cause detectable damage, while an exposure of 500 R will be fatal. The definition of the roentgen is technical, as it refers to the amount of ion pairs formed in 1 cc (cubic centimeter) of air; the definition can be found, if you are interested, in any reference book on radioactivity or in any college physics textbook.

Absorbed Dose

The rad (for *radiation absorbed dose*), is different from the roentgen in that the roentgen is a measure of exposure while the rad is the amount of radiation absorbed. For gamma radiation, the numerical difference between 1 roentgen and 1 rad is very small. The rate of absorbed dose is given in rads per hour.

Dose Equivalent

The dose equivalent is calculated by converting the dose in rads to a dose in rems (radiation *e*quivalent *m*an) by using a factor of one for beta and gamma radiation, and a factor of ten for alpha radiation. A more convenient measurement of dose is the millirem, or one-thousandth of a rem. The dose equivalent or rem is used to determine the biological damage done by the exposure.

Curie

The curie is the unit of activity, which is defined as the number of nuclei in a sample that disintegrate every second. It defines the rate of decay of a radioactive material. A sample of low activity is not as dangerous as the same-sized sample of a radioactive material with a higher activity. With the size of the sample known, you will know how dangerous it is by knowing its specific activity, which is defined as the number of curies per gram of radioactive material.

Summary

The handling of a radioactive material or any incident involving radioactivity is probably the most hazardous situation in which a firefighter can be placed. The first thing you must do when you respond to this type of incident is to notify the *experts*, the Nuclear Regulatory Commission. Let them decide what to do if you have gained control of the situation—and especially if you have not.

Second, you must remember that there is *no* chemical difference between radioactive materials and their stable counterparts. You cannot tell the difference between radioactive carbon and the stable form of carbon. More important, when radioactive carbon burns, the combustion products will be *radioactive carbon monoxide and radioactive carbon dioxide*. A radioactive material is just as soluble in water as the stable form. In other words, however a stable form of an element will act to form compounds in chemical reactions, or in any other reactions, the radioactive form of what material will react *chemically* in the same way.

Third, whenever you do handle a radioactive incident, remember the only protection you will have is time, distance, and shielding. Someone must keep track of exposure time; you must stay as far away from the source of radioactivity as you can (remember the inverse-square law), and you must always keep something dense between you

Table 30 / Radioactive Materials Commonly Transported in the U.S.A.

plutonium nitrate solution
radioactive device, n.o.s.
radioactive fissile, n.o.s.
radioactive material, limited quantity, n.o.s.
radioactive material, low specific activity (LSA), n.o.s.
radioactive material, n.o.s.
radioactive material, special form, n.o.s.
thorium metal, pyrophoric
thorium nitrate
uranium hexafluoride, fissile
uranium hexafluoride, LSA
uranium metal, pyrophoric
uranyl acetate
uranyl nitrate hexahydrate solution
uranyl nitrate, solid

and the radioactive source, especially if that is emitting gamma radiation.

Table 30 is a list of those radioactive materials commonly transported in the United States.

Glossary

Alpha Particle: The largest of the common radioactive particles, having a mass of 4 a.m.u.; it is identical to the nucleus of the helium atom, travels only three or four inches, and is stopped by a sheet of paper.

Beta Particle: A particle that is the same size as an electron, but may be charged positively or negatively, can travel up to 100 feet, and requires special protection in addition to turnout gear.

Contamination: The depositing of radioactive material upon the surfaces of people, structures, and objects.

Curie: A measure of radioactivity.

Dose: The total quantity of ionizing radiation received by a person.

Dose Rate: The amount of ionizing radiation to which a person has been exposed over a period of time.

Dosimeter: An instrument used for measuring total accumulated exposure to ionizing radiation.

Fissile: Radioactive material that may be able to undergo a nuclear fission reaction.

Gamma Ray: The most dangerous form of common radiation particle because of the speed at which it moves, and the great distances it can cover.

Geiger Counter: A device that may be used to detect low levels of nuclear radiation.

Half-Life: The time required for a radioactive material to undergo enough disintegrations to reduce its mass to one-half its original mass.

Inverse-Square Law: A rule which states that the amount of radiation absorbed falls off as the inverse square of the distance from the source of radiation.

Ionization: The production of an electrically charged particle from a neutral atom or molecule.

Isotope: A form of the same element having identical chemical properties but a different number of neutrons in the nuclei of its atoms.

Positron: A positively charged electron.

Rad: A unit of absorbed dose of radiation.

Radioactivity: The spontaneous disintegration of unstable nuclei accompanied by emission of nuclear radiation.

RBE: Acronym for *Relative Biological Effectiveness:* The conversion of the number of rads of gamma radiation to the number of rads of a different radiation that will produce the same biological effect.

Rem: A unit of biological dose of radiation.

Roentgen: A unit of exposure to gamma or X-ray radiation.

Shielding: Any material or obstruction that absorbs radiation.

Transmutation: The changing of one element into another by a nuclear reaction.

Unstable Elements: Elements that emit particles and decay to form other elements (radioactive elements).

X-Rays: Electromagnetic radiation of high energy possessing wavelengths lower than those in the ultraviolet region.

17

Explosives

Introduction

An explosive, as defined by DOT, is any chemical compound, mixture, or device, the primary or common purpose of which is to function by explosion, i.e., with substantially instantaneous release of gas and heat, unless such compound, mixture, or device is otherwise specifically classified in 49CFR, Parts 170-189 (Sec. 173.50). (See figure 49.) Stated another way, an explosive is a material or device that is created specifically for the purpose of exploding, which it will do upon demand. This definition eliminates all materials that will explode but were not created or designed to function that way. An example would be a mixture of natural gas in air; there is no doubt that in the right combination with air, and with the proper ignition source, there *will* be an explosion. Natural gas is collected and piped to consumers to be used as a *fuel*, however, and not as an explosive; therefore, even though natural gas will explode under the proper conditions, it is not classed as an explosive. There are many more examples of materials that will explode under certain conditions, such as ammonium nitrate, organic peroxides, monomers, all flammable gases, the vapors of all flammable and combustible liquids, coal and grain dusts, metal and organic powders, and blasing agents. The explosions caused by these materials will be devastatingly real, but none of the materials that can explode are classified as explosives.

Perhaps at this time we should look at explosions, classify them, and then move on to the discussion of explosives, simply because there

Figure 49
Placards for Explosives.

are many materials which are not explosives that will explode. By looking at the various types of explosions, we may be better able to define explosives.

Types of Explosions

An explosion is defined as a sudden, violent release of mechanical, chemical, or nuclear energy from a confined region. Another technical definition of an explosion is a process of rapid physical and/or chemical transformation of a system into mechanical work, accompanied by a change of its potential energy. Another author said that an explosion is a very loud "bang," followed immediately by the rapid going away of everything that was there. Regardless of the definition, whether serious or facetious, the effect is what concerns us. In every explosion, there is a loud noise of some sort, accompanied by the physical disturbance and/or destruction of the surrounding environment. The energy released by the explosion is usually very destructive to life and property, especially if the explosion is of an accidental nature. Of course, during time of war, or in criminal or terrorist activity, the explosion will not be accidental, but specifically designed to cause death and destruction.

There are four basic types of explosions; they are:

1. pressure relief
2. rapid oxidation
3. runaway polymerization
4. molecular decomposition

Pressure Relief

The pressure-relief explosion is typical of catastrophic container failure, and only on specific occasions is it associated with another explosion. The pressure-relief explosion is really a mechanical operation, the result of the failure of a container due to the rise of pressure above the container's design strength. A typical pressure-relief explosion is the failure of a boiler, or some other hot-water heating system, caused by overheating and/or the failure of a pressure-relief device to operate properly. It can also be caused by the weakening of the container fabric by corrosion, metal fatigue caused by excessive heat, or physical damage. One of the most devastating types of explosions is the *Boiling Liquid, Expanding Vapor Explosion*, or BLEVE; this particular phenomenon is really two explosions. The first is a pressure-relief explosion caused by the concurrent rise in internal pressure caused in turn by applied heat energy from another source and the weakening of the metal container by impinging flame; the second explosion is a rapid oxidation explosion caused by the ignition of the flammable gases that are instantaneously produced by the catastrophic failure of the container. You must remember that the BLEVE is *not* a pressure-relief explosion; it is caused by the conditions brought about by the pressure-relief explosion and subsequent ignition of flammable gases that result in a rapid oxidation explosion.

Rapid Oxidation

Examples of the rapid-oxidation explosion have been given in the preceding paragraph. The principle is the same as in any combustion reaction. Any time that fuel and an oxidizer (for example, the oxygen in the air) and the proper energy are brought together, there will be a fire. In this case, we have gases, vapors, or fine mists of flammable liquids intimately mixed with the air. In such a situation, when the proper energy is brought to this intimate mixture, the result is instantaneous combustion or an explosion. It might also be defined as a deflagration, which is extremely rapid burning.

Runaway Polymerization

In the runaway polymerization type of explosion, the instantaneous release of energy is the energy that results from a chemical reaction, which is different from the rapid oxidation reaction. In this case, there is a tremendous amount of energy released instantaneously, but not from an oxidation reaction. You will recall that monomers (the unstable materials that polymerize to form polymers) are extremely small molecules which are unstable *because* they are small *and* contain double bonds. When heat and/or pressure is applied to these materials outside the control of the chemical engineer (that is, outside the reaction vessel in which the polymerization reaction normally occurs), the double bonds present in these monomers break all at once. This instantaneous release of energy is the classic explosion known as runaway polymerization. As the pressure rises from this reaction, the container is ruptured, and the released energy is sufficient to vaporize and ignite whatever remaining monomer exists. This results in another explosion that resembles the *Boiling Liquid, Expanding Vapor Explosion,* or BLEVE. Whatever it resembles, however, the resulting explosion will cause serious destruction of property and will kill anyone within the danger zone. The runaway polymerization explosion is typical of monomers; it might be advisable for you to review Chapter Fourteen and become familiar with the names of these hazardous materials.

A runaway polymerization explosion becomes possible when heat and pressure override the safety of the inhibitor placed in the monomer to prevent this very occurrence. You must be aware of the fact that, even if the inhibitor is not driven out of the monomer, the monomers, most of which are easily liquifiable gases, are indeed subject to the BLEVE. No matter which explosion takes place, death and destruction will be the result. Regardless of how long the reaction takes, or the fact that a pressure-relief explosion may take place first as a result of the energy released in the runaway polymerization, and a rapid oxidation explosion may follow, although you may argue that the runaway polymerization explosion does not exist at all, be advised that there will be an explosion in this situation.

Molecular Decomposition

There are some materials which, rather than burning, undergo an instantaneous decomposition of their molecules, releasing enormous amounts of energy. This explosion is known as the molecular decomposition explosion. A classic example of a hazardous material that is not designed to be an explosive but which reacts in this way is acetylene. Presence of the triple bond in the acetylene molecule makes it extremely unstable. When burned in a controlled manner, acetylene will give off tremendous amounts of energy in a flame so hot it will melt any common metal. When acetylene is under pressure and shocked, however, all of the energy is released in the instantaneous decomposition of its molecule, resulting in a spectacular explosion that includes the generation of shrapnel from its container. Any explosive that operates by molecular decomposition will be a very effective explosive and, under most conditions, highly dangerous. A typical explosive that operates in this manner is nitroglycerine.

Other Types of Explosions

1. Nuclear Fission

The nuclear fission explosion is the detonation of an atomic weapon. Nuclear fission is the splitting of an atomic nucleus induced by bombardment with neutrons from an external source and propagated by the neutrons so released. A nuclear fission explosion will not take place until there is a critical mass of fissionable material present. The critical mass is the minimum amount of fissionable material needed to sustain the chain reaction. There is a vast amount of energy released with a nuclear fission explosion, and, under controlled conditions, that is, with the partial absorption of neutrons in a reactor, the fission can be controlled so that the energy released can be used to heat water and generate electrical power. Gamma radiation is a byproduct of nuclear fission.

An example of a nuclear fission explosion is the explosion of an atomic bomb. Another nuclear fission explosion may take place in a nuclear reactor *if* a breakdown occurs and a critical mass is achieved.

2. Nuclear Fusion

A nuclear fusion reaction is one in which the nuclei of light atoms (the hydrogen isotopes, deuterium and tritium) fuse together to form helium. The nuclear fusion explosion is represented in the detonation of the hydrogen bomb, which is usually referred to as uncontrolled fusion. The energy released from a nuclear fusion explosion is greater than that from a nuclear fission explosion. A controlled form of nuclear fusion is very "clean," that is, there is no great radioactive hazard. Scientists are working hard to develop a nuclear-fusion energy plant, to replace the current nuclear fission plants, because more energy is liberated in the fusion reaction, and there is no radioactive waste. A further example of nuclear fusion is the reaction that constantly takes place on the surface of the sun, from which the Earth receives its energy.

Types of Explosives

Since the pressure-relief explosion is a mechanical operation, runaway polymerization involves hazardous materials that are not explosives, and the nuclear fission and nuclear fusion bombs are meant for use in wartime, attention will be centered on explosives that operate by rapid oxidation and molecular decomposition. Nothing will be said about military explosives, since there are few transportation accidents which occur that involve them. One or two will be mentioned in passing, but the majority of our time will be spent on the commercial explosives, some of which, of course, are used as, or in making, military explosives.

High Explosives

A definition of high explosive is a chemical compound or mixture of compounds, usually containing nitrogen, that detonates as a result of shock or heat. It may or may not be combustible, but it may be relatively insensitive to heat and shock, even though heat and shock cause its detonation. The high explosives are set apart from low explosives in that detonation is the method by which high explosives explode, while low explosives usually deflagrate (burn very, very rapidly). An explosion, according to some experts, proceeds at speeds below the speed of sound, and its method of propagation is a heat wave, while a detonation proceeds at supersonic speed and is propagated by a shock wave. Some go so far as to suggest that an explosion and a deflagration are the same, while the detonations differ because of the shock wave. High explosives are usually divided into classes according to their sensitivity. These classes are: the main charge, primary explosives, and secondary explosives.

1. Main Charges

A main charge (so named because it is one of the powerful explosives that do the work for which explosives are intended) is usually a high explosive that is *relatively* insensitive to heat or shock. These explosives usually are the most powerful of the explosives in their *brisance*; that is, the sharp, shattering effect an explosive has on its surroundings. They are so insensitive to heat and shock, however, that it usually takes another explosive to cause them to detonate.

Probably the most famous of the high explosives is 2,4,6-trinitrotoluene, better known as TNT. It exists as yellow crystals, and it is toxic by inhalation, ingestion, or by skin absorption. It is flammable and will detonate only if violently shocked or heated to 450°F. It is shipped in wooden casks or paper bags. TNT is relatively safe to handle, and it requires a strong initiator to cause detonation. Another name for TNT is methyltrinitrobenzene.

Another high explosive that is closely related to TNT is picric acid. It also appears as yellow crystals and is toxic by skin absorption. It is shipped in one- and five-pound bottles, 25-pound boxes, 100-pound kegs, and 300-pound barrels. Other names for picric acid are picronitric acid, trinitrophenol, nitroxanthic acid, carbazoic acid, and phenoltrinitrate.

Ammonium picrate is another high explosive shaped as yellow crystals. It is a hazard only when dry, but it is flammable when wet. Ammonium picrate is the explosive used in armor-piercing military shells; it is also used in pyrotechnics. Other names for ammonium picrate are ammonium carbazoate and ammonium picronitrate.

One of the most famous of all explosives is nitroglycerine, a yellow, viscous liquid. When pure, nitroglycerine is relatively safe to handle. If it is contaminated or hot, it then becomes the extremely shock-sensitive liquid you have read about or seen in movies. It is toxic by ingestion, inhalation, and skin absorption. It is seldom used alone as an explosive because of the danger of its becoming sensitized. When nitroglycerine is mixed with an inert substance, its sensitivity is overcome; as long as it remains mixed with the inert material, it will be safe to handle. In this form, it is known as dynamite. Alone, it is also known as trinitroglycerine and glyceryl trinitrate.

Dynamite was invented in 1888 by the Swedish researcher Alfred Nobel, as a safe way to handle nitroglycerine. He mixed it with an inert mineral, diatomaceous earth, and found that the result could be handled quite safely. He also discovered that if you wrap the dynamite in waxed paper, you can form it into "sticks" of predetermined amounts, resulting in an ability to judge the power of each stick. Nobel's invention was used in modern warfare, as the answer to the dangerous nitroglycerine, and he became so guilt-ridden that he established a series of prizes to be awarded annually to the greatest scholars in predetermined topics, and used this means also to establish his Peace Prize. Today, dynamites still use nitroglycerine, but the overall use of nitroglycerine dynamites is decreasing. When nitroglycerine *is* used, it is mixed with a combustible dope, wood pulp and calcium carbonate.

Other high explosives that are relatively stable are Composition B, HBX-1, EDNA (ethylenedinitramine), gelatin dynamite, and the "plastic" explosives, C-3 and C-4.

2. Primary Explosives

Primary explosives are those high explosives that are *extremely* sensitive to heat and shock. For this reason they are among the most dangerous of the high explosives. Because of this sensitivity they are absolutely useless as the main charge, but they *are* useful for *initiating* the detonation of high explosives. Furthermore, they do not have the brisance, or explosive power, that the high explosives do, but they are powerful enough to detonate the main charge. They are known as initiators or detonators. The principal members of this group are mercury fulminate, silver fulminate, and lead azide.

3. Secondary Explosives

Secondary explosives are those high explosives that are less sensitive to heat and shock than the primary explosives but more sensitive than the main-charge explosives. They are also more powerful than the initiators but less powerful than the main-charge explosives. Secondary explosives are used in those situations where the sensitivity of the main-charge explosives is such that the power of the primary explosive is not enough to cause detonation of the main charge. You can always use more of the primary explosive, but its detonation may just scatter the main charge rather than detonate it; thus the secondary explosive will be detonated and in turn it will detonate the main charge. In boosting the *relatively* weak power of the primary explosive, the secondary explosives have become known as "booster" explosives. When the primary explosive, the secondary explosive, and the main charge are arranged so that the primary explosive detonates the secondary explosive, which in turn detonates the main charge (all of which occurs, of course, in milliseconds), this sequence of events is known as an *explosive train*. Some secondary explosives are RDX (cyclotrimethylenenitramine), tetryl (trinitrophenylmethylnitramine), lead styphnate (lead trinitroresourcinate), and PETN (pentaerythritoltetranitrate).

Do not be misled by the terms *relatively insensitive* or *less powerful* into thinking that the primary and secondary explosives are not powerful. It does not take much of these materials to kill you when they detonate. As a matter of fact, a blasting cap contains these materials, may be half the thickness of a pencil and only three inches long, and may effectively blow off the hand that is holding one when it detonates. Also, the property that sets high explosives apart from low explosives is the speed with which the respective reactions take place. You will be killed by the shock of either explosion; if the shock does not get you, the flying debris or the collapsing building will.

Low Explosives

Low explosives are those explosives that deflagrate rather than detonate, unless they are confined. The low explosives are also called propellants, since that is, or originally was, their main function. To deflagrate is to burn very, very fast; if the deflagrating material is confined at any given point, the deflagration will transit to an explosion. The term low explosive does not indicate a low hazard. As a matter of fact, the presence of one particular explosive in the classification is responsible for the fact that low explosives are considered more hazardous than high explosives. There are only three principal explosives in this class: black powder, smokeless powder, and cordite. Of these three, black powder is the explosive that is so very dangerous because of its sensitivity to heat, shock, or energy of *any* kind.

1. Black Powder

Black powder is one of the oldest explosives known and probably the easiest to make. The ingredients are so readily obtainable that they will not be mentioned here for fear that the reader will experiment to see if black powder really is so easily made. There is no definite ratio of its ingredients, since it is *so* sensitive over such a wide range of proportions of the ingredients. Because it is so easy to make, and the ingredients are so available, black powder has become the favorite explosive of terrorists. Since it is so sensitive, it is also the explosive that kills the most terrorists, usually during its manufacture. When the "right" proportions are chanced upon, it is said that black powder is sufficiently sensitive that the friction of a fly landing on the material will cause it to explode. Whether this is true or not is scarcely important, since the material *is* so sensitive that you *can* cause it to explode by gently pouring it from one container to another. Needless to say, the grinding together of the ingredients to get an intimate mixture will end the mixing with a bang! Unfortunately, black powder is no laughing matter, since many well-meaning people (people who are not terrorists and/or do not want to hurt others) are seriously injured or killed trying to make their own black powder. *Don't try it!*

2. Smokeless Powder

Smokeless powder is a low explosive developed to replace black powder as the propellant in gun cartridges. The development was not necessarily to make guns safer, but to eliminate the telltale puff of smoke emitted when a gun was fired, thus giving away the gunner's position, and allowing someone else with a gun to shoot back at the puff of smoke, thus revealing the opponent's position. Nevertheless, smokeless powder is safer than homemade black powder by several orders of magnitude. The implication here is that commercial black powder is safe; it is *still* a low explosive, regardless of the label on the can, but it is very much safer than homemade black powder. Smokeless powder is made from guncotton (nitrocellulose) and nitroglycerine.

3. Cordite

Cordite is a form of smokeless powder also made from nitroglycerine and nitrocellulose, but it contains an extra ingredient to thicken and stabilize it. It is then rolled and cut into cords to be used as the propellant for military artillery shells.

Other Explosives

There are many other types of commercial explosives, each developed and designed to do a specific job. They are usually organic compounds containing the azide, bromate, chlorate, chlorite, iodate, nitrate, nitrite, perchlorate, and picrate functional groups; these are *covalent* compounds, not *ionic* compounds, with the exception of the ammonia compounds, and an occasional lead or silver compound. Beware of any compound that has an organic name (with -methyl-, -ethyl-, -phenyl-, and so on, somewhere in the name) and one of the above-mentioned functional groups in it. You *must* treat that as an explosive until you prove it otherwise. Another hint is the presence of the nitro group in the name. Not all the organic nitro compounds are explosives, but only a few are not!

You must also contend with brand names and generic names and initials. There are many new explosives made with ammonium nitrate and powdered aluminum; others use liquid oxygen with some organic material soaked into it to form an explosive gel to be used right where it is made. To keep current with the state of the art in explosives and their manufacture, you should arrange to have an expert in explosives (an expert associated with the manufacture of explosives, not their detonation) train you annually in how to identify explosives and what to do in an incident involving them.

Blasting Agents

A blasting agent is a material that is designed to explode but is not classified as an explosive (see figure 50). It is a mixture of materials that is intended to have enough power to shatter or loosen something, usually in a construction project, but is so insensitive to shock that it will take a powerful blasting cap (which *is* an explosive) to make it explode. Most blasting agents have similar compositions, of ammonium nitrate plus a liquid fuel, usually a fuel oil (a combustible liquid). To be classified as a blasting agent, the mixture must not contain any material that can be classified as an explosive. There are other materials, such as ground nut hulls and other combustibles, that may be added to the mixture. These blasting agents are also known as nitro-carbo-nitrates and may be listed on a bill of lading as NCN.

Many blasting agents are transported by truck, although many are mixed at the location where they are to be used. The blaster (the person mixing and using the blasting agent) supposedly knows all the safety regulations necessary to keep him from blowing himself and others to bits.

A special word of caution is required here, concerning an ingredient of blasting agents, ammonium nitrate. Ammonium nitrate is manufactured principally as a *fertilizer*. Far and away the largest amount of ammonium nitrate manufactured is for agricultural purposes and not as an ingredient in blasting agents. This does not imply that the ammonium nitrate that a farmer uses as a fertilizer will not explode, because it will.

Ammonium nitrate will not explode every time it is shocked, or every time it burns, or every time it is contaminated and *then* is shocked or burned. Ammonium nitrate *will* explode, however, as proved by the Texas City disaster. When will it explode? No one is

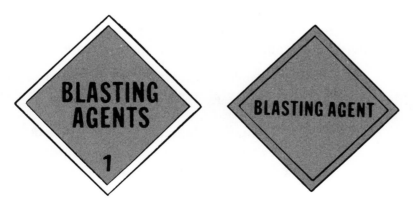

**Figure 50
Placard and Label
for Blasting Agents.**

exactly sure *when* the conditions are just right. Are you willing to bet your life that it will *not* explode the next time you are handling an emergency involving ammonium nitrate?

Other Definitions

There are many other classifications into which explosives may be placed, and other phenomena of explosives of which you should be aware. There will be no further classification (other than DOT's), but you should know the definitions that accompany these phrases.

1. A forbidden explosive is an explosive that may not be transported by common carrier; that is, these explosives may not be shipped by the regular means of transportation by which all other goods are shipped. These particular explosives must be shipped by the manufacturer in his own vehicles or by firms which specialize in the transport of explosives and carry no other goods.

2. An acceptable explosive, on the other hand, may be shipped by common carrier.

3. A sympathetic explosion occurs when one material is detonated by the shock wave of another explosion.

4. A permissible explosive is an explosive that may be used in underground mines, on the premise that the heat generated by the explosion will not ignite any flammable gases or vapors present.

5. A magazine is a special storage area for explosives.

DOT Classification of Explosives

Class A explosives (see figure 51) are explosives that are detonating or otherwise of maximum hazard. They are as follows:

Type 1. Solids that deflagrate on contact with flame but cannot be detonated by a number 8 blasting cap. An example is black powder.

Type 2. Solids containing liquid explosive ingredients that can be detonated by a number 8 blasting cap when unconfined. An example is dynamite made from nitroglycerine.

**Figure 51
Placard and Label for
Class A Explosives.**

Type 3. Solids with no liquid ingredients but which can be detonated with a number 8 blasting cap when unconfined. Examples are amatol, picric acid, tetryl, and TNT.

Type 4. Solids that can be caused to detonate, when unconfined, by contact with sparks or flame. Examples are mercury fulminate and lead azide.

Type 5. Desensitized nitroglycerine.

Type 6. Liquids that can be exploded when an eight-pound weight is dropped on them from less than ten inches. An example is nitroglycerine.

Type 7. Blasting caps, where the total of explosives per unit does not exceed 150 grains.

Type 8. Any solid or liquid compound not included above; an example is a commercial "shaped" charge.

Type 9. Certain propellant explosives such as black powder.

Class B explosives (see figure 52) are those explosives that, in general, function by rapid combustion rather than detonation and include some explosive devices such as special fireworks and flash powders.

Class C explosives (see figure 53) are certain types of manufactured articles containing Class A or Class B explosives, or both, as components, but in restricted quantities, and certain types of fireworks.

Emergencies Involving Explosives

Regardless of any statement as to the *relative* safety of any explosive, or its insensitivity to heat or shock, or the fact that it might be a relatively "weak" explosive, all these so-called safeguards disappear

**Figure 52
Placard and Label for
Class B Explosives.**

**Figure 53
Placard and Label for
Class C Explosives.**

when the material is exposed to heat, has already been shocked, or has been contaminated. If any of these things happen to explosives—*any* explosives—anything can happen. Many of the explosives are flammable and, while burning, of course are subjected to high heat; this situation can cause the unburned portion of the explosive to detonate, even when the explosive normally deflagrates and does not *ordinarily* detonate. In other words, you should be as careful in handling incidents involving explosives as you would be with any other hazardous material capable of taking your life in an instant.

In a fire situation, your first order of business is to identify the explosive involved. If the fire is in the cargo area, in a transportation incident, or in a storage area, you should not fight the fire unless it can be done from behind explosion-proof barriers—as if such barriers truly existed. Remember, you must protect yourself and others from the blast which will surely come, but you must also be aware of the shrapnel and the collapse of the structure which may follow the explosion, in addition to the many other fires that may be started by the heat of the explosion.

If the fire is approaching the storage or cargo area, as in a tire fire, do what you can to extinguish the fire from a protected area; that is, use flooding amounts of water from as far away as possible. Do not approach the explosives, because they may have already been sensitized by the heat of the approaching fire. Treat any other fire involving the body or engine of the truck in the same manner. Get help as soon as you know explosives are involved; contact the manufacturer *and* the shipper, and let them know what is happening. Listen to their advice before you make your decisions; above all, evacuate all unnecessary personnel to a safe distance. Protection of property should be considered only *after* all people, including firefighters, are in a safe location. Any time you bring an incident involving explosives to a safe conclusion, it means you were either extremely lucky or extremely conservative. Do *not* let anyone with property at stake talk you into exposing human life to almost certain destruction.

Other Explosives

Many materials are regulated by the U.S. Department of Transportation according to the function of the chemical, compound, or device. Table 31 is a list of such materials and devices, broken down by the DOT's classification.

Table 31 / Other Explosives

Class A Explosives
ammunition, chemical, explosive, with Poison A material
ammunition, chemical, explosive, with Poison B material
ammunition, chemical, explosive, with irritant
ammunition for cannon with explosive projectile
ammunition for cannon with gas projectile
ammunition for cannon with illuminating projectile
ammunition for cannon with incendiary projectile
ammunition for cannon with smoke projectile
ammunition for cannon with tear gas projectile
ammunition for small arms with explosive projectile
ammunition for small arms with incendiary projectile
booster explosive
burster, explosive
charged oil well jet perforating gun (total 20 percent or more explosive)

Table 31 / Other Explosives (cont.)

Class A Explosives (cont.)

commercial shaped charge
detonating fuze, Class A explosive
detonating primers, Class A explosives
detonators, Class A explosive
explosive bomb
explosive mine
explosive projectile
explosive torpedo
fuze, detonating
fuze, detonating, radioactive
grenade, hand or rifle, explosive
high explosive
high explosive, liquid
igniter, jet thrust (jato)
igniter, rocket motor
initiating explosives
jet thrust unit (jato)
low explosive
propellant explosive
rocket ammunition with explosive projectile
rocket ammunition with gas projectile
rocket ammunition with illuminating projectile
rocket ammunition with incendiary projectile
rocket ammunition with smoke projectile
rocket motor
shaped charge
supplementary charge, explosive

Class B Explosives

ammunition for cannon with empty projectile
ammunition for cannon with inert loaded projectile
ammunition for cannon without projectile
ammunition for cannon with solid projectile
ammunition for cannon with tear gas projectile
explosive power device
fireworks, special
grenade, without bursting charge (with incendiary material)
igniter, jet thrust (jato)
igniter, rocket motor

Table 31 / Other Explosives (cont.)

Class B Explosives (cont.)

jet thrust unit (jato)
propellant explosive in water (smokeless powder)
propellant explosive in water, unstable, condemned, or deteriorated (smokeless powder)
propellant explosive, liquid
propellant explosive, solid
rocket ammunition with empty projectile
rocket ammunition with inert loaded projectile
rocket ammunition with solid projectile
rocket engine, liquid
rocket motor
starter cartridge
torpedo, railway

Class C Explosives

actuating cartridge, explosive
black powder igniter with empty cartridge bag
cannon primer
cartridge bags, empty, with black powder igniter
cartridge cases, empty, primed
cartridge, practice ammunition
charged oil well jet perforating gun (less than 20 percent explosive)
cigarette load
combination fuze
combination primer
cordeau detonant fuse
delay electric igniter
detonating fuze, Class C explosive
detonating primers, Class C explosives
detonators, Class C explosives
electric squib
empty cartridge bag with black powder igniter
explosive auto alarm
explosive cable cutter
explosive power device
explosive relief device
explosive rivet
fireworks, common
flexible linear shaped charge, metal-clad
fuse igniter

Table 31 / Other Explosives (cont.)

Class C Explosives (cont.)

fuse, instantaneous
fuse lighter
fuse, mild detonating, metal-clad
fuse, safety
fuze, combination
fuze, detonating, Class C explosive
fuze, percussion
fuze, time
fuze, tracer
grenade without bursting charge (with smoke charge)
grenade, empty, primed
hand signal device
igniter
igniter cord
igniter fuse, metal-clad
oil-well cartridge
percussion cap
percussion fuze
safety squib
signal flare
small arms ammunition
small arms ammunition, irritating (tear gas) cartridge
small arms primer
smoke candle
smoke grenade
smoke pot
smoke signal
starter carter cartridge
toy caps
toy propellant device
toy smoke device
tracer
tracer fuze
trick matches
trick noise maker, explosive
Very signal cartridge

Table 32 / Explosives and Their Classifications

Explosive	Classification
ammonium fulminate	forbidden
ammonium nitrate	high explosive
ammonium perchlorate	high explosive
ammonium picrate	high explosive
black powder	low explosive
C-3	high explosive
C-4	high explosive
chlorate powder	high explosive
Composition B	high explosive
cordite	low explosive
diazodinitrophenol	initiating explosive
dinitroglycol	rocket fuel
dinitromethane	forbidden
dynamite	high explosive
ethylenedinitramine (EDNA)	high explosive
fulminate of mercury (dry)	forbidden
fulminate of mercury (wet)	initiating explosive
guanyl nitrosoamino guanylidene hydrazine	initiating explosive
guanyl nitrosoamino guanyladene tetrazine	initiating explosive
lead azide (dextrinated type only)	initiating explosive
lead mononitroresorcinate (wet)	initiating explosive
lead mononitroresorcinate (dry)	forbidden
lead styphnate	initiating explosive
methyl nitrate	rocket propellant
nitrocellulose	low explosive
nitroethane	propellant
nitroguanidine	high explosive
nitroglycerine	high explosive
nitromannite (dry)	initiating explosive
nitromannite (wet)	initiating explosive
nitrosoguanidine	initiating explosive
nitrourea	high explosive
pentaerythritol tetranitrate	secondary explosive
picric acid	high explosive
silver acetylide	initiating explosive
tetranitroaniline	primary explosive
tetranitromethane	rocket fuel
tetrazene	initiating explosive
tetryl	high explosive
trinitroanaline (picramide)	high explosive

Table 32 / Explosives and Their Classifications (cont.)

Explosive	Classification
trinitroanisol (methyl picrate)	high explosive
trinitrobenzene	high explosive
trinitrobenzoic acid	high explosive
trinitrocresol (cresolite)	high explosive
trinitromethylenetriamine (cyclonite) (RDX)	high explosive
trinitronaphthalene	high explosive
trinitroresorcinol (styphnic acid)	primary explosive

Table 32 lists specific chemical compounds and mixtures classified as explosives by DOT. They are all shipped with an Explosive A placard and label. The exceptions are those explosives marked forbidden, which means that they may not be shipped by common carrier.

Glossary

Blasting Agent: A mixture of materials, usually ammonium nitrate and fuel oil, that is used instead of explosives in certain applications. It does not contain any explosives and is so insensitive to heat and shock that special procedures are required to cause it to explode.

Blasting Cap: A device containing a primary and/or a secondary explosive, used to detonate the main charge.

Booster Explosive: *See* secondary explosive.

Brisance: The sharp shattering effect an explosive has on its surroundings; also referred to as the explosive power of an explosive.

Deflagrate: To burn very, very rapidly; the speed of reaction of deflagration is much faster than ordinary combustion but slower than detonation.

Detonation: Same definition as an explosion in many cases; technically, the major difference is that a detonation is propagated by a shock wave and travels at supersonic speeds.

Explosion: A sudden, violent release of mechanical, chemical, or nuclear energy from a confined region; an explosion is propagated by a heat wave and travels at subsonic speeds; often used interchangeably with detonation.

Explosive: Any material or device that is created specifically for the purpose of exploding, which it will do on demand.

Explosive Train: The building up of sufficient brisance to detonate the main charge; usually uses a primary explosive to detonate a secondary explosive, which detonates the main charge.

Forbidden Explosive: An explosive that may not be transported by common carrier.

High Explosive: An explosive that is *relatively* insensitive to heat and shock, and usually detonates rather than deflagrates; usually divided into primary or initiating, secondary, and main-charge explosives.

Initiating Explosive: *See* primary explosive.

Low Explosive: An explosive that is sensitive to heat and shock and usually deflagrates rather than detonates, unless confined; also called propellants.

Main Charge: The high explosive that does the most damage to the surroundings.

Permissible Explosive: An explosive that may be used in mines.

Primary Explosive: A very sensitive high explosive that is used to detonate a secondary explosive in an explosive train; also called an initiating explosive.

Propellant: *See* low explosive.

Secondary Explosive: A sensitive high explosive that is used to detonate the main charge in an explosive train, if the brisance of the primary explosive is not high enough to cause the detonation; also called "booster explosives"; may also be used as an initiating explosive.

18

Water- and Air-Reactive Materials

Water-Reactive Materials

There is a classification of hazardous materials that pose a new and different hazard for firefighters, who must be aware of them when responding to an incident, whether fire is present or not. Firefighters rely on water for more than extinguishment of fires—for "sweeping" water-soluble gases and vapors from the air or dispersing of flammable gases or vapors by the use of fine sprays or fogs. There are, however, some hazardous materials that will react in a highly detrimental manner when they come into contact with water. In some cases, these materials will break into flame, produce toxic or flammable gases or vapors, tremendous amounts of heat, or some other adverse effect. As a class, they are known as water-reactive materials or, more simply, as water reactives (see figure 54). Knowing the adverse effects of applying water to these materials, you may still choose to apply water, simply because it is all you have with which to fight, or because the dangerous conditions which will result from the application of water are not as bad the failure to apply water. Nevertheless, you should be aware of what these materials are, how to handle them without water, and what will result if water does contact them (see figure 55). We will look at each group of water-reactive materials and discuss what happens in each case.

Some of the materials mentioned will be classified by DOT as flammable solids, even if the materials themselves do not burn. This is

Figure 54
Placard and Labels for Water-Reactive Materials.

Figure 55
Fighting a Fire of Water-Reactive Materials (red fuming nitric acid spill)

David M. Lesak

Chapter 18 / Water- and Air-Reactive Materials

because when such a material comes into contact with water, it will liberate a flammable gas that will usually be ignited by the heat also liberated by the reaction, or simply because extremely flammable gases are liberated. The "Flammable Solid" designation is usually accompanied by a "Dangerous When Wet" placard or label.

Other water-reactive materials are classified by DOT as "Corrosive Liquids," since this is the major hazard. When water is used to control these substances, however, toxic, irritating, or flammable gases may be evolved.

You may not always be able to tell from the placard or label if the material is water-reactive or not. It is important for you to know which materials will react with water, how violent that reaction will be, what is evolved when water is used, and whether the results of the spill will be worsened if you do choose to use water. Sometimes, you may have no choice but to use water. Know the consequences of your actions!

The Alkali Metals

The alkali metals (the elements of group IA of the Periodic Table), lithium (Li), sodium (Na), potassium (K), rubidium (Rb), and cesium (Cs), will all react violently when they come into contact with water. The alkali metals are so reactive that they do not exist as metals in nature; they want to react so badly that they will seek out the oxygen in the water molecule and literally rip it from its chemical (covalent) bonding with hydrogen. The action is *exothermic*; that is, heat is liberated in the reaction. The exothermic reaction produces so much heat that it is sufficient to ignite the hydrogen released in the same reaction. Furthermore, the reaction is so violent that, if the piece of metal is very large (the size of the last joint of your thumb, or larger), the metal will burst apart into smaller pieces, scattering these about and causing many more reactions. As the pieces of metal shrink in size during the reaction, they assume the shape of tiny ball bearings, rolling around the surface of the water (lithium, sodium, and potassium all have specific gravities of less than 1.0), spitting and cracking as they react with the water. Remember hydrogen burns very hot, and that most of its energy of combustion is liberated as heat and very little as light; you will not be able to see the flames unless the surrounding light is not very bright.

When the alkali metals react with water, the corresponding hydroxide is formed. These hydroxides (LiOH, NaOH, and KOH) are all strong caustics; that is, they are highly corrosive. If the amount of water in which the alkali metals are reacting is small, the resulting solution may be corrosive.

When the alkali metals are shipped, they are shipped either in vacuum-packed metal cans or in jars that contain enough kerosene to cover the metal. Only liquids that do not have oxygen in their molecules may be used to cover the alkali metals during shipment or storage, since the metals will react with a liquid that does. Kerosene is used rather than gasoline since the former has a much higher flash point and therefore is safer. Kerosene also has a higher specific gravity. As a matter of fact, if a piece of sodium or potassium is removed from its protective cover of kerosene and begins to react in moist air, it can be replaced in its container to stop its reaction. This will, of course, cause the kerosene to ignite, but it is much easier to extinguish a small fire involving a combustible liquid than a burning metal.

Let us pause here to expand upon the statement that the alkali metals must be shipped in vacuum-packed cans, or *under a liquid that has no oxygen in its molecule*. The alkali metals are so reactive that they are never found free in nature in the elemental form. They want to achieve stability so badly that they seek to react with anything that will give them that stability. The most common element with which they can react is oxygen, and they will do so whenever they come into contact with it, *even if it is tied up chemically in another compound*. This is why the alkali metals are water-reactive. They want the oxygen in the water molecule, and they want it so badly that they will tear apart the quite stable water molecule to get it, liberating the extremely flammable hydrogen molecules formed as hydrogen is left behind in the reaction.

This molecular attack is not limited to water. The alkali metals will seek oxygen in whatever molecule it exists (with one exception). This fact means that if you try to substitute an alcohol, an aldehyde, a ketone, an ether, an organic acid, an ester, or any other liquid that contains oxygen in its molecular makeup (including mixtures of liquids), the metal will react with the liquid to get to the oxygen. The exception mentioned above is pure, dry oxygen. The alkali metals will not react with the oxygen in the surrounding air unless that air contains a lot of moisture. The reaction that occurs in moist air is with the water in the air, not with the oxygen. Once the metal begins burning, however, its combustion *will* be supported by atmospheric oxygen.

Even this exception has an exception. Beware of powdered alkali metals! The increase of surface area by the reduction in size to powder increases the reactivity of the metal so that it appears to react with the oxygen in the surrounding air. Whether powered alkali metals are pyrophoric or not is immaterial; you must be aware that powdered alkali metals will act as if they are and will be extremely hazardous.

The implication of the reactivity of the alkali metals with oxygen-containing molecules means that carbon dioxide may not be used as an extinguishing agent on burning alkali metals (nor indeed, on other burning metals). There may also be reactions with chlorinated hydrocarbons, and so these should not be used to extinguish burning metals.

The only other possible extinguishing agents are the Class D dry powders. Even these materials may be ineffective. The moister the air, the higher the probability that an alkali metal may react with it. An effective coating of the dry powder over the alkali metal must be made to have any chance of sealing off moist air from the metal. You should become aware of any manufacturing process in your jurisdiction that uses any of the alkali metals.

The Alkaline Earth Metals

The alkaline earth metals are the elements of group IIA on the Periodic Table. They are less reactive than the alkali metals, but they too do not exist free in nature. There are also very few uses for the pure metals (with the exception of magnesium), so you will stand little chance of encountering them. They are beryllium (Be), magnesium (Mg), calcium (Ca), strontium (Sr), barium (Ba), and radium (Ra). Radium is radioactive and is not shipped in significant quantities to be considered a water-reactive hazardous material.

Magnesium is a useful metal, combining strength with lightness, but the only way it can exist as a metal is with a very fine coating of magnesium oxide over it to prevent further oxidation. Magnesium, like the other alkaline earth metals, is water-reactive, and even though the fine coating of magnesium oxide usually prevents unexpected reactions with water, there are documented instances of damp magnesium detonating under certain conditions. Those conditions always include the metal's state as small particles, such as turnings, chips, shavings, and particularly as a powder. In fact, if you reduce the alkaline earth metals (as well as the alkali metals) to a fine enough powder, they will not only be water-reactive, but pyrophoric; that is, they will react upon contact with air, even dry air. This phenomenon is due to the tremendous increase in surface area of the metal when it is reduced in size. As this surface area is increased, more and more of the metal is exposed to the oxygen in the air, and, finally, when the ratio of exposed surface to air is perfect, a spontaneous reaction will occur, involving every particle of the metal instanteously, with a tremendous release of energy. This detonation can also be brought about by shocking larger pieces of the damp metal (still as small as chips and powders, but not so fine as to be pyrophoric), such as those small pieces of metal produced in a grinding operation. *All* powdered metals are dangerous, but those of the alkaline earth metals and the alkali metals are the most dangerous.

Again, if large chunks of a metal are ignited from another reaction, your only choice of an extinguishing agent may be the dry powders. These may be ineffective against burning metallic powders because they may scatter the burning metal as they are applied.

The Hydrides

The hydrides are a group of inorganic, water-reactive materials that are made up of a metal and hydrogen. They are all irritants, toxic, and highly flammable. Whenever they come into contact with water, they liberate hydrogen and the corresponding hydroxide. The explosiveness and flammability of hydrogen is well known, and the hydroxides of the alkali metals are extremely caustic (corrosive).

Lithium hydride, LiH, is a white crystalline powder that might tend toward bluish gray. It has a specific gravity of 0.82 and is soluble in ether. Upon contact with water, lithium hydride will liberate enormous amounts of hydrogen, and whatever water is left after the reaction will be a solution of lithium hydroxide, LiOH. Lithium hydride is used to ship hydrogen around the country because it is so light. It is the most common of the hydrides, is shipped in cans, drums, and cases, and is so water-reactive that it will ignite in moist air. Dry chemical may be the only effective fire-extinguishing agent in the case of all the hydrides. Then again, you may only be able to protect exposures.

Sodium hydride, NaH, is a white powder, while potassium hydride, KH, is a grayish powder dispersed in oil. Both will decompose in moist air to liberate hydrogen. Sodium aluminum hydride is a white, crystalline powder with a specific gravity of 1.24; it decomposes at 361°F., with the evolution of hydrogen; it also will liberate hydrogen upon contact with water; its molecular formula is $NaAlH_4$. Lithium aluminum hydride, $LiAlH_4$, is a white powder with a specific gravity of 0.917; it sometimes turns gray, due to the slow decomposition of the material, leaving some colloidal lithium in the mixture. Lithium aluminum hydride is very sensitive to moist air and will liberate hydrogen in moist air or when water is applied. At 257°F., it will decompose into lithium hydride, aluminum, and hydrogen.

The boron hydrides are a series of compounds of boron and hydrogen, the simplest of which is borane, BH_3. Borane is highly unstable at normal temperature and pressure and will decompose to become the gas, diborane, B_2H_6. Pentaborane, B_5H_9, is a colorless liquid with a pungent odor, highly flammable, and highly toxic by inhalation and ingestion; it is not as water-reactive as the rest of the boron hydrides. Decaborane, $B_{10}H_{14}$, is a colorless, crystalline solid that decomposes slowly at elevated temperatures; it is highly toxic, and may explode in contact with heat or flame. The higher boron hydrides, which go up to $B_{20}H_{26}$, must all be examined individually for their water sensitivity, as well as some of the shorter-chain boron hydrides.

The Carbides

The carbides are a family of binary compounds that include carbon and another element. They are not all water-reactive; we will mention only those that are.

Probably the most familiar of the carbides is calcium carbide, CaC_2. It is a grayish black solid with a specific gravity of 2.22 and has a faint odor of garlic. When it comes into contact with water it liberates the highly flammable and unstable gas, acetylene, C_2H_2, and calcium hydroxide, $Ca(OH)_2$, a caustic material. Acetylene is so unstable that it is never shipped very far from the place where it is generated by the adding of water to calcium carbide in a controlled manner and in a special container. Therefore, acetylene is shipped around the country as calcium carbide (just as hydrogen is shipped as lithium hydride), in sift-proof, air-tight containers that are delivered to those people who generate the acetylene and bottle it in cylinders. Calcium carbide is shipped as a "Flammable Solid" and "Dangerous When Wet" material. The only use for calcium carbide is to generate acetylene.

Aluminum carbide, Al_4C_3, is a yellow solid that has a specific gravity of 2.36 and is a quite stable compound until it comes into contact with water. It will liberate methane gas, CH_4, in the same manner that calcium carbide liberates acetylene; the caustic aluminum hydroxide, $Al(OH)_3$, will also be formed. Aluminum carbide is used as a drying agent, a reducing agent, and a catalyst, as well as a generator of methane. It is shipped as a "Flammable Solid" and "Dangerous When Wet" material.

Beryllium carbide, Be_2C, is a crystalline solid used in the reactor cores of nuclear power plants. When it contacts water, it too will liberate methane and produce the corresponding beryllium hydroxide, $Be(OH)_2$.

Magnesium carbide, Mg_2C_3, upon contact with water, will liberate propyne gas, C_3H_4, which is the next compound after acetylene in the alkyne series. Propyne contains a triple bond between the first and second carbons and therefore possesses a great deal of energy. Propyne is also known as methyl acetylene or allylene; when it is mixed with propadiene, propane, and butane, it becomes stabilized and is known as MAPP gas.

Once water has reached the carbides, your choice of extinguishing agents to use is dictated by what gas has evolved. The standard procedure in a flammable gas fire is to stop the flow of gas! In such a situation this action may not be possible. If you can keep additional water from reaching the hazardous material, you may be able to accomplish this aim. Otherwise, you may only be able to protect exposures.

The carbides that are *not* hazardous are usually extremely hard and are consequently used as grinding materials. They are silicon carbide, SiC, tungsten carbide, W_2C, titanium carbide, TiC, and tantalum carbide, TaC.

The Nitrides

The *nitrides* are binary compounds containing the nitride ion, N^{-3}. They are different from the *azides*, which contain the azide ion, N_3^{-1}. The nitrides and the azides are both hazardous materials, with the nitrides classified as water-reactive materials, while the azides are explosives.

Magnesium nitride, Mg_3N_2, a greenish yellow powder, reacts with water to liberate ammonia, NH_3, and the caustic, magnesium hydroxide, $Mg(OH)_2$.

Lithium nitride, Li_3N, is a brownish red solid that is both water-reactive and *pyrophoric*, which means it will react in air. Upon contact with water lithium nitride will release ammonia and lithium hydroxide, LiOH. It is used to nitride metals, a process which results in a special, very hard surface being formed on the metal; it also has additional uses in manufacturing other chemicals.

The nitrides listed here are solids and may be controlled more easily than liquids. They may be swept up or otherwise confined.

Ammonia is the evolved gas common to water contact with the nitrides. Ammonia is less hazardous than many other evolved gases with other materials and may pose fewer problems for you. Ammonia, an irritant, will not burn until it reaches its lower flammable limit of 16 percent in air. It is highly soluble in water and may be "swept" from the air with a fine fog, if you are careful to keep additional water from reaching the spilled materials and careful of the run-off, which will be a solution of ammonium hydroxide, NH_4OH.

The Phosphides

The phosphides are ionic compounds of phosphorus and metals, with the phosphide ion being P^{-3}. When the phosphides react with water, they form the flammable, toxic gas, phosphine, PH_3, and the corresponding hydroxide. Phosphine is an analogous compound to ammonia, NH_3, with the same ratio of hydrogens to the non-metal.

Aluminum phosphide, AIP, is a dark gray or dark yellow solid, with a specific gravity of 2.85. It is toxic and is used as an insecticide and a fumigant. It is shipped as a "Flammable Solid" and "Dangerous When Wet" material.

Calcium phosphide Ca_3P_2, is a reddish brown solid with a specific gravity of 2.51 and is toxic. It is used in signal fires, torpedoes, pyrotechnics, and rodenticides. It too is shipped as a "Flammable Solid" and "Dangerous When Wet" material.

The phosphides are solids and are easier to control in an incident that liquids; they may be swept up or otherwise contained to prevent contact with water. The phosphine that is generated is highly toxic by inhalation, and it is flammable; evacuation will be a prime concern in large spills. A controlled fire of burning phosphine *may* be preferable to the uncontrolled spread of toxic vapors. Before you choose actually to *start* a fire, be prepared for the consequences if it were to get out of control.

The Inorganic Chlorides

There are many inorganic chlorides that are stable in contact with water. As a matter of fact, all that they usually do is dissolve in water, to form neutral salt solutions. The only stable inorganic chlorides, however, are the chlorides of the alkali metals and the alkaline earth metals. Many of the others are water-reactive.

Aluminum chloride, $AlCl_3$, usually referred to as *anhydrous* (no water present) aluminum chloride, is a yellowish white solid with a specific gravity of 2.44, which sublimes at 352°F. It is moderately toxic by ingestion and inhalation and a strong irritant. Aluminum chloride reacts violently with water, liberating hydrogen chloride (HCl) gas. If water must be used to control an emergency involving aluminum chloride or other water-reactive inorganic chloride, it should be used in flooding amounts. Small amounts of water will liberate the acid gas, while large amounts of water will retain the gas in solution, which, of course, is hydrochloric acid (also HCl). Always be mindful of the run-off water! As a solid, aluminum chloride may be swept up to be contained, so long as protection is maintained against breathing the dust or permitting it contact with the skin.

Antimony pentachloride, SbCl₅, is a reddish yellow, oily liquid with a pungent odor. It is toxic and corrosive, and it also yields hydrogen chloride in contact with water. It is shipped as a corrosive solid. Antimony trichloride, SbCl₃, is corrosive and a strong irritant, but it fumes just slightly in moist air, and it is not considered water-reactive.

Boron trichloride, BCl₃, is a colorless, fuming liquid that is a strong tissue irritant. Its fumes are both corrosive and toxic. It is shipped as a corrosive liquid in cylinders and tank cars, and it liberates HCl, as do the other water-reactive chlorides.

Stannic chloride, SnCL₄, also known as stannic tetrachloride, is a colorless, fuming, caustic liquid, which is also toxic. It is valuable in many chemical manufacturing operations and will liberate heat along with the HCl. It is shipped as a corrosive liquid.

Titanium tetrachloride, TiCl₄, is a colorless liquid that is toxic by inhalation and a strong skin irritant that liberates HCl on contact with water. It is shipped in glass bottles, steel drums, and tank cars as a corrosive liquid.

Phosphorus oxychloride, POCl₃, is a colorless, fuming liquid with a strong, pungent odor. It is toxic by inhalation and ingestion and a strong irritant. It liberates large amounts of heat in addition to HCl; phosphoric acid is also formed. Phosphorus oxychloride is shipped in steel-jacketed lead cylinders and carboys and nickel-lined drums and tank cars as a corrosive liquid.

Phosphorus pentachloride, PCl₅, also known as phosphoryl chloride, is a yellowish solid that fumes in moist air. It sublimes at 320°F. and has a specific gravity of 3.60. It is flammable and corrosive to eyes and skin. Phosphoric acid is formed on contact with water, along with HCl. Although it is a solid, care must be taken in handling phosphorus pentachloride. If the surrounding temperature rises to 320°F., the material will sublime and generate toxic and corrosive vapors.

Phosphorus trichloride, PCl₃, is a colorless, fuming liquid that decomposes rapidly in moist air, liberating great amounts of HCl. It is highly toxic and is corrosive to skin and tissue. It is shipped in carboys, cylinders, and tank cars as a corrosive liquid.

Silicon tetrachloride, SiCl₄, is a colorless fuming liquid whose fumes cause a feeling of suffocation. It will produce enough HCl on contact with water not only to be corrosive to skin and tissue, but also to any metal nearby. Silicon tetrachloride is shipped in iron drums, glass bottles, and steel tanks as a corrosive liquid. It is toxic by ingestion and inhalation.

Sulfur monochloride, S₂Cl₂, also known as sulfur chloride, is an amber to yellowish red, oily, fuming liquid, which decomposes on contact with water, yielding corrosive fumes, including HCl. It is shipped as a corrosive liquid.

Sulfuryl chloride, SO_2CL_2, is a colorless liquid with a pungent odor which liberates HCl on contact with water. It is shipped in five-gallon carboys, 55-gallon drums, and 725-gallon drums as a corrosive liquid.

Thionyl chloride, $SOCl_2$, is a reddish yellow, fuming liquid whose fumes cause a feeling of suffocation. It decomposes at 284°F, is toxic, and is a strong tissue irritant. It is used as a pesticide and in the manufacture of some engineering plastics; it is shipped in glass carboys and steel drums as a corrosive liquid.

With the exception of two materials, the inorganic chlorides mentioned are all liquids, which present a containment problem. If you can dig a containment pit that will hold all the spilled material, you will have a better chance of controlling the fuming and protecting the liquid from contact with water. Remember that while you may have some control over your own application of water, it may rain! If a pit is not possible, try to dike the liquid with dry sand or dry soil.

All the materials mentioned are corrosive, and a preferred tactic with corrosives is to neutralize them. No one material is best with all corrosive liquids, but soda ash (sodium carbonate) is probably the cheapest and most effective neutralizer for the greatest number of liquids. Calcium carbonate may be as cheap, and more readily available, but it works more slowly. Sodium bicarbonate may be the most effective, but it is probably the most expensive.

Whatever you select as a neutralizing agent, *always* experiment first by getting some of the spilled liquid into a container and adding some of the neutralizer. This test is done to see if the neutralizer really works, and, if it does, does it also make the fuming worse for a period of time? Be prepared for accelerated fuming no matter what steps you take, and be prepared for evacuation.

Remember, you may have to use water, regardless of the consequences. Just be prepared to handle the problems that arise as you use the water, and, if possible, contain the run-off. Your actions may save lives at the time of the spill, but you may be held responsible for damage to the environment after the incident has been concluded.

Peroxides

The inorganic peroxides, sodium peroxide (Na_2O_2), lithium peroxide (Li_2O_2) and potassium peroxide (K_2O_2), will react with water to liberate either hydrogen peroxide, H_2O_2, and the corresponding inorganic hydroxide or will liberate the hydroxide plus oxygen, depending upon the temperature of the water. The peroxides are liberated with cold water; oxygen, with warm water. The peroxides of the alkaline earth metals may also have the same reactions. Just remember that the principal hazard of the inorganic peroxides in their oxidizing power, and that water reactivity is secondary.

Other Compounds

There are many other materials that react with water in some adverse way. Table 33 lists some of them; four of them and the substances liberated are described in the following paragraphs. You will notice that sometimes only heat is liberated. This result is hazardous because in many cases the heat liberated is enough to ignite ordinary combustibles.

Calcium oxide, CaO, a white solid, also known as lime or quicklime, liberates tremendous amounts of heat when in contact with water. The resulting solution will contain the caustic calcium hydroxide, $Ca(OH)_2$, also known as slaked lime.

Sulfuric acid, H_2SO_4, a powerful corrosive liquid, also liberates tremendous amounts of heat in contact with water, and it does so with such speed that the boiling points of both liquids are reached almost instantaneously. Because of this fact, water is never deliberately added to sulfuric acid, but the acid is always added to water. Of course, this action is not always possible in a spill or other accidental-release situation.

Acetic anhydride, $(CH_3CO_2)_2O$, a colorless liquid, liberates acetic acid, CH_3COOH, upon contact with water to produce a great deal of strong vapors of the acid.

Acetyl chloride, CH_3COCl, a fuming liquid, liberates HCl and acetic acid on contact with water.

Pyrophoric Materials

Pyrophoric materials are those substances that react spontaneously in air. Some scientists make a distinction between those that react in clean, dry air and those that react in moist air. Technically, those substances that react in moist air are water-reactive materials, and those that react upon contact with dry air are pyrophoric. Regardless of just how the reaction takes place, there *will* be a reaction, and it will be hazardous.

Some of these materials may break into flame, some may decompose slightly less violently into some noxious components, and

326

Table 33 / Water-Reactive Materials

acetic anhydride
acetyl chloride
alkylaluminums
aluminum borohydride
aluminum carbide
aluminum chloride
aluminum hydride
aluminum isopropoxide
aluminum nitride
aluminum phenoxide
aluminum phosphide
aluminum propoxide
aluminum selenide
ammonium bromoselenate
ammonuim copper (I) iodide
antimony bromide
antimony chloride
antimony hydride
antimony iodide
antimony pentachloride
arsenic hydride
arsine
barium
barium amide
barium arsenide
barium dithionate
barium hydride
barium selenate
benzoyl chloride
beryllium borohydride
beryllium carbide
beryllium hydride
beryllium nitride
bismuth trichloride
bismuth nitrate
bismuth oxolate
borane
boron hydrides
boron pentasulfide
boron phosphide
boron tribromide

boron trichloride
boron trisulfide
bromine fluorosulfate
bromine monofluoride
bromine trifuoride
bromine pentafluoride
butyllithium
calcium
calcium arsenide
calcium carbide
calcium cyanamide
calcium cyanide
calcium hydride
calcium hypochlorite
calcium oxide
calcium phosphide
cerium hydride
cesium
cesium aluminum sulfate
cesium hydride
chlorine trifluoride
chlorodiethylaluminum
chlorosulfuric acid
chromium trioxide
copper hydride
diethyl aluminum chloride
diethyl beryllium
diethyl zinc
dimethyl cadmium
dimethyl zinc
ethyl sodium
fluorine
iodine pentafluoride
lead hydride
lithium
lithium borohydride
lithium nitride
lithium peroxide
magnesium (small particles)
magnesium aluminum hydride
magnesium borohydride

magnesium carbide
magnesium hydride
magnesium nitride
methyldichlorosilane
methyltrichlorosilane
nitryl chloride
nitryl fluoride
oleum (fuming sulfuric acid)
pentaborane
phosphorus pentachloride
phosphorus pentafluoride
phosphorus pentasulfide
phosphorus oxychloride
   (phosphoryl chloride)
phosphorus tribromide
phosphorus trichloride
phosphorus trifluoride
phosphorus trisulfide
phosphoryl chloride
potassium
potassium amide
potassium borohydride
potassium hydride
potassium nitride
potassium peroxide
rubidium
rubidium hydride
silane compounds
silane (silicon hydride)
silicon tetrachloride
sodium
sodium amide (sodamide)
sodium borohydride
sodium carbide
sodium hydride
sodium hydrosulfide
sodium methylate
sodium peroxide
sodium phosphide
sodium-potassium alloys
sodium selenide

Table 33   /   Water-Reactive Materials (cont.)

stannic chloride (stannic tetrachloride)	sulfuryl chloride	trichlorosilane
strontium	tellurium hydride	triethyl aluminum
sulfur dibromide	thionyl chloride	triethyl aluminum ethereate
sulfur difluoride	titanium bromide	trimethyl aluminum
sulfuric acid	titanium tetrachloride	uranium hydride
sulfur monochloride (sulfur chloride)	tri(iso)butylaluminum	vanadium tetrachloride

others may detonate. It is not possible to classify every material into what it does in contact with air, but some space will be given to those pyrophoric materials that you are likely to encounter.

Phosphorus

Elemental phosphorus in its white or yellow allotropic form is truly pyrophoric. It is usually kept under water to prevent its reaction, and, once it dries, it begins to burn violently. It is used in military weapons as tracer material for bullets, and in incendiary bombs. There the phosphorus is packed around an explosive, which, when it explodes, will spread the phosphorus in all directions, starting one fire for each of the countless pieces that have been strewn about. Such a bomb is a terrible anti-personnel weapon, since phosphorus will quickly burn a hole through human skin, tissue, and bones.

Red phosphorus is an allotrope of phosphorus and is not pyrophoric. Chemical engineers can change phosphorus back and forth between the different allotropes and have thus devised the safe way to transport and store it, until it is needed as the white or yellow form.

Organic Compounds

Many of the organo-metallic compounds are pyrophoric. Organo-metallic compounds are substances with a lot of strange chemistry in their manufacture. These compounds all have an organic portion or radical to which a metal is attached directly. They all have strange-sounding names, and all of them ought to be easy to recognize *because* of their names. Such compounds as tetraethyl lead, $Pb(C_2H_5)_4$, and nickel carbonyl, $Ni(CO)_4$, are two of the most common organo-metallic compounds, but neither is pyrophoric. Tetraethyl lead, the anti-knock liquid added to "leaded" gasolines, is a Class B poison, and nickel carbonyl is a flammable liquid.

Dimethyl arsine, (CH₃)AsH, a colorless, poisonous liquid, *is* pyrophoric, however, and will break into flame whenever air reaches it. The alkylaluminums, triethyl aluminum, trimethyl aluminum, trimethyl aluminum ethereate, chlorodiethyl aluminum, and tri(iso)-butyl aluminum are all colorless liquids which ignite in air *and* decompose explosively if contacted by cold water. Other pyrophoric materials include dimethyl cadmium and diethyl zinc.

Other Pyrophoric Substances

Even though the alkali metals and the alkaline earth metals are not truly pyrophoric, there is evidence that when these metals have been reduced to a very fine powder, they *will* react in air. Even if this reaction is due to the moisture in the air, we must consider them as pyrophoric, simply because we cannot control the humidity whenever there is a release of these materials.

Emergencies Involving Pyrophoric Materials

Since the very essence of the hazard of pyrophoric materials is their exposure to air, any extinguishing agent which functions by the exclusion of air *should* be effective; water works very well with phosphorus. You must be careful, however, not to add something to the situation that will make it worse. The pyrophoric substances are so reactive that they *might* react with the fire-extinguishing agent you select. Use your reference materials to look up the best materials for control of *each* pyrophoric substance individually. Remember, too, if we agree to list the powdered alkali metals as pyrophoric, they are *still* water-reactive. Refer to table 34 for a list of pyrophoric materials.

Table 34 / Pyrophoric Materials

cesium (powdered)	rubidium (powdered)
chlorodiethyl aluminum	sodium (powdered)
diethyl cadmium	titanium dichloride
dimethyl arsine	triethyl aluminum
dimethyl zinc	triisobutyl aluminum
lithium (powdered)	trimethyl aluminum
phosphorus (white or yellow)	trimethyl aluminum ethereate
potassium (powdered)	

Glossary_____

Allotrope: One of several possible forms of a substance.

Binary Compound: A compound containing only two elements.

Pyrophoric: Possessing the ability to react in air.

Sublime: To change from the solid state directly to the vapor state, bypassing the liquid state.

Appendix A

Chemical Structure of Monomers and Polymers: Molecular Structure of Certain Plastics

1. ABS

ABS is a terpolymer, whose repeating units are from the monomers acrylonitrile, butadiene, and styrene.

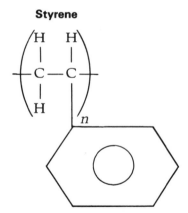

Acrylonitrile **Butadiene** **Styrene**

2. Acetal

3. Acrylic

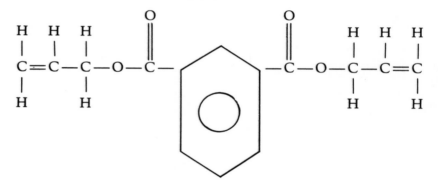

4. Allyl

Allyl monomers are diallyl phthalate and diallyl isophthalate.

Diallyl phthalate

Diallyl isophthalate

5. Amino

Amino monomers are formaldehyde, urea, and melamine.

Formaldehyde

$$H-\overset{\overset{\displaystyle H}{|}}{C}=O$$

Urea

$$\underset{\overset{\displaystyle |}{H}}{\overset{\overset{\displaystyle H}{|}}{N}}-\overset{\overset{\displaystyle O}{||}}{C}-\underset{\overset{\displaystyle |}{H}}{\overset{\overset{\displaystyle H}{|}}{N}}$$

Melamine

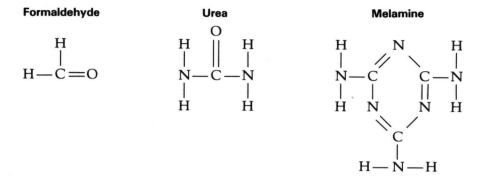

6. Cellulosics

Cellulosics are based on esters that chemically modify cellulose, which is a natural polymer.

7. Epoxy

Epoxy resins contain a highly reactive, three-membered ring referred to as the epoxide group, which is usually at the polymerization terminal point.

Epoxide group (or epoxy radical)

8. Ethylene-vinyl acetate

9. Fluoroplastics

Fluoroplastics is a catch-all name for all carbon-based plastics containing fluorine rather than chlorine. The most common is polytetrafluoroethylene (PTFE).

10. Ionomer

Ionomer is a generic name for polymers whose interchain bonding is ionic rather than covalent. These polymers based on sodium or zinc salts of ethylene-methacrylic acid copolymers.

Ethylene   Methacrylic acid

11. Nitrile

Nitriles are based on the acrylonitrile monomer.

$$\left(\begin{array}{c} \text{H} \quad \text{H} \\ | \qquad | \\ \text{C}-\text{C} \\ | \qquad | \\ \text{H} \quad \text{C}\equiv\text{N} \end{array}\right)_n$$

12. Phenolics

Phenolics are reaction products of phenol and formaldehyde.

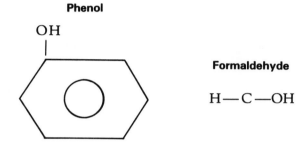

Phenol

OH
|

Formaldehyde

H — C — OH

13. Polyamide

$$\left(\text{NH (CH}_2)_6\text{ NHOC (CH}_2)_4\text{ CO}\right)_n$$

14. Polybutylene

Polybutylene is polymerized from the butene-1 monomer.

$$\begin{array}{cccc} \text{H} & \text{H} & \text{H} & \text{H} \\ | & | & | & | \\ \text{C}=\text{C}-\text{C}-\text{C}-\text{H} \\ | & & | & | \\ \text{H} & & \text{H} & \text{H} \end{array}$$

15. Polycarbonate

Polycarbonate is based on the bisphenol-A monomer.

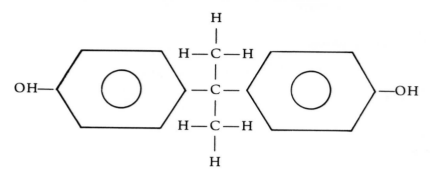

16. Polyesters

Polyesters are made by the reaction of terephthalic acid and ethylene glycol.

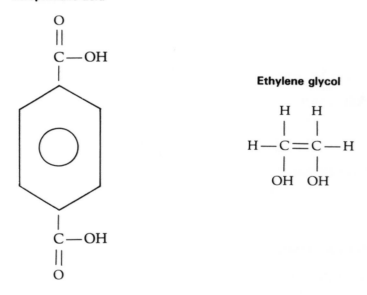

17. Polyethylene

The repeating unit of polyethylene is ethylene.

18. Polyphenylene oxide

These resins are produced by a patented process for oxidative coupling of phenolic monomers.

19. Polypropylene

These resins are produced by polymerizing propylene.

$$\left(\begin{array}{ccc} H & H & H \\ | & | & | \\ C & C & C \\ | & | & | \\ H & H & H \end{array}\right)_n$$

20. Polystyrene

The repeating unit of polystyrene is styrene.

$$\left(\begin{array}{cc} H & H \\ | & | \\ C & C \\ | & | \\ H & \end{array}\right)_n$$

21. Polyurethane

Polyurethanes are block copolymers composed of the extender-isocyanate reaction product and a polymeric polyester or polyether.

22. Polyvinyl chloride

The repeating unit of polyvinyl chloride is vinyl chloride.

$$\left(\begin{array}{cc} H & H \\ | & | \\ C & C \\ | & | \\ H & \\ & Cl \end{array}\right)_n$$

338

23. Silicone

Silicon replaces carbon in the polymer backbone of silicone.

$$
\left(
\begin{array}{ccccccc}
& H & & H & & H & \\
& | & & | & & | & \\
-Si & -O- & Si & -O- & Si & - \\
& | & & | & & | & \\
& H & & H & & H &
\end{array}
\right)_n
$$

24. Styrene-acrylonitrile

Styrene-acrylonitrile is a copolymer of styrene and acrylonitrile.

Styrene

$$
\begin{array}{cc}
H & H \\
| & | \\
C & = C \\
| & | \\
H & \\
\end{array}
$$

Acrylonitrile

$$
\begin{array}{cc}
H & H \\
| & | \\
C & = C \\
| & | \\
H & C\equiv N \\
\end{array}
$$

25. Sulfones

Polysulfone and polyethersulfone are examples of sulfones.

Polysulfone

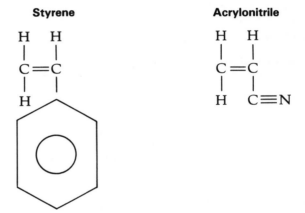

Polyethersulfone

$$\left(O - \left\langle \bigcirc \right\rangle - \underset{\underset{O}{\overset{\overset{O}{\|}}{\|}}}{S} - \left\langle \bigcirc \right\rangle \right)_n$$

26. Vinylidene chloride

$$\left(\begin{array}{cc} H & Cl \\ | & | \\ C & - C \\ | & | \\ H & Cl \end{array} \right)_n$$

Hazardous Chemicals Forbidden to Be Carried by Common Carriers

Table 35 / Hazardous Chemicals Not to Be Carried by Common Carriers (U.S. Department of Transportation)

acetyl acetone peroxide
acetyl benzoyl peroxide
acetyl cyclohexanesulfonyl peroxide
acetylene, liquid
acetylene silver nitrate
acetyl peroxide
aluminum dross, wet or hot
ammonium azide
ammonium bromate
ammonium chlorate
ammonium fulminate
ammonium nitrite
antimony sulfide mixture with a chlorate
arsenic sulfide mixture with a chlorate
ascaridole (organic peroxide)
azaurolic acid (salt of), dry
3-azido-1,2-propylene glycol nitrate
5-azido-1-hydroxy tetrazole
azidodithiocarbonic acid
azidoethyl nitrate
azido guanidine picrate, dry
azido hydroxy tetrazole (mercury and silver salts) oxide

azotetrazole, dry
benzene diazonium chloride, dry
benzene diazonium nitrate
benzene triozonide
benzoxidiazoles, dry
benzoyl azide
biphenyl triozonide
bromine azide
4-bromo-1,2-dinitrobenzene (unstable at 59°C.)
4-bromo-2-nitrobenzene (unstable at 59°C.)
bromosilane
1,2,4-butanetriol trinitrate
tert-butoxycarbonyl azide
tert-butyl hydroperoxide
tert-butyl peroxyacetate
n-butyl peroxydicarbonate
tert-butyl peroxyisobutyrate
cabazide
charcoal, wet
chlorine azide
chloroprene, uninhibited
coal briquettes, hot

Table 35 / Hazardous Chemicals Not to Be Carried by Common Carriers (U.S. Department of Transportation) (cont.)

coke, hot
copper acetylide
copper amine azide
copper tetramine nitrate
cyclotetramethylene tetranitramine, dry
2,2-di-(4,4-di-tertbutyl-
    peroxycyclohexyl)propane, more than
    42 percent with inert solid
diacetone alcohol peroxides
p-diazidobenzene
1,2-diazidoethane
1,1'-diazoaminonaphthalene
diazoaminotetrazole, dry
diazodinitrophenol: *see* initiating explosive
diazdiphenylmethane
diazomethane
diazonium nitrates, dry
diazonium perchlorates, dry
1,3-diazopropane
dibenzyl peroxydicarbonate
di-(beta-nitroxyethyl)ammonium nitrate
dibromoacetylene
2,2-di-(tert-butylperoxy)butane
di-(tert-butylperoxy)phthalate
N,N'-dichloroazodicarbonamidine
dichloroacetylene
2,4-dichlorobenzoyl peroxide
diethylene glycol dinitrate
diethylgold bromide
diethyl peroxydicarbonate
di-(1-hydroxytetrazole), dry
di-(1-naphthoyl)peroxide
1,8-dihydroxy-2,4,5,7-tetranitroanthraquinone
    (chrysamminic acid)
diisopropylbenzene hydroperoxide
diisotridecyl peroxydicarbonate
2,5-dimethyl-2,5-di(tert-butylperoxy)hexane,
    more than 82 percent with water
dimethylhexane dihydroperoxide
dinitolmide

1,4-dinitro-1,1,4,4-tetramethylolbutane-
    tetranitrate
3,5-dinitro-o-toluamide: *see* dinitolmide
2,4-dinitro-1,3,5-trimethylbenzene
1,3-dinitro-4,5-dinitrosobenzene
1,3-dinitro-5,5-dimethyl hydantoin
dinitro-7,8-dimethylglycoluril, dry
1,1-dinitroethane, dry
1,2-dinitroethane
dinitroglycoluril
dinitromethane
dinitropropylene glycol
2,4-dinitroresorcinol (heavy metal salts of)
4,6-dinitroresorcinol (heavy metal salts of)
3,5-dinitrosalicylic acid (lead salt), dry
dinitrosobenzylamidine and salts of, dry
2,2-dinitrostilbene
a,a'-di-(nitroxy)methylether
1,9-dinitroxy penta-
    methylene-2,4,6,8-tetramine, dry
ethanol amine dinitrate
ethylene diamine diperchlorate
ethylene glycol dinitrate
ethylhydroperoxide
ethyl perchlorate
explosive, forbidden
forbidden explosives
forbidden materials
fulminate of mercury, dry
fulminating gold
fulminating mercury
fulminating platinum
fulminating silver
fulminic acid
galactsan trinitrate
glycerol-1,3-dinitrate
glycerol monoglutinate trinitrate
glycerol monolactate trinitrate
guanyl nitrosamino guanylidene
    hydrazine, dry

Table 35 / Hazardous Chemicals Not to Be Carried by Common Carriers (U.S. Department of Transportation) (cont.)

hexamethylene triperoxide diamine, dry
hexamethylol benzene hexanitrate
2,2',3',4,4',6,6'-hexanitro-3,3'-dihydroxyazobenzene, dry
hexanitroazoxy benzene
2,2',3',4,4',6-hexanitrodiphenylamine
2,3',4,4',6,6'-hexanitrodiphenylether
N,N'-(hexanitrodiphenyl)ethylene dinitramine, dry
hexanitrodiphenyl urea
hexanotroethane
hexanitrooxanilide
hydrazine azide
hydrazine chloride
hydrazine dicarbonic acid diazide
hydrazine perchlorate
hydrazine selenate
hydrocyanic acid
hydroxyl amine iodide
hyponitrous acid
initiating explosives, dry
inositol hexanitrate, dry
inulin trinitrate, dry
iodine azide, dry
iodoxy compounds, dry
iridium nitratopentamine iridium nitrate
isothiocyanic acid
lead azide, dry
lead mononitroresorcinate
lead picrate, dry
lead styphnate, dry
magnesium dross, wet or hot
mannatan tetranitrate
mercurous azide
mercury acetylide
mercury iodide aquabasic ammonobasic (iodide of Millian's base)
mercury nitride
mercury oxycyanide
metal salts of methyl nitramine, dry

methazoic acid
methylamine dinitramine and dry salts thereof
methylamine nitroform
methylamine perchlorate, dry
methylene glycol dinitrate
methyl ethyl ketone peroxide, in solution with more than 9 percent by weight active oxygen
a-methylglucoside tetranitrate
a-methylglycerol trinitrate
methyl isobutyl ketone peroxide, in solution with more than 9 percent by weight active oxygen
methyl nitrate
methyl picric acid (heavy metal salts of)
methyl trimethylol methane trinitrate
monochloroacetone, unstabilized
naphthyl amineperchlorate
nickel picrate
nitrated paper, unstable
nitrates of diazonium compounds
2-nitro-2-methylpropanol nitrate
6-nitro-4-diazotouluene-3-sulfonic acid, dry
N-nitroaniline
m-nitrobenzene diazonium perchlorate
nitroethylene polymer
nitroethyl nitrate
nitrogen trichloride
nitrogen triiodide
nitrogen triiodide monoamine
nitroglycerine, liquid, not desensitized
nitroguanidine nitrate
1-nitro hydantoin
nitro isobutane triol trinitrate
nitromannite
N-nitro-N-methylglycolamide nitrate
m-nitrophenyldinitro methane
nitrosugars, dry
1,7-octadiene-3,5-diyne-1,3-dimethoxy-9-octadecynoic acid

Table 35 / Hazardous Chemicals Not to Be Carried by Common Carriers (U.S. Department of Transportation) (cont.)

pentaerythrite tetranitrate, dry

pentanitroaniline, dry

perchloric acid exceeding 72 percent strength

peroxyacetic acid, more than 43 percent, and with more than 6 percent hydrogen peroxide

m-phenylene diaminediperchlorate, dry

potassium carbonyl

propionyl peroxide

pyridine perchlorate

quebrachitol pentanitrate

selenium nitride

shaped charges (commercial), containing more than eight ounces of explosives

silver azide

silver chlorite

silver fulminate

silver oxalate

silver picrate

sodium picryl peroxide

sodium tetranitride

sucrose octanitrate

sulfur and chlorate, loose mixtures of

tetraazido benzene quinone

tetraethylammonium perchlorate, dry

tetramethylene diperoxide dicarbamide

2,3,4,6-tetranitrophenol

2,3,4,6-tetranitrophenyl methyl nitramine

2,3,4,6-tetranitrophenylnitramine

tetranitroresorcinol

2,3,5,6-tetranitroso-1,4-dinitrobenzene

2,3,5,6-tetranitroso nitrobenzene

tetrazene

tetrazolyl azide, dry

tri-(b-nitroxyethyl)ammonium nitrate

trichloromethyl perchlorate

triformaxime trinitrate

1,3,5-trimethyl-2,4,6-trinitrobenzene

trimethylene glycol diperchlorate

trimethylol nitromethane trinitrate

2,4,6-trinitro-1,3,5-triazido benzene

2,4,6-trinitro-1,3-diazobenzene

trinitroacetic acid

trinitroacetonitrile

trinitroamine cobalt

trinitroethylnitrate

trinitromethane

1,3,5-trinitrophthalene

2,4,6-trinitrophenyl guanidine, dry

2,4,6-trinitrophenyl nitramine

2,4,6-trinitrophenyl trimetylol methyl nitramine trinitrate

2,4,6-trinitroso-3-methyl nitraminoanisole

trinitrotetramine cobalt nitrate

vinyl nitrate polymer

p-xylyl diazide

National Fire Protection Association NFPA 704 Standard System for the Identification of the Fire Hazards of Materials

Identification of Materials by Hazard Signal System

6-1. One of the systems delineated in the following illustrations shall be used for the implementation of this standard.

Identification of Materials by Hazard Signal Dimensions

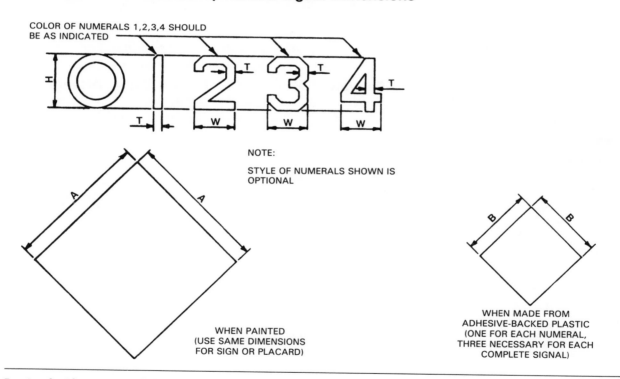

COLOR OF NUMERALS 1,2,3,4 SHOULD BE AS INDICATED

NOTE:

STYLE OF NUMERALS SHOWN IS OPTIONAL

WHEN PAINTED
(USE SAME DIMENSIONS
FOR SIGN OR PLACARD)

WHEN MADE FROM
ADHESIVE-BACKED PLASTIC
(ONE FOR EACH NUMERAL,
THREE NECESSARY FOR EACH
COMPLETE SIGNAL)

Minimum Dimensions of White Background
for Signals
(White Background is Optional)

Size of Signals H	W	T	A	B
1	0.7	5/32	2½	1¼
2	1.4	5/16	5	2½
3	2.1	15/32	7½	3¾
4	2.8	5/8	10	5
6	4.2	15/16	15	7½

All Dimensions Given in Inches

Exception: For containers with a capacity of one gallon or less, symbols may be reduced in size, provided:

1. This reduction is proportionate.
2. The color coding is retained.
3. The vertical and horizontal dimensions of the diamond are not less than 1 in. (2.5 cm).
4. The individual numbers are no smaller than ⅛ in. tall.

Identification of Materials by Hazard Signal Arrangement

ADHESIVE-BACKED PLASTIC BACKGROUND PIECES — ONE NEEDED FOR EACH NUMERAL, THREE NEEDED FOR EACH COMPLETE SIGNAL.

FIGURE C1. FOR USE WHERE SPECIFIED COLOR BACKGROUND IS USED WITH NUMERALS OF CONTRASTING COLORS.

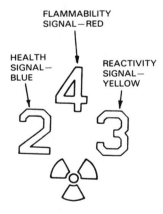

FLAMMABILITY SIGNAL—RED

HEALTH SIGNAL—BLUE

REACTIVITY SIGNAL—YELLOW

FIGURE C2. FOR USE WHERE WHITE BACKGROUND IS NECESSARY.

WHITE PAINTED BACKGROUND, OR, WHITE PAPER OR CARD STOCK

FIGURE C3. FOR USE WHERE WHITE BACKGROUND IS USED WITH PAINTED NUMERALS, OR, FOR USE WHEN SIGNAL IS IN THE FORM OF SIGN OR PLACARD.

ARRANGEMENT AND ORDER OF SIGNALS
— OPTIONAL FORM OF APPLICATION

Distance at Which Signals Must be Legible	Minimum Size of Signals Required
50 feet	1″
75 feet	2″
100 feet	3″
200 feet	4″
300 feet	6″

NOTE:

This shows the correct spatial arrangement and order of signals used for identification of materials by hazard.

FIGURE C4. STORAGE TANK

IDENTIFICATION OF THE FIRE HAZARDS OF MATERIALS   NFPA 704

	Identification of Health Hazard Color Code: BLUE		Identification of Flammability Color Code: RED		Identification of Reactivity (Stability) Color Code: YELLOW
	Type of Possible Injury		Susceptibility of Materials to Burning		Susceptibility to Release of Energy
Signal		Signal		Signal	
4	Materials which on very short exposure could cause death or major residual injury even though prompt medical treatment were given.	**4**	Materials which will rapidly or completely vaporize at atmospheric pressure and normal ambient temperature, or which are readily dispersed in air and which will burn readily.	**4**	Materials which in themselves are readily capable of detonation or of explosive decomposition or reaction at normal temperatures and pressures.
3	Materials which on short exposure could cause serious temporary or residual injury even though prompt medical treatment were given.	**3**	Liquids and solids that can be ignited under almost all ambient temperature conditions.	**3**	Materials which in themselves are capable of detonation or explosive reaction but require a strong initiating source or which must be heated under confinement before initiation or which react explosively with water.
2	Materials which on intense or continued exposure could cause temporary incapacitation or possible residual injury unless prompt medical treatment is given.	**2**	Materials that must be moderately heated or exposed to relatively high ambient temperatures before ignition can occur.	**2**	Materials which in themselves are normally unstable and readily undergo violent chemical change but do not detonate. Also materials which may react violently with water or which may form potentially explosive mixtures with water.
1	Materials which on exposure would cause irritation but only minor residual injury even if no treatment is given.	**1**	Materials that must be pre-heated before ignition can occur.	**1**	Materials which in themselves are normally stable, but which can become unstable at elevated temperatures and pressures or which may react with water with some release of energy but not violently.
0	Materials which on exposure under fire conditions would offer no hazard beyond that of ordinary combustible material.	**0**	Materials that will not burn.	**0**	Materials which in themselves are normally stable, even under fire exposure conditions, and which are not reactive with water.

Appendix D

International Maritime Dangerous Goods Code

The International Maritime Dangerous Goods (IMDG) Code is published by The International Maritime Organization (IMO). The IMO was established by the United Nations in 1948 as the first agency devoted exclusively to maritime matters. Its most important objectives are maritime safety and the prevention of marine pollution. The IMDG Code concerns itself with the shipping of dangerous goods by sea. It is designed to assist compliance with and to supplement the requirements of Chapter VII of the Safety of Life at Sea Convention. It is also intended to help bring into agreement different national requirements.

Reproduced here are just the Class Definitions of the International Maritime Dangerous Goods Code. The complete code including amendments is available in five looseleaf volumes at .103.50, including shipping to the U.S., from The Publications Section, International Maritime Organization, 4 Albert Embankment, London SE1 7SR, England.

Class Definitions of the IMDG Code

CLASS 1—EXPLOSIVES

Class 1 comprises:

(a) explosive substances,[1] except those which are too dangerous to transport or those where the predominant hazard is one appropriate to another class;

(b) explosive articles, except devices containing explosive substances in such quantity or of such a character that their inadvertent or accidental ignition or initiation during transport shall not cause any manifestation external to the device either by projection, fire, smoke, heat or loud noise: and

(c) substances and articles not mentioned under (a) and (b) above which are manufactured with a view to producing a practical, explosive or pyrotechnic effect.

Transport of explosive substances which are unduly sensitive or so reactive as to be subject to spontaneous reaction is prohibited.

For the purpose of this Code the following definitions apply:

(a) An explosive substance is a solid or liquid substance (or a mixture of substances) which is in itself capable by chemical reaction of producing gas at such a temperature and pressure and at such speed as to cause damage to the surroundings. Included are pyrotechnic substances even when they do not evolve gases.

(b) A pyrotechnic substance is a substance or a mixture of substances designed to produce an effect by heat, light, sound, gas or smoke or a combination of these as the result of non-detonative self-sustaining exothermic chemical reactions.

(c) An explosive article is an article containing one or more explosive substances.

Class 1 is divided into five divisions:

Division 1.1 Substances and articles which have a mass explosion hazard[2]

Division 1.2 Substances and articles which have a projection hazard but not a mass explosion hazard

---

[1] A substance which is not itself an explosive but which can form an explosive atmosphere of gas, vapour or dust is not included in Class 1.

[2] A mass explosion is one which affects virtually the entire load practically instantaneously.

Division 1.3 Substances and articles which have a fire hazard and either a minor blast hazard or a minor projection hazard or both, but not a mass explosion hazard

This division comprises substances and articles:

(a) which give to rise to considerable radiant heat, or

(b) which burn one after another, producing minor blast or projection effects or both

Division 1.4 Substances and articles which present no significant hazard

This division comprises substances and articles which present only a small hazard in the event of ignition or initiation during transport. The effects are largely confined to the package and no projection of fragments of appreciable size or range is to be expected. An external fire must not cause practically instantaneous explosion of virtually the entire contents of the package.

NOTE: Substances and articles in this division so packaged or designed that any hazardous effects arising from accidental functioning are confined within the package unless the package has been degraded by fire, in which case all blast or protection effects are limited to the extent that they do not significantly hinder firefighting or other emergency response efforts in the immediate vicinity of the package, are in Compatibility Group S.

Division 1.5 Very insensitive substances which have a mass explosion hazard

This division comprises explosive substances which are so insensitive that there is very little probability of initiation or of transition from burning to detonation under normal conditions of transport. As a minimum requirement they must not explode in the external fire test.

NOTE: The probability of transition from burning to detonation is greater when large quantities are carried in a ship.

Class 1 is unique in that the type of packaging frequently has a decisive effect on the hazard and therefore on the assignment to a particular division. Where multiple hazard classifications have been assigned, they are listed on the individual schedule.

CLASS 2—GASES: COMPRESSED, LIQUIFIED OR DISSOLVED UNDER PRESSURE

Because of the difficulty in reconciling the various main systems of regulation, definitions in this Class are of a general nature to cover all such systems. Moreover, since it has not been found possible to reconcile two main systems of regulation in respect of the differentiation between a liquified gas exerting a low pressure at a certain temperature and an inflammable liquid, this criterion has been omitted; both methods of differentiation are recognized.

This Class comprises:

(a) Permanent gases
     Gases which cannot be liquefied at ambient temperatures;
(b) Liquefied gases
     Gases which can become liquid under pressure at ambient temperatures;
(c) Dissolved gases
     Gases dissolved under pressure in a solvent, which may be absorbed in a porous material;
(d) Deeply refrigerated permanent gases - e.g., liquid air, oxygen, etc.

In the cases (a), (b) and (c) above, the gases are normally under pressure.

For stowage and segregation purposes Class 2 is subdivided further, namely:

Class 2.1—Inflammable gases[3]
Class 2.2—Non-inflammable gases
Class 2.3—Poisonous gases[4]

CLASS 3—INFLAMMABLE LIQUIDS[5]

These are liquids, or mixtures of liquids, or liquids containing solids in solution or suspension (e.g., paints, varnishes, lacquers, etc., but not including substance which, on account of their other dangerous characteristics, have been included in other classes) which give off an inflammable vapour at or below 61°C (141°F) closed cup test (corresponding to 65.6°C (150°F) open cup test).

---

[3] "Inflammable" has the same meaning as flammable.
[4] Poisonous gases which are also inflammable should be segregated as Class 2.1 gases.
[5] "Inflammable" has the same meaning as flammable.

In this Code, Class 3 is subdivided further, namely:

Class 3.1   Low flashpoint group of liquids having a flashpoint below −18°C (0°F), closed cup test;

Class 3.2   Intermediate flashpoint group of liquids having a flashpoint of −18°C (0°F), up to, but not including 23°C (73°F), closed cup test;

Class 3.3   High flashpoint group of liquids having a flashpoint of 23°C (73°F) up to, and including, 61°C (141°F), closed cup test.

Substances which have a flashpoint above 61°C (141°F), closed cup test, are not considered to be dangerous by virtue of their fire hazard. Where the flashpoint is indicated for a volatile liquid it may be followed by the symbol "c.c.", representing determination by a closed cup test, or by a symbol "o.c.", representing an open cup test.

CLASS 4—INFLAMMABLE SOLIDS OR SUBSTANCES

In this Code, Class 4 deals with substances other than those classed as explosives, which, under conditions of transportation, are readily combustible, or may cause or contribute to fires.

Class 4 is subdivided further, namely:

Class 4.1   Inflammable solids. The substances in this Class are solids possessing the properties of being easily ignited by external sources, such as sparks and flames, and of being readily combustible, or of being liable to cause or contribute to fire through friction.

Class 4.2   Substances liable to spontaneous combustion. The substances in this Class are either solids or liquids possessing the common property of being liable spontaneously to heat and to ignite.

Class 4.3   Substances emitting inflammable gases when wet. The substance in this Class are either solids or liquids possessing the common property, when in contact with water, of evolving inflammable gases. In some cases these gases are liable to spontaneous ignition.

CLASS 5—OXIDIZING SUBSTANCES (AGENTS) AND ORGANIC PEROXIDES

In this Code, Class 5 deals with oxidizing substances (agents) and organic peroxides.

Class 5 is subdivided further, namely:

Class 5.1   Oxidizing substances (agents). These are substances which, although in themselves not necessarily combustible, may, either by yielding oxygen or by similar processes, increase the risk and intensity of fire in other materials with which they come into contact.

Class 5.2   Organic peroxides. Most substances in this Class are combustible. They may act as oxidizing substances, and are liable to explosive decomposition. In either liquid or solid form they may react dangerously with other substances. Most will burn rapidly and are sensitive to impact or friction.

CLASS 6—POISONOUS (TOXIC) AND INFECTIOUS SUBSTANCES

In this Code, Class 6 is subdivided further, namely:

Class 6.1   Poisonous (toxic) substances. These are substances liable either to cause death or serious injury or to harm human health if swallowed or inhaled, or by skin contact.

Class 6.2   Infectious substances[6]. These are substances containing disease-producing micro-organisms.

CLASS 7—RADIOACTIVE SUBSTANCES

In this Code, Class 7 comprises substances which spontaneously emit a significant radiation and of which the specific activity is greater than 0.002 microcurie per gramme.

---

[6] For the reasons given in the Introduction to Class 6, detailed recommendations for Class 6.2 have not been included in this Code.

CLASS 8—CORROSIVES

In this Code, Class 8 comprises substances which are solids or liquids possessing, in their original state, the common property of being able more or less severely to damage living tissue. The escape of such a substance from its packaging may also cause damage to other cargo or to the ship.

CLASS 9—MISCELLANEOUS DANGEROUS SUBSTANCES

Any other substances which experience has shown, or may show to be of such a dangerous character that the provisions of this Class should apply to it. In this Code, in particular, Class 9 includes a number of substances and articles which cannot be properly covered by the provisions for the other classes, or which present a relatively low transportation hazard.

Periodic Table

Group I A

1 **H** 1.008

II A

3 **Li** 6.941	4 **Be** 9.01

11 **Na** 22.99	12 **Mg** 24.305

		III B	IV B	V B	VI B	VII B	VIII
19 **K** 39.102	20 **Ca** 40.08	21 **Sc** 44.956	22 **Ti** 47.9	23 **V** 50.941	24 **Cr** 51.996	25 **Mn** 54.938	26 **Fe** 55.847	27 **Co** 58.933
37 **Rb** 85.468	38 **Sr** 87.62	39 **Y** 88.906	40 **Zr** 91.22	41 **Nb** 92.906	42 **Mo** 95.94	43 **Tc** 98.906	44 **Ru** 101.07	45 **Rh** 102.91
55 **Cs** 132.91	56 **Ba** 137.34	57 **La** 138.91	72 **Hf** 178.49	73 **Ta** 180.95	74 **W** 183.85	75 **Re** 186.2	76 **Os** 190.2	77 **Ir** 192.22
87 **Fr** (223)	88 **Ra** 226.03	89 **Ac** (227)	104					

Lanthanide series

58 **Ce** 140.12	59 **Pr** 140.91	60 **Nd** 144.24	61 **Pm** (147)	62 **Sm** 150.4	63 **Eu** 151.96	64 **Gd** 157.25

Actinide series

90 **Th** 232.04	91 **Pa** 231.04	92 **U** 238.03	93 **Np** 237.05	94 **Pu** 239.05	95 **Am** (243)	96 **Cm** (247)